Physics and Applications of
Negative Refractive Index Materials

Physics and Applications of

Negative Refractive Index Materials

S. Anantha Ramakrishna
Tomasz M. Grzegorczyk

SPIE
PRESS

Bellingham, Washington USA

CRC Press
Taylor & Francis Group
Boca Raton London New York

CRC Press is an imprint of the
Taylor & Francis Group, an **informa** business

Co-published by SPIE
P.O. Box 10
Bellingham, Washington 98227-0010 USA
Tel.: +1 360-676-3290
Fax: +1 360-647-1445
Email: Books@SPIE.org
spie.org
ISBN: 9780819473998

CRC Press
Taylor & Francis Group
6000 Broken Sound Parkway NW, Suite 300
Boca Raton, FL 33487-2742

First issued in paperback 2020

ISBN-13: 978-0-367-57748-3 (pbk)
ISBN-13: 978-1-4200-6875-7 (hbk)

To Kanchan, Kartik, and Kanishka

To Alessandra, Eva, Davide, and to my parents

In memoriam: Jin Au Kong

Jin Au Kong, Professor in the Department of Electrical Engineering and Computer Science at the Massachusetts Institute of Technology (MIT), contributed tremendously to the development of left-handed media from the very beginning, in 2001. His influence can be felt in all areas of this field, from theory and numerical simulations for which he was internationally renowned, to experiments that he helped conceive and carry through all around the world, notably within MIT and the MIT Lincoln Laboratory, as well as in Asia and Europe. The conference he created, *Progress in Electromagnetic Research Symposium*, was one of the first, if not the first, to promote technical sessions on left-handed media, thus contributing to the cross-fertilization of ideas among researchers worldwide. His journals, the *Journal of Electromagnetic Waves and Applications* and the *Progress in Electromagnetic Research*, constantly call for innovative papers on all aspects of left-handed media, and have been well regarded and well cited. Finally, his textbooks, written and published years before the advent of left-handed media, often contain remarkable ideas and concepts that were later rediscovered and accepted as pillars in this new field.

Prof. Kong passed away unexpectedly on March 12, 2008, of complications from pneumonia, before this book went to press. Many of the ideas and concepts presented in the following pages have been directly inspired by him and discussed at length with him, often late into the night in his office at MIT. The course of events made it such that my efforts toward the realization of this book have become a tribute to his work during the last 7 years. The international community has lost one of its giants, which is in addition a personal loss for me.

T. M. Grzegorczyk

Foreword

The past ten years have seen an astonishing explosion of interest in negative refractive index materials. First explored systematically by Veselago in 1968 from a theoretical point of view these materials remained without an experimental realisation for more than 30 years. That had to await development of suitable metamaterials, materials whose function is due as much to their internal sub wavelength structure as to their chemical composition. The added flexibility to create new materials enables properties unavailable in nature to be realised in practice. That opened the floodgates to a host of new experiments.

Why the great interest? From its rebirth at the beginning of this century negative refraction has provoked controversy. To be consistent with the laws of causality a material has to do much more than refract negatively. For example, it must necessarily be dispersive. Thus did many misunderstandings arise and pioneers had to endure some testing assaults. Yet even that aspect can now be seen as positive because controversy drew attention to the fledgling subject and showed that negative refraction contains subtleties that even experienced scientists did not at first appreciate. Even now we as a community are learning from our errors and discovering new aspects of this long hidden subject. As work progressed and news of amazing results spread beyond the scientific community into the popular press, a broader excitement has been generated. Some of the more extraordinary results such as the prescription for a perfect lens, and particularly the possibility of making objects invisible, had already been foreseen in science fiction and fed a ready-made appetite in the popular imagination. Thus the ancient subject of classical optics has brought us new discoveries and excitement.

This book, written by two leading practitioners of negative refraction, arrives at an opportune time because there is a substantial body of results available in the field that need to be gathered together in a systematic fashion sparing new arrivals hours of wasted time trawling through the very many papers in the literature. And yet new discoveries are continually reported. This is work in progress and the authors must steel themselves eventually to write a second edition!

Sir John B. Pendry
Imperial College London

Preface

Rarely in the history of science does one have the opportunity to witness an explosion of interest for a given topic, to participate in its development from its beginning, and to witness its growth at a pace almost exponential over a period of about a decade. Yet, we believe that this is precisely what has happened to us, with regard to the new development of materials that are now called metamaterials, left-handed media, or negative refractive media. Fundamentally rooted in the electromagnetic theory and governed by the equations proposed by the Scottish physicist James Clerk Maxwell at the end of the 19th century, the development of these structured composite materials that we call *metamaterials* could have been another incremental step in the more general research in electromagnetics and optics. Yet, the scientific community quickly realized that the implications and applications opened by the study of metamaterials are unprecedented, potentially revolutionary, and scientifically as well as technologically highly interesting and challenging. A new paradigm of electromagnetic and optical materials has evolved today from these studies.

The study of metamaterials is often thought of as being associated with negative refraction. It is much more than that. Over the past decade, scientists have shown how to manipulate the macroscopic properties of matter at a level unachieved before. For decades, our world was limited to materials with primarily positive permittivities and permeabilities, with some exceptions such as plasmas, for example, whose permittivities can be negative. The research in metamaterials coupled with the rapid advancements in micro- and nano-fabrication technology has totally lifted this limitation, and has opened the door to almost arbitrary material properties with some extraordinary consequences across the electromagnetic spectrum, from radio frequencies to optical frequencies. This book is devoted to a discussion of these consequences as well as their theoretical implications and practical applications.

It is inevitable that such a growing field has attracted much attention in the scientific as well as in the more popular literature: the number of scientific articles has been in constant and almost exponential growth since about the year 2000, many popular articles have been published in scientific as well as nonscientific journals, while technical reviews and a few books have already been devoted to this field. It therefore appears ambitious at best and risky at worst to attempt the publication of an additional reference in this arena.

Nonetheless, we think that such an addition is necessary and was, in fact, missing. The extremely large number of scientific papers published is certainly vivid proof of the rapid evolution of this research area, but getting familiar with and appreciating so much information also represent a daunting task for the student or researcher who is new to this field. In addition, the large number of new articles appearing on a weekly basis may also appear difficult to track, even by the expert researcher. It is with this spirit that we have targeted this book at as vast an audience as possible: the reader unfamiliar, but interested in this field, will find in the following pages the synthesis and organization of what we believe to be the most important and influential papers related to metamaterials, whereas the expert reader will hopefully find a useful viewpoint and detailed explanations of some of the most recent papers at the time of this writing, touching on as many aspects of this field as possible.

An additional motivation to undertake the writing of this book was our feeling that a coherent reference presenting the history, development, and main achievements of metamaterials was missing. Although some excellent books are already available to the reader, they are usually focused on either a very specific aspect of this field, or a compilation of chapters written by renowned scientists. In the present book, we have tried to remedy what we believe are limitations of the previous two formats by offering a book covering a wide variety of topics, yet having a coherence across chapters that enables the reader to cross-reference similar topics and, hence, to delve deeper into their presentation and explanation.

Naturally, it is impossible to present in a short book all aspects of a given scientific field, all the more when this field has become so vast and complex as the one the present book is devoted to. In addition, and despite our best efforts, our grasp of the field is also incomplete and is being refined by the day. We would therefore like to apologize upfront to those authors who may feel that their work is misrepresented or underrepresented in the following pages. May they put it on the account of our limited knowledge and not on our judgment of the quality of their contributions.

Finally, we must remark that it has been very difficult to write a book on an emerging area: it has almost been like writing about the personality of a growing teenager. New topics of today might disappear tomorrow or, instead, might reveal unexpected promises and become the front-runners of this research field. Metamaterials of the future will necessarily be robust and reliable, multifunctional, and reconfigurable to perform satisfactorily in various demanding environments. Today's metamaterials are quite primitive by these standards and developments are happening at breathtaking speeds. These have been the reasons why we decided not to have a concluding chapter – this book is an ongoing account of metamaterials.

S. A. Ramakrishna
Kanpur, India

T. M. Grzegorczyk
Cambridge, Massachusetts, USA

Acknowledgments

This book came about not only because of our privilege to have witnessed the birth of this field, but more importantly because of our privilege to have actively participated in its development from a very early date. The research we have carried out over almost an entire decade brought us in contact with many researchers and students who, in many ways, have helped us discover and learn about this exciting topic. We would like in particular to acknowledge the contributions of our most closely related colleagues: Benjamin E. Barrowes, Sangeeta Chakrabarti, Hongsheng Chen, Jianbing J. Chen, Xudong Chen (with a special thanks for proofreading parts of the manuscript), Sebastien Guenneau, Brandon A. Kemp (with a special thanks for proofreading parts of the manuscript), Jin Au Kong, Narendra Kumar, Akhlesh Lakhtakia, Jie Lu, Olivier J. F. Martin, Christopher Moss, Lipsa Nanda, Stephen O'Brien, Joe Pacheco, Jr., Sir John Pendry, Lixin Ran, Zachary Thomas, Harshawardhan Wanare, Bae-Ian Wu, and Yan Zhang.

We specifically thank L. Nanda and S. Chakrabarti for their help in making some of the figures and compiling the bibliography. We thank our colleagues from across the world who have given us permission to reuse or reproduce their figures and data which, at times, might have even been original and unpublished.

SAR acknowledges the support of the Centre for Development of Technical Education, IIT Kanpur via a book-writing grant and encouragement from his colleagues in the Physics Department at IIT Kanpur.

Finally, for their constant support and encouragements, we would like to thank our respective families to whom we dedicate this book.

About the authors

S. Anantha Ramakrishna received his M.Sc. in physics from the Indian Institute of Technology, Kanpur, and his Ph.D. in 2001 for his research work on wave propagation in random media at the Raman Research Institute, Bangalore. During 2001–2003 he worked with Sir John Pendry at Imperial College London on the theory of perfect lenses made of the newly discovered negative refractive index materials. In 2003, he joined the Indian Institute of Technology, Kanpur as an assistant professor and is presently an associate professor of physics there. His research interests concern complex wave phenomena in optics and condensed matter physics. He published the first comprehensive, technical review on the development of negative refractive index materials in 2005. He is a Young Associate of the Indian Academy of Science, Bangalore, a recipient of the Young Scientist Medal for 2007 of the Indian National Science Academy, Delhi, and was selected as an affiliate of the Third World Academy of Science, Trieste, in 2007. He was an invited professor at the Institut Fresnel, Université Aix–Marseille I in May 2006, and a visiting professor at the Nanophotonics and Metrology Laboratory at the Ecole Polytechnique Federale de Lausanne during June–July 2006. He is a member of SPIE and a life member of the Indian Physics Association.

Tomasz M. Grzegorczyk received his Ph.D. from the Swiss Federal Institute of Technology, Lausanne, in December 2000. In January 2001, he joined the Research Laboratory of Electronics (RLE), Massachusetts Institute of Technology (MIT), U.S.A., where he was a research scientist until July 2007. Since then, he has been a research affiliate at the RLE-MIT, and founder and president of Delpsi, LLC, a company devoted to research in electromagnetics and optics. His research interests include the study of wave propagation in complex media and left-handed metamaterials, electromagnetic induction from spheroidal and ellipsoidal objects for unexploded ordnances modeling, optical binding and trapping phenomena, and microwave imaging. He is a senior member of IEEE, a member of the OSA, and was a visiting scientist at the Institute of Mathematical Studies at the National University of Singapore in December 2002 and January 2003. He was appointed adjunct professor of The Electromagnetics Academy at Zhejiang University in Hangzhou, China, in July 2004. From 2001 to 2007, he was part of the Technical Program Committee of the *Progress in Electromagnetics Research Symposium* and a member of the Editorial Board of the *Journal of Electromagnetic Waves and Applications* and *Progress in Electromagnetics Research*.

Contents

1

Introduction

This book is devoted to the description of metamaterials, their origins and physical principles, their electromagnetic and optical properties, as well as to their potential applications. This field has witnessed an immense gain of interest over the past few years, gathering communities as diverse as those from optics, electromagnetics, materials science, mathematics, condensed matter physics, microwave engineering, and many more. The field of metamaterials being therefore potentially extremely vast, we have limited the scope of this book to those composite materials whose structures are substantially smaller, or at the least smaller, than the wavelength of the operating radiation. Such structured materials have been called metamaterials in order to refer to the unusual properties they exhibit, while at the same time being describable as effective media and characterized by a few effective medium parameters insofar as their interaction with electromagnetic radiation is concerned. We also include in this book a chapter on photonic crystals, which work on a very different principle than metamaterials, but which have been closely connected to them and have been shown to exhibit many similar properties.

The metamaterials discussed in this book are designer structures that can result in effective medium parameters unattainable in natural materials, with correspondingly enhanced performance. Much of the novel properties and phenomena of the materials discussed in this book emanate from the possibility that the effective medium parameters (such as the electric permittivity and the magnetic permeability) can become negative. A medium whose dielectric permittivity and magnetic permeability are negative at a given frequency of radiation is called a *negative refractive index medium* or, equivalently, a *left-handed medium*, for reasons that will become clear shortly. In this book, we do not, however, discuss another important and powerful manner of attaining extraordinary material properties – that of *coherent control* whereby atomic and molecular systems are driven into coherence by strong and coherent electromagnetic fields (Scully and Zubairy 1997). Due to the extremely coherent nature of the excitation and response, the quantum mechanical nature of the atoms and molecules is strongly manifested in these cases and the description of the atomic systems relies necessarily on quantum mechanics. In contrast, we remark that since the sizes of the metamaterial structures we are interested in are microscopically large (compared to atomic sizes) and the resonances reasonably broad, it is the classical electromagnetic properties that are apparent. Hence, we ignore the quantum mechanical nature of light and matter throughout our discussions.

This chapter offers a general introduction to the topic this book is devoted to, starting with a brief description of the historical development of the subject. We give first a general account of the development of optics, electromagnetism, and the characterization of their effects by effective constitutive parameters. A more specific account of the development of the ideas surrounding metamaterials and negative effective medium parameters follows. We then clarify mathematically our definitions, and discuss the Lorentz model for the dispersion of the dielectric permittivity of a dispersive medium and the basic definitions for the description of negative refractive index media. In the course of this chapter, we hope to set out the basic foundations that will allow the reader to follow the book without much confusion.

1.1   General historical perspective

The study of optical phenomena has accompanied the evolution of mankind from almost its origins. Astronomy, which is often said to be the oldest of all science, has led humans through an incredibly vast journey of discoveries, turning philosophers into scientists and passive observations into active research. During the last couple of centuries, man has achieved an unprecedented understanding and control over light thanks to one fundamental property: light exhibits just the right amount of interaction with matter. This interaction is intense enough compared to that between other particles or matter waves such as neutrons or neutrinos, and yet weak enough compared to the interaction between charged particles such as electrons or protons. The fact that light, or the *photon*, is one of the fundamental particles of nature and that its propagation velocity sets the ultimate limit on the speed of any signal further underlines the significance of this control.

Despite being one of the oldest topics in science, Optics has remained a very fundamental area of physics and engineering because of the simplicity of its theoretical grounds. It is, for example, formidable to realize that optical properties of many materials can be characterized by a single number called the refractive index, n. This number allows one to understand refraction processes and enables the design of lenses and prisms that led to the understanding of colors and dispersion. For a long time, this refractive index was a number that represented the *optical density* of a medium, a notion reasonably supported by the definition of the refractive index as

$$n = \frac{c}{v}, \tag{1.1}$$

where c is the speed of light in vacuum and v is the speed of light in the medium.

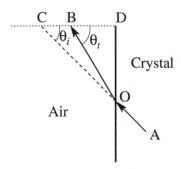

Figure 1.1 An adaptation of Ibn Sahl's original drawing showing refraction at a planar interface. AO is the incident ray from inside the crystal and OB is the refracted ray. Ibn Sahl obtained the reciprocal of the refractive index as $1/n = \frac{OB}{OC}$ $(= \frac{OB}{OD}\frac{OD}{OC} = \frac{\sin\theta_i}{\sin\theta_t}$ in today's terminology).

The roots of Optics as a science go as far as the ancient Greek civilization, where Aristotle, upon studying visual perception, recognized the importance of the medium in-between the eye and an object. Another Greek astronomer, Ptolemy, performed several experiments on the effects of refraction on visual perception of objects in the 2nd century AD. Despite these early works, the real credit for the association of a number to the refraction effects of a transparent medium is probably due to Ibn Sahl, an Arabic scholar of Catalonian origin. Ibn Sahl, who lived in Baghdad around 984 AD, wrote a treatise on *Burning Instruments* where he clearly stated a law of refraction for light passing across a plane interface from a material medium into air. This law, completely identical to what we now call the *Snell law's* of refraction, defined the refractive index n in terms of the incident and refracted rays as shown in Fig. 1.1. Ibn Sahl further used this refractive index for a crystal to study the focusing properties of a biconvex lens and several other focusing instruments.[*]

For a long time all optically transparent crystals were mainly characterized by the refractive index. Based on the experimental findings of Willebrord Snellius in 1621, the French philosopher René Descartes (1596–1650) published in his *"Dioptrique"* the law of refraction in the form we know it today. The refractive index was considered to be a quantification of the resistance offered by a medium to the passage of light. Based on this idea, Fermat enunciated his famous *Principle of Least Action*, which proved invaluable for studies of light propagation in media with spatially varying refractive index. Erasmus Bartholinus had discovered the double refraction in calc-spar in 1669 which led to the realization that there was a polarization associated with light. Malus also discovered polarization, and the rotation of polarization of light

---

[*]The reader is referred to Rashed (1990) for a lucid description of Ibn Sahl's work.

upon passage through an anisotropic medium in 1808, again with a calc-spar crystal. This was one of the first cases when the optical material could not be characterized by a single number, but the description necessarily depended on the propagation direction and the relative orientation of the crystal.

In parallel to the development of optics, the 19th century also witnessed the emergence of the theories of electricity and magnetism. A plethora of experimental observations challenged the physicists to look for underlying explanations and gave birth to fundamental laws such as Ampère's, Gauss, or Faraday's laws. Yet, electricity, magnetism, and optics were seen as independent fields, ruled by independent laws and yielding independent applications. It took the incredible insight and genius of James Clerk Maxwell (see Fig. 1.2) to first unify the former two, and then all the three fields under a uniquely simple and complete theory. With his work, Maxwell showed that electricity and magnetism are entangled phenomena, inseparable, and self-sustaining, ruled by four simple equations known today as the Maxwell equations. The concepts of dielectric permittivity and magnetic permeability, denoted by the letters ε and μ, respectively, became fundamental for the description of media and their response to electric and magnetic fields, and were called constitutive parameters. Moreover, upon studying the self-sustaining solutions of the electromagnetic field in vacuum, Maxwell discovered electromagnetic waves, effectively revolutionizing the field for a second time with descriptions of frequency, wavelength, and propagation speed, with all their fundamental and technological impacts. Finally, upon calculating the propagation speed of the newly discovered electromagnetic waves, Maxwell realized that it was very close to that of light in vacuum, which led him to bridge the two independent fields by declaring that light is an electromagnetic wave. The independent demonstrations of radio frequency waves and their propagation in vacuum by H. R. Hertz, N. Tesla, J. C. Bose, [†] and G. Marconi, as well as the theoretical work of Einstein obviating the need for the all permeating "aether," made quick developments in optics possible by utilizing the Maxwell equations.

The connection between the two fields, optics and electromagnetics, was summarized by the very simple equation (also known as the Maxwell relation)

$$n^2 = \varepsilon\mu, \tag{1.2}$$

relating the index of refraction, an optical quantity, to the permittivity and permeability of media, two electromagnetic quantities. It was also then realized that all media could be described by the concepts of permittivity and permeability, whose definitions had to be properly generalized. Hence, absorption of light in materials was described by complex valued ε and μ, whereas many anisotropic crystals (where all directions are not equivalent) were described by second-rank tensors $\bar{\bar{\varepsilon}}$ and $\bar{\bar{\mu}}$, effectively yielding different values in different propagation directions or for different polarization states.

[†] J. C. Bose is credited with the discovery of millimeter waves.

Figure 1.2 Two giants of electromagnetism: J. C. Maxwell (*left*) mathematically unified Electricity, Magnetism, and Optics through his equations. The image is taken from the Wikipedia project, http://www.wikipedia.org. H. A. Lorentz (*right*) gave a microscopic model for the dispersion of the dielectric permittivity. (Courtesy of C. W. J. Beenakker, from the "Collection Instituut-Lorentz, Leiden University.")

Although dispersion of the refractive index with frequency was a well-known empirical fact by then, it was Fresnel (of the diffraction fame) who first tried to explain it in terms of the molecular structure of matter. This was also supported by Cauchy who gave the well-known dispersion formula which goes by his name. But it was essentially H. A. Lorentz (see Fig. 1.2, right) who gave a reasonably robust theory of dispersion in terms of the polarization of the basic molecules constituting a material. This Lorentz theory of dispersion (described in Section 1.3.2) has been very successful at describing the variation of the dielectric permittivity with frequency and is used as a workhorse model for describing the dispersion in resonant systems. At frequencies well away from an absorption resonance, the Lorentz theory easily approximates into the Cauchy dispersion formula. In dense media (high pressure gases, liquids, and solids), it had to be corrected for local field effects – effects of other neighboring polarized molecules, which yielded the Lorentz-Lorenz model, akin to the Clausius-Mossotti relations for the electrostatic case (Jackson 1999).

Interestingly, although there was no *a priori* bound on the values of the constitutive parameters, all known transparent media were described by a refractive index between about 1.2 and 1.9 only at optical frequencies.[‡] These bounds were broken for the first time when it was realized that stratified

---

[‡]Excluding semiconductors where it could be as large as 4 in the infrared regions.

materials, where layers of transparent materials with different refractive indices are stacked together, could exhibit very different optical properties due to well-controlled interference phenomena of the multiply scattered waves at the interfaces between the different media. The most striking examples of such technology are the quarter wavelength anti-reflection coatings and high reflection thin film coatings. The theory of periodic media was later generalized to higher dimensions, making the layered medium a special case of structures later to be called photonic crystals, where strong modifications of the properties of electromagnetic radiation come from multiple scattering or Bragg scattering within the structure. A drastic example is the realization of structures in which light is not able to propagate at all in any direction in a band of frequencies (bandgap) because of the proper interplay of scattering and destructive interference. Actually the realization of a one-dimensional stop-band structure should be credited to Lord Rayleigh who was probably the first to systematically investigate the wave propagation in layered materials (Rayleigh 1887). Lord Rayleigh had already realized the existence of a stop-band and the fact that a layered medium would cause complete reflection of the incident light for frequencies within this band. For further reading on these topics, the reader is referred to Joannopoulos et al. (1995) and Sakoda (2005).

By the middle of the 20th century, the optics of layered media had been well established, benefiting from the thrust in military requirements during World War II. Improvements were demanded in all areas of optical instrumentation, from binoculars to periscopes, and provided the impetus for industrial activity in this area. The strong modification of light propagation in such systems resulted in a variety of optical properties, ranging from highly reflecting multilayer coatings to their opposite, the anti-reflective coatings. The reader is referred to details in the classic book by Born and Wolf (1999) for further reading on these topics.

In 1987, the generalization from one-dimensional periodic media (i.e., layered media) to three-dimensional periodic media was independently proposed by Yablonovitch (1987) and John (1987) who also discussed the strong modification of the density of photon states in such systems. Thus, even the spontaneous emission probability for an atom within the photonic crystal, emitting at a frequency in the forbidden band (called the bandgap) was shown to be possibly strongly modulated. Yablonovitch et al. (1991) pursued this work with the demonstration of a face-centered cubic photonic crystal at microwave frequencies. It was demonstrated by calculations (Ho et al. 1990) that a diamond-like lattice structure with a strong enough refractive index contrast could result in a complete bandgap for light propagating in any direction (a three-dimensional bandgap). For the last decade or so, photonic crystals with negligible absorption have become one of the most promising avenues for the development of all-optical circuits. For example, Akahane et al. (2003) have reported optical cavities using two-dimensional photonic crystals with some of the highest ever reported Q-factors ($\sim 10^6$) at optical frequencies.

An important limitation in controlling the propagation of light in matter came from the fact that the index of refraction could still take positive values only. In fact, a negative refractive index was often seen as being incompatible with the definition of optical density, and hence was often viewed as unphysical. However, careful theoretical considerations showed that a negative refraction could indeed be physical, provided that the medium exhibits other fundamental and necessary properties. The two most important ones were shown to be frequency dispersion (where the permittivity and permeability are not constant with frequency) and dissipation, the two not being independent but related to one another by the necessity of causality. Despite these additional constraints, materials with a negative refractive index had no further reasons to remain hypothetical and the scientific community began a quest for their physical realization.

The germs of the possibility of negative refraction probably first appeared in 1904 during discussions between Sir Arthur Schuster and Sir Horace Lamb regarding the relationship between the group velocity and the phase velocity of waves (see Boardman et al. (2005) for a detailed discussion). The negative group velocity that is possible due to anomalous dispersion at frequencies close to an absorption resonance was the point in contention. For the case of negative refraction, Schuster believed that the group velocity should have a component away from the interface while the phase velocity vector should point inward to the interface. Although Schuster's conclusion came about from a confusion regarding negative group velocity (the energy flow need not coincide with the group velocity direction in the vicinity of a resonance), it was probably the first consideration of negative phase velocity vectors. In 1944, Mandelshtam considered the possibility of oppositely oriented phase and group velocities (Mandelshtam 1950). He noted that Snell's law for refraction between two media admitted the mathematical solution of refraction at an angle of $(\pi - \theta_t)$ in addition to the usual angle of refraction at θ_t, and reconciled it with the fact that the phase velocity still tells nothing about the direction of energy flow. Mandelshtam then also presented examples of negative group velocity structures in spatially periodic dielectric media (Mandelshtam 1945) with the periodicity at wavelength scales. Sivukhin was probably the first to notice the possibility of a medium with negative ε and μ, but rejected it since the possibility of their existence was yet to be clarified.

Viktor G. Veselago first formally considered media with simultaneous negative ε and μ from a theoretical point of view (Veselago 1968), and concluded that the phase velocity and the energy flow in such media would point in opposite directions. Thus, the media could be considered as having a negative refractive index. He systematically investigated several effects resulting from his conclusions, including the negative refraction at an interface, the negative Doppler shifts, an obtuse angle for Čerenkov radiation, and the possibility of momentum reversal. He also considered the behavior of convex and concave lenses made of such media and also showed that a flat slab of material with $n = -1$ could image a point source located on one side of the slab onto two other points, one inside the slab and one on the other side of it (provided that the thickness of the slab was sufficient). His results, however, did not spark much interest at the time and remained an academic curiosity for many subsequent years, primarily because there were no media available at the time which had both ε and μ negative at a given frequency. The realization of these media had to wait for another 30 years for the development of ideas allowing their experimental realization.

Metamaterials have been the most recent development in this quest for control over light via material parameters, with the recognition that engineered materials, structured in specific manners, can exhibit resonances unique to the structure at certain frequencies. The structures are engineered such that at these frequencies, the wavelength of the electromagnetic radiation is much larger than the structural unit sizes, and thus can excite these resonances while still failing to resolve the details of the structure (shape, size, etc.). Consequently, an array of these structural units (periodic or otherwise) appears to be effectively homogeneous to the radiation and can be well described by effective medium parameters such as a dielectric permittivity ε and a magnetic permeability μ.[§]

1.2   The concept of metamaterials

Interestingly, the tremendous interest surrounding media with simultaneously $\varepsilon < 0$ and $\mu < 0$ arose despite the fact that no natural materials have been, and still are, known to exhibit these properties and all known such media today are artificially structured metamaterials. Although Veselago speculated in his landmark paper (Veselago 1968) that some "gyrotropic substances possessing both plasma and magnetic properties" could be anisotropic examples of left-handed media, to date there is no report of a natural medium with such properties. Therefore, their realization took the path of engineered structures that have been called *metamaterials*.[¶]

The word *"meta"* implies "beyond" (as in "metaphysics") and the terminology "metamaterials" today implies composite materials consisting of structural units much smaller that the wavelength of the incident radiation and displaying properties not usually found in natural materials. Although many

---

[§]It is important to note that the effective medium parameters might have little to do with the bulk material parameters of the medium making up the structures as is discussed in Chapter 3.

[¶]The origin of the term Metamaterial has been attributed to R. M. Walser who defined them as "Macroscopic composites having a manmade, three-dimensional, periodic cellular architecture designed to produce an optimized combination, not available in nature, of two or more responses to specific excitation" in 1999 (Walser 2003).

of the ideas of metamaterials have their origin in the theories of homogenization of composites (see for example Milton (2002)), metamaterials differ from those in that they are crucially dependent on resonances for their properties and the nature of the bulk material of the structural units is often of marginal importance in determining the effective medium parameters in the relevant frequency bandwidth. Typically, the resonances in metamaterials can induce large amounts of dispersion (large changes with frequency) in the effective medium parameters at frequencies close to resonance. By properly driving and enhancing these resonances, one can cause the materials parameters ε or μ to become negative in a frequency band slightly above the resonance frequency.

Pendry et al. (1996) first theoretically suggested and later experimentally demonstrated (Pendry et al. 1998) that a composite medium of periodically placed thin metallic wires can behave as an effective plasma medium for radiation with wavelength much larger than the spatial periodicity of the structure. For frequencies lower than a particular (plasma) frequency, the thin wire structure therefore exhibits a negative permittivity ε. Although dense wire media had been considered with much interest as artificial impedance surfaces by electrical engineers (Brown 1960, Rotman 1962, King et al. 1983), they were usually considered when the wavelength was comparable to the period of the lattice and were therefore not really metamaterials per se, for which effective medium parameters can be defined.

In 1999, Pendry et al. described how one could tailor a medium whose effective magnetic permeability could display a resonant Lorentz behavior and therefore achieve negative values of the permeability within a frequency band above the resonant frequency (Pendry et al. 1999). Again, although similar structures consisting of loops, helices, spirals or Omega-shaped metallic particles had been considered earlier by the electrical engineering community (Saadoun and Engheta 1992, Lindell et al. 1994) as the basis of artificial chiral and bianisotropic media, the work reported in Pendry et al. (1999) was the first to consider them as magnetizable particles that could lead to an effective negative μ.

In light of the connection between (ε, μ) and the index of refraction n expressed in Eq. (1.2), one should immediately wonder what happens to n when both ε and μ are negative. While in usual materials with positive constitutive parameters it is natural to take the positive square root in Eq. (1.2), $n = \sqrt{\varepsilon\mu}$, physical and mathematical considerations lead into choosing the negative square root $n = -\sqrt{\varepsilon\mu}$ when $\varepsilon < 0$ and $\mu < 0$. More arguments in favor of this conclusion are provided subsequently in this chapter and within the body of this book.

With the basis for a negative permittivity and a negative permeability having been laid out, researchers went on to actually experimentally demonstrate the reality of a negative index medium in a prism experiment at microwave frequencies (Smith et al. 2000, Shelby et al. 2001b). A photograph of one of the original metamaterial structures possessing a negative index of refrac-

Figure 1.3 One of the world's first negative refractive index medium at microwave frequencies reported in Shelby et al. (2001b). The system has negative refractive index for wave propagating in the horizontal plane with the electric field along the vertical direction. The ring-like metallic structures printed on a circuit board provide the negative magnetic permeability while metal wires make the composite acquire a negative dielectric permittivity. (Reproduced with permission from Shelby et al. (2001b). © 2001 by the American Association for the Advancement of Science.)

tion is reproduced in Fig. 1.3 and illustrates how the proposals for a negative permittivity and a negative permeability were put together in a single configuration. Although these initial experiments were met with some criticism, they were quickly confirmed by free-space experiments (Greegor et al. 2003, Parazzoli et al. 2003) with large sample sizes. As a consequence, at the time of the present writing, negative refractive index materials are well accepted and have become available at frequencies spanning a wide portion of the electromagnetic spectrum, from static to microwave to optical frequencies, although the extent of homogenization and description as a homogeneous material is often questionable at the higher end of the spectrum.

The realization that engineered structures can exhibit a negative index of refraction opens up several conceptual frontiers in electromagnetics and optics: several new properties become realizable while most known electromagnetic effects have to be revisited. Even pedestrian effects like refraction between two media, one of them with a negative index, are modified whereby the wave refracts on to the same side of the normal. Several other phenomena were shown to be modified, such as the Doppler shift, the Čerenkov radiation, the Goos-Hänchen shift for reflection of a beam, the radiation pressure, etc. In addition, media with negative permittivity and permeability have the ability to support surface electromagnetic modes, which has given an impetus to the new field of plasmonics (Barnes et al. 2003). The surface plasmon excitations on a metal surface (i.e., at an interface with a medium exhibiting a nega-

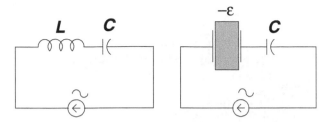

Figure 1.4 A capacitor and an inductor form a resonant circuit that can oscillate at $\omega_0 = 1/\sqrt{LC}$. A capacitor filled with a negative dielectric has negative capacitance, acts as an inductor and can resonate with another usual capacitor. (Reproduced with permission from Ramakrishna (2005). © 2005, Institute of Physics Publishing, U.K.)

tive permittivity) have been well known (Raether 1986), whereas materials with negative magnetic permeability are totally novel and can be expected to support the analogous surface plasmon but of a magnetic nature. These surface plasmons on a structured metallic surface can resonantly interact with radiation and give rise to a host of novel electromagnetic effects.

The origin of the surface plasmon can be simply understood as a resonance effect at the interface between two media. Let us consider, for example, the simple case of a capacitor: it is well known that a capacitor can be formed by two parallel conducting plates with an insulating dielectric placed in-between. Filling the gap with a negative dielectric material instead would lead to a capacitor with negative capacitance, which is equivalent to an inductor. Thus two capacitors in a circuit, one filled with a positive dielectric (ε_p) and the other filled with a negative dielectric (ε_m), can become resonant (see Fig. 1.4). The condition for resonance with two such capacitors turns out to be simply $\varepsilon_m = -\varepsilon_p$, which is exactly the condition for the excitation of a surface plasmon at the interface between a semi-infinite positive medium and a semi-infinite negative medium in the static limit. Including negative dielectric materials within regular structures of positive dielectrics can therefore yield media in which a variety of resonances can be excited and the structured media would then display many novel phenomena. The excitation of surface plasmons on small implanted metal particles has been exploited for several centuries in Europe to make brilliantly colored glass windows, and it was explained only at the beginning of the 20th century by the Mie theory of light scattering (Bohren and Huffman 1983).

A direct and very novel application of these surface plasmon modes is the perfect lens, which is an imaging device that can preserve subwavelength details in the image and thus overcome the classic *diffraction limit* (Born and Wolf 1999). It was demonstrated that not only could such a slab of negative refractive medium image a point source in the sense already pointed out in Veselago (1968) for the propagating modes, but that this reconstruction

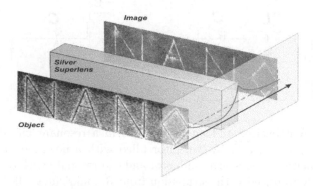

Figure 1.5 Imaging of an arbitrary object "NANO" by a slab of silver that acts as a super lens. The line width of the "NANO" object is 40 nm. The developed image is found to reproduce subwavelength features of the object to the extent of $\lambda/6$. The figure shows the FIB image of the actual object used at the object plane and the AFM image of the developed image on a photoresist. (Figure kindly supplied by Prof. X. Zhang and based on work published in Fang et al. (2005).)

also holds for the non-propagating near-field modes of the source (Pendry 2000). Thus the imaging action is not limited by the diffraction limit and, in principle, the image can be perfect with infinite resolution. However, the conditions for a perfect resolution were shown to be highly theoretical and unphysical, and the resolution is, in fact, limited by other processes, primarily dissipation in the negative refractive index material (Ramakrishna et al. 2002, Smith et al. 2003). Nonetheless, even if perfect resolution is out of reach, subwavelength image resolution is still achievable and is used in optical lithography with subwavelength details as illustrated in Fig. 1.5 (Fang et al. 2005). This lensing effect has been generalized to the idea of complementary media (Pendry and Ramakrishna 2003), which brings in a new view point on negative refractive index media as *electromagnetic anti-matter* that annihilates the effects of ordinary electromagnetic matter on radiation.

In parallel to the development of resonant metamaterials, (Eleftheriades and Grbic 2002, Eleftheriades et al. 2002) and (Caloz and Itoh 2005) independently developed a transmission line approach with lumped circuit elements for planar metamaterials (see Fig. 1.6 for an implementation) which could support backward waves, or, in other words, an effectively planar negative refractive index medium. A host of effects predicted in negative refractive index materials, such as the negative refraction effect, the obtuse angle for Čerenkov radiation, and the subwavelength image resolution, were quickly realized in these transmission line systems, primarily due to the ease in implementing these designs with lumped circuit elements.

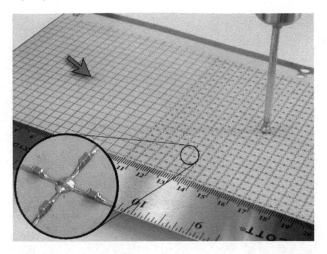

Figure 1.6 A two-dimensional transmission line system that displays a negative refractive index. The transmission line has been implemented using lumped circuit elements: essentially it is a microstrip grid loaded with surface-mounted capacitors and an inductor embedded into the substrate at the central node. The figure also shows a probe to detect the near-field radiation. The inset shows the expanded unit cell of the metamaterial. (Reproduced with permission from Iyer et al. (2003). © 2003, Optical Society of America.)

As the field of metamaterials grew rapidly, various communities were drawn into this research field, bringing a variety of viewpoints, expertise, and interesting ideas. This cross-fertilization between so many different fields of physics, mathematics, and engineering is reflected for example in the development of metamaterial antennae (Ziolkowski and Erentok 2006), optical nano-antennae for plasmonics (Muhlschlegel et al. 2005), and a new circuit element approach to the optics or plasmonics of nanosized metallic particles (Alù et al. 2006a). The emerging area of plasmonics quickly became fundamentally related to metamaterials, particularly at optical frequencies. In fact, the very mechanism and designs of negative refractive index media at optical frequencies are, in one way, intimately related to the excitation of these plasmons in the nano-metallic particles making up the structures (Alù and Engheta 2007, Ramakrishna et al. 2007a). Surface plasmon excitations have been shown to be crucial in the mechanisms of several novel optical phenomena such as the extraordinary transmission of light (Ebbesen et al. 1998, Krishnan et al. 2001) through subwavelength-sized hole arrays in metallic films (see Fig. 1.7), large non-linearities due to local field enhancements on rough metal surfaces, single photon tunneling through subwavelength-sized holes (Smolyaninov et al. 2002), etc.

HFW	HV	det	mag	WD
8.00 µm	15.00 kV	TLD	16 000 x	5.0 mm

Figure 1.7 A 2-D array of holes (190 nm diameter and 415 nm periodicity) etched by focused ion beam technology in a film of gold deposited on fused silica made at IIT Kanpur. This sample shows a resonantly enhanced transmission peak for light with a wavelength of about 540 nm and 620 nm.

1.3   Modeling the material response

This section reviews some fundamental concepts of continuum electromagnetism that are essential to the ideas of metamaterials. For more in-depth discussions and theoretical details, which are beyond the scope of this book, the reader is referred to standard textbooks of electromagnetic theory such as Landau et al. (1984), Jackson (1999), Kong (2000).

1.3.1   Basic equations

The Maxwell equations are the fundamental equations for the understanding of all electromagnetic and optical phenomena. In their differential form, these equations are written as

$$\nabla \cdot \mathcal{E} = \frac{\varrho}{\varepsilon_0}, \tag{1.3a}$$

$$\nabla \cdot \mathcal{B} = 0, \tag{1.3b}$$

$$\nabla \times \mathcal{E} = -\frac{\partial \mathcal{B}}{\partial t}, \tag{1.3c}$$

$$\nabla \times \mathcal{B} = \mu_0 \mathcal{J} + \varepsilon_0 \mu_0 \frac{\partial \mathcal{E}}{\partial t}, \tag{1.3d}$$

where $\mathcal{E}$ and $\mathcal{B}$ are the electric field and the magnetic induction, respectively, and ϱ and $\mathcal{J}$ are the volume charge and current densities, respectively. These equations are called the microscopic Maxwell equations because ϱ and $\mathcal{J}$ here represent the actual microscopic charge and current densities. In a material medium, for example, ϱ would describe the electronic and nuclear charge distributions. Thus ϱ and $\mathcal{J}$ would necessarily be complicated and vary extremely fast on very small length scales. Most often, however, we are not interested in the correspondingly fast variations of the electric and magnetic fields over atomic length scales and a macroscopic description is sufficiently accurate.[||] The fundamental Maxwell equations are therefore rewritten at the macroscopic level as

$$\nabla \cdot \mathbf{D} = \rho, \tag{1.4a}$$

$$\nabla \cdot \mathbf{B} = 0, \tag{1.4b}$$

$$\nabla \times \mathbf{E} = -\frac{\partial \mathbf{B}}{\partial t}, \tag{1.4c}$$

$$\nabla \times \mathbf{H} = J + \frac{\partial \mathbf{D}}{\partial t}, \tag{1.4d}$$

where $\mathbf{E}$ and $\mathbf{H}$ are the macroscopic electric and magnetic fields, $\mathbf{D}$ is the displacement field, and $\mathbf{B}$ is the macroscopic magnetic induction. Similarly, ρ and $\mathbf{J}$ are the macroscopic net charge and current densities. Here the microscopic fields are averaged over sufficiently large volumes to yield the macroscopic field quantities wherein the fast variations over small length scales are not observable. Thus, the underlying medium appears homogeneous and shows a homogeneous response to the applied fields. We refer the reader to Jackson (1999) for an insightful derivation of these equations from the microscopic Maxwell equations.

In most materials, the time domain displacement field $\mathbf{D}$ is directly and linearly proportional to the applied electric field $\mathbf{E}$, and is a function of the material in which the field propagates. Due to the mass of the electrons in

---

[||] Note that the wavelength of electromagnetic radiation is of the order of 10^{-2} m at microwave frequencies and about 10^{-7} m for optical (visible) radiation. In addition, the time period of the oscillations are of the order of 10^{-9} seconds to 10^{-15} seconds, respectively. Therefore, one usually seeks only spatially averaged and time-averaged information, averaged over much longer length scales and time scales.

the medium that introduce a certain inertia in the response, $\mathbf{D}$ does not vary instantaneously with $\mathbf{E}$, but instead is a function of the entire time history of how $\mathbf{E}$ excited the medium. A somewhat general form for $\mathbf{D}$ can therefore be written in the following form:

$$\mathbf{D}(\mathbf{r}, t) = \int_{-\infty}^{t} \mathrm{d}t' \; \phi(\mathbf{r}; t, t') \mathbf{E}(\mathbf{r}, t'), \qquad (1.5)$$

where $\phi(\mathbf{r}; t, t')$ is called the local response function. We assume here that the polarization that sets in a medium depends on the local fields – an assumption that can be violated at small lengthscales due to correlations in the polarization over a given volume of the material. For stationary processes, $\phi(\mathbf{r}; t, t') = \phi(\mathbf{r}; t - t')$, i.e., all physical quantities depend only on the elapsed time intervals and the above integral becomes a convolution. Frequency domain displacement field and electric field can be defined such as

$$\mathbf{E}(\mathbf{r}, t) = \int_{-\infty}^{+\infty} \mathrm{d}\omega \mathbf{E}(\mathbf{r}, \omega) \, e^{-i\omega t}, \qquad (1.6a)$$

$$\mathbf{D}(\mathbf{r}, t) = \int_{-\infty}^{+\infty} \mathrm{d}\omega \mathbf{D}(\mathbf{r}, \omega) \, e^{-i\omega t}. \qquad (1.6b)$$

Introducing these definitions into Eq. (1.5) and using the convolution theorem of Fourier transforms (Arfken 1985), it can immediately be seen that the frequency domains $\mathbf{E}(\mathbf{r}, \omega)$ and $\mathbf{D}(\mathbf{r}, \omega)$ are related by the simple linear relation

$$\mathbf{D}(\mathbf{r}, \omega) = \varepsilon_0 \varepsilon(\mathbf{r}, \omega) \mathbf{E}(\mathbf{r}, \omega), \qquad (1.7)$$

where $\varepsilon(\mathbf{r}, \omega)$ is the frequency-dependent dielectric function given by

$$\varepsilon(\mathbf{r}, \omega) = \frac{1}{\varepsilon_0} \int_{-\infty}^{\infty} \mathrm{d}\tau \; \phi(\mathbf{r}; \tau) e^{i\omega\tau}. \qquad (1.8)$$

This relation indicates that ε is dispersive, i.e., function of the frequency ω. The dispersive nature arises from the inertia of the dipoles in a causal medium (due to the mass of the electrons), which defines a material polarization that does not respond instantaneously to the applied fields, but depends on its time history as we have seen. At extremely high frequencies, for example x-rays or γ-rays, the matter cannot even respond and the "electronic" matter is almost transparent leading to the limit

$$\lim_{\omega \to \infty} \varepsilon(\omega) \to 1.$$

We shall see some examples of frequency-dependent dielectric functions in the next section. A similar analysis also holds true for the magnetic permeability $\mu(\mathbf{r}, \omega)$, which can be space and frequency dependent.

The expression of $\phi(\mathbf{r}, \tau)$ can be obtained from an inverse Fourier transform of Eq. (1.8), and subsequently introduced in Eq. (1.5). Supposing that the orders of integration can be interchanged, it can be shown that the polarization is related to the electric field via the Fourier transform of $[\varepsilon(\omega)/\varepsilon_0 - 1]$. The analyticity of this latter function in the upper ω plane allows the application of the Cauchy theorem over a contour extending over the real axis, jumping the pole, and closing itself at infinity in the upper plane. This direct complex plane integration provides two relations between the real and imaginary parts of $\varepsilon(\omega)$, known as the Kramers-Kronig relations, and expressed as (Jackson 1999)

$$\mathrm{Re}(\varepsilon(\omega)) - 1 = \frac{1}{\pi}\mathrm{PV}\int_{-\infty}^{\infty}\mathrm{d}\omega'\frac{\mathrm{Im}(\varepsilon(\omega'))}{\omega' - \omega}, \tag{1.9a}$$

$$\mathrm{Im}(\varepsilon(\omega)) = -\frac{1}{\pi}\mathrm{PV}\int_{-\infty}^{\infty}\mathrm{d}\omega'\frac{\mathrm{Re}(\varepsilon(\omega')) - 1}{\omega' - \omega}, \tag{1.9b}$$

where PV denotes the Cauchy principal value. Similar relations hold for the real and imaginary parts of the magnetic permeability μ. Consequently, in addition to being frequency dispersive, ε and μ are also required to be complex functions on the account of causality. The imaginary parts account for absorption of radiation in the medium and the total absorbed energy in a volume V is given by (Landau et al. 1984)

$$\int_V \mathrm{d}^3 r \int_{-\infty}^{\infty} \omega \left[\mathrm{Im}(\varepsilon(\omega))|\mathbf{E}(\mathbf{r},\omega)|^2 + \mathrm{Im}(\mu(\omega))|\mathbf{H}(\mathbf{r},\omega)|^2\right] \frac{\mathrm{d}\omega}{2\pi}. \tag{1.10}$$

For example, consider a time harmonic plane wave $\exp[i(kz - \omega t)]$ propagating along the z-axis in a dissipative medium with $\mu = 1$ and a complex ε where $\mathrm{Im}(\varepsilon) > 0$. It is clear that the amplitude of the wave decays exponentially due to absorption of the wave as it propagates, which clearly implies that $\mathrm{Im}(k) > 0$. This complex wave-vector can be obtained from the Maxwell equations as $k^2 = \varepsilon\omega^2/c^2$.

Eqs. (1.9) indicate that the real and imaginary parts of the permittivity (and similarly the permeability) are Hilbert transforms of each other, as illustrated in Fig. 1.8. These relations are derived for material media in thermodynamic equilibrium solely on the grounds of causality. The restriction that they provide on the variation in the real and imaginary parts of material parameters should be regarded as very fundamental. The Kramers-Kronig relations allow an experimentalist, for example, to measure the imaginary part of the permittivity easily by absorption experiments at various frequencies and deduce the real parts of the dielectric permittivity from the imaginary part. An example of this procedure is shown in Fig. 1.8 where the imaginary part of the permittivity is calculated from the real parts by a Hilbert transform with different frequency ranges for the integration. Note that the integrals in the Kramers-Kronig relations involve frequencies all the way up to infinity, whereas it is clear that the effective medium theories break down at high frequencies. However, this does not really affect us in the case of usual optical media since the macroscopic material response functions hold almost down to

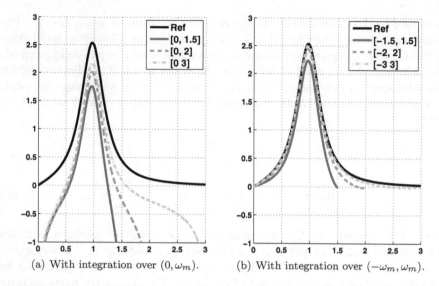

Figure 1.8 Comparison between the analytic imaginary part of the permittivity (thick black curve on both graphs) and the imaginary part obtained via a Hilbert transform. The analytic expression is $\varepsilon_r = 1-(\omega_p^2-\omega_o^2)/(\omega^2-\omega_o^2+i\gamma\omega)$ with $\omega_o = 2\pi \times 10$ GHz, $\omega_p = 2\pi \times 15$ GHz, and $\gamma = \omega_0/2$. The labels 1, 2, and 3 refer to the subscript m of ω_m and correspond to $\omega_1 = 2\omega_p$, $\omega_2 = 4\omega_p$, and $\omega_3 = 6\omega_p$ respectively.

the level of few atomic distances. Thus, the very high frequency limit is never really probed. In the case of metamaterials, the wavelength is usually larger than the periodicity by only one or two orders of magnitude and this high frequency cutoff, when the homogenization becomes invalid, is easily accessed. Thus these relations should be applied cautiously to metamaterials keeping this in mind: if the effective medium theory itself cannot describe the system, the effective medium parameters obtained from the Kramers-Kronig relations are not meaningful.

1.3.2   Dispersive model for the dielectric permittivity

This section briefly presents a dispersive model for the dielectric permittivity that is due to H. A. Lorentz. The resulting expression is very general and it has been found that many metamaterials exhibit effective constitutive parameters in agreement with this law. As another example, amplifying media such as laser gain media, whose imaginary part of the permittivity is negative, are often modeled by a generalized Lorentz model where the oscillator strength is taken to be negative. The Lorentz model is also a good approximation to the density matrix equations of a weakly perturbed two-level quantum system.

Within this approximation, having a negative oscillator strength corresponds to a population inversion. Note that a similar discussion would hold for magnetizable media and the corresponding magnetic fields. The dispersion of the magnetic permeability in many magnetic materials also exhibits a Lorentz-like dispersion although the resonances usually occur at radio and microwave frequencies. Because of this fundamental importance and relevance to the specific field of metamaterials, we shall introduce the derivation of the Lorentz dispersion law here in order to make it familiar to the reader as well as to bring out its generic features.

The frequency dispersive nature of a medium is related to the polarizability of its basic units, *viz.*, the atoms and molecules. Although one can give an adequate description of dispersion only by a quantum mechanical treatment, a simplified description is possible by using only a few basic results concerning the properties of atoms and molecules. One starts by noting that an applied electric field causes charge separation of the positively charged nuclei and the negatively charged electrons in an atom or molecules. Thus a dipole moment is generated and to a good approximation dominates over the other multipole moments. The induced dipole moments can be determined by the displacements of the charges from their equilibrium positions. The atoms or molecules may additionally have a permanent dipole moment in which case the equilibrium positions of the positive and negative charges do not coincide (we may ignore the motion of the nuclei due to their comparatively large mass). The force on the electrons is given by the Lorentz force:

$$\mathbf{F} = -e(\mathbf{E} + \mathbf{v} \times \mathbf{B}), \tag{1.11}$$

where $\mathbf{v}$ is the velocity of the electrons. We can usually neglect the magnetic field effects as $|\mathbf{B}|/|\mathbf{E}| \sim 1/c$ and most of the speeds involved are non-relativistic.

The electron in an atom or molecule can be assumed to be bound to the equilibrium position through an elastic restoring force. Thus, if m is the mass of the electron, the equation of motion becomes

$$m\ddot{\mathbf{r}} + m\gamma\dot{\mathbf{r}} + m\omega_0^2\mathbf{r} = -e\mathbf{E_0}\exp(-i\omega t), \tag{1.12}$$

where $\mathbf{r}$ is the displacement vector, ω_0 is the resonance angular frequency characterizing the harmonic potential trapping the electron to the equilibrium position, and ω is the angular frequency of the light. Here $m\gamma\dot{\mathbf{r}}$ is a phenomenological damping (viscous) force on the electron due to all inelastic processes. This damping term is extremely important as the oscillating electrons radiate electromagnetic waves and energy, although they can also lose energy in several other manners including collisions.

Using a trial solution $\mathbf{r} = \mathbf{r_0}\exp(-i\omega t)$, the displacement of the electron is obtained as

$$\mathbf{r_0} = \frac{-e\mathbf{E_0}/m}{\omega_0^2 - \omega(\omega + i\gamma)}. \tag{1.13}$$

The dipole moment due to each electron is $\mathbf{p} = -e\mathbf{r}$ and the polarization, defined as the total dipole moment per unit volume, $\mathbf{P}$, is given by the vectorial sum of all the dipoles in the unit volume. Assuming one dipole per molecule and an average number density of N molecules per unit volume, one obtains

$$\mathbf{P} = N\mathbf{p} = \frac{Ne^2\mathbf{E}/m}{\omega_0^2 - \omega(\omega + i\gamma)} = \varepsilon_0 \chi_e \mathbf{E}, \qquad (1.14)$$

where χ_e is the dielectric susceptibility. Hence one can write for the dielectric permittivity

$$\varepsilon(\omega) = 1 + \chi_e(\omega) = 1 + \frac{Ne^2/m\varepsilon_0}{\omega_0^2 - \omega(\omega + i\gamma)}. \qquad (1.15)$$

The quantity $f^2 = Ne^2/m\varepsilon_0$ is often called the oscillator strength.

Eq. (1.15) is called the Lorentz formula for the dispersion of ε whose real and imaginary parts are plotted in Fig. 1.9. The imaginary part, $\text{Im}(\varepsilon)$, is seen to strongly peak at ω_0 and the full width at half maximum is determined by the levels of the dissipation parameter γ. The real part, $\text{Re}(\varepsilon)$, changes in a characteristic manner near ω_0 which is consistent with the Kramers-Kronig relations given by Eqs. (1.9).

One should note that the above discussion strictly holds only for a dilute gas of the polarizable objects. In a dense material medium with a much larger concentration, the fields that arise due to nearby polarized objects affect the polarization at any given point. These fields are known as *local fields* and the polarization that sets in the medium is proportional to the effective field, which is the vectorial sum of the applied field and the local fields.

Needless to say, the actual description of the local fields would be very complicated. On the other hand, in the spirit of homogenization, we can a think of each polarizable object to be within a small sphere surrounded by a uniformly polarized medium rather than being a set of discrete dipoles at various locations. Assuming the polarization outside to be a constant, $\mathbf{P}$, one obtains the effective field as**

$$\mathbf{E}' = \mathbf{E_{appl}} + \frac{\mathbf{P}}{3\varepsilon_0}. \qquad (1.16)$$

Thus, we would have to replace the applied electric field in Eq. (1.14) with the effective field. Note that the polarizability (α) of the polarizable object is defined by $\mathbf{p} = \varepsilon_0 \alpha \mathbf{E}'$, where $\mathbf{p}$ is the induced dipole moment so that the net polarization is expressed as $P = N\varepsilon_0 \alpha E'$. From Eq. (1.14) we can write

$$\alpha = \frac{e^2/\varepsilon_0 m}{\omega_0^2 - \omega(\omega + \gamma)}, \qquad (1.17)$$

**One uses the result that the field in a uniformly polarized sphere is $E = P/3\varepsilon_0$ in the quasi-static limit.

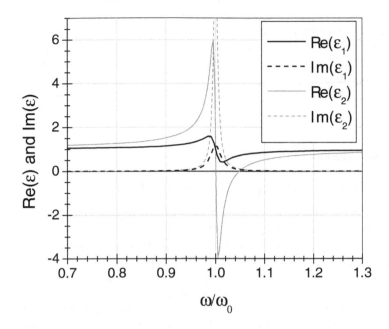

Figure 1.9 Real and imaginary parts of the dielectric permittivity predicted by the Lorentz model. The parameters for $\varepsilon_1(\omega)$ are $f_1^2 = 0.03\omega_0^2$ and $\gamma_1 = 0.025\omega_0$ and those for $\varepsilon_2(\omega)$ are $f_2^2 = 0.1\omega_0^2$ and $\gamma_1 = 0.01\omega_0$. Note that if the oscillator strength is strong and the dissipation is small enough, the real part of the permittivity can become negative at frequencies just above the resonance frequency as in the case of ε_2.

and the dielectric susceptibility that relates the polarization and the applied fields as

$$\chi_e = \frac{N\alpha}{1 - \frac{N\alpha}{3}}. \tag{1.18}$$

The dielectric permittivity thus takes the form

$$\varepsilon = 1 + \chi_e = \frac{1 + \frac{2N\alpha}{3}}{1 - \frac{N\alpha}{3}}, \tag{1.19}$$

where the local field corrections have been incorporated. This formula is known as the Lorentz-Lorenz formula after the two scientists who came to these conclusions independently and almost simultaneously. For static fields, an analogous result holds and is known as the Clausius-Mossotti relation for dielectrics.

Finally, we should point out that a crucial approximation made here is that the size of the polarizable objects (atoms and molecules) is very much smaller than the wavelength of radiation. This enabled us to treat all the polarizable objects in the volume as if subjected to the same field with no spatial variation (limit of infinite wavelength). The discussion, however, holds true even for more complicated but small polarizable objects, not just atoms and molecules, which is discussed subsequently.

1.4   Phase velocity and group velocity

Shortly after Maxwell introduced the concept of electromagnetic waves, he immediately went about calculating the velocity of these waves and realized that, for a single frequency and in vacuum, they were propagating at the velocity of light (which allowed him to make the connection between the field of electromagnetics and the field of optics). The concept of velocity is fundamental in the study of waves and signals since it provides information on how the wave evolves in space and time, and how fast information can be transfered from one point to another. Yet, one needs to be careful when assigning a physical significance to the various velocities that can be defined.

Let us first take the case of a monochromatic plane wave propagating in the $\hat{z}$ direction. In the time domain, the field is written as $E_y = E_0 \cos(kz - \omega t)$, where E_0 is the amplitude of the wave. For a propagating wave, we can track a point of constant phase and realize that it is traveling at a velocity

$$v_p = \frac{dz}{dt} = \frac{\omega}{k}. \tag{1.20}$$

Because of this definition, v_p is called the phase velocity. In the case of free-space, $k = \omega/c$ so that the phase front propagates at the velocity of light. In the case of a more general lossless non-dispersive medium, $k = \omega\sqrt{\varepsilon\mu}/c$ which is a linear function of frequency: the phase velocity is constant, typically the velocity of light in the medium. For yet more general dispersive media, the phase velocity is not a constant with frequency and the phase velocity can be typically larger than the speed of light in the medium. As we shall see subsequently, this does not violate the principle of special relativity since the phase velocity is not associated with a transport of energy, or more strictly, transmission of a signal. Nonetheless, in such a case, various components of a multi-frequency signal propagate at different velocities and cause a phase distortion.

All physical signals are composed of multiple frequencies, i.e., are spread over a certain frequency band. The spectrum of such a wave is never just a Dirac delta function. The assumption of monochromatic plane waves is therefore a theoretical idealization, whereas in the real world, the signal is typically composed of a slowly varying envelope confining a rapidly oscillating wave. The simplest multi-frequency signal is composed of two closely separated frequencies $\omega_0 \pm \Delta\omega$, where $\Delta\omega << \omega_0$, to which correspond the wave-numbers $k \pm \Delta k$. The superposition of the two waves is simply written as

$$
\begin{aligned}
E_y &= \cos\left[(k + \Delta k)z - (\omega + \Delta\omega)t\right] + \cos\left[(k - \Delta k)z - (\omega - \Delta\omega)t\right], \\
&= 2\cos(\Delta k z - \Delta\omega t)\cos(kz - \omega t).
\end{aligned}
\tag{1.21}
$$

Tracking the constant fronts of the two terms yields two velocities:

1. $kz - \omega t = $ constant yields the velocity of the rapidly oscillating wave, which is similar to the monochromatic case discussed previously:

$$
v_p = \frac{dz}{dt} = \frac{\omega}{k}.
\tag{1.22a}
$$

2. $\Delta k z - \Delta\omega t = $ constant yields the velocity of the envelope, called the group velocity:

$$
v_g = \frac{dz}{dt} = \frac{\Delta\omega}{\Delta k}.
\tag{1.22b}
$$

   Intuitively, the group velocity is seen to correspond to the velocity of the envelope or the packet, and corresponds to the velocity of propagation of the energy in many cases.

In the limit of a very narrow-band signal, $\Delta\omega \to 0$ and the group velocity is expressed as

$$
v_g = \frac{1}{\partial k/\partial\omega}.
\tag{1.23}
$$

We can also express the group velocity in terms of the phase velocity:

$$
\frac{1}{v_g} = \frac{1}{v_p} + \omega\frac{\partial}{\partial\omega}\left(\frac{1}{v_p}\right),
\tag{1.24}
$$

which indicates that if there is no frequency dispersion, $v_g = v_p$. In the case of normal dispersion, $\frac{\partial}{\partial\omega}(1/v_p) > 0$ so that $v_g < v_p$. We had mentioned above that v_p can be larger than the velocity of light inside the medium. It can easily be shown that v_g is in fact lower than this limit. Since v_g corresponds to the velocity at which information is carried, it is in compliance with the principle of relativity.

   In the case of anomalous dispersion relation, $\frac{\partial}{\partial\omega}(1/v_p) < 0$ so that $v_g > v_p$: the group velocity can be even larger than the speed of light in vacuum. In this case, however, the group velocity loses its meaning as signal velocity, which has to be defined in terms of the electromagnetic energy flow. This issue is discussed in greater detail in Section 6.5.

Finally, let us mention that the definition of the group velocity can be generalized to a vectorial relation as

$$\mathbf{v}_g = \nabla_k \omega. \tag{1.25}$$

This gradient relationship indicates that the direction of the group velocity is normal to the iso-frequency contour in the spectral domain. This property is extensively used in Sections 5.1 and 5.2 for example.

1.5   Metamaterials and homogenization procedure

1.5.1   General concepts

One of the crucial ideas in a homogenization procedure is that the wavelength of radiation is several times, preferably several orders of magnitude, larger than the underlying polarizable objects (such as atoms and molecules). In this case, the radiation is sufficiently *myopic* so as to not resolve the spatially fast varying structural details, but only responds to the macroscopic charge and current densities. Upon averaging in macroscopic measurements, the only remaining important parameters are the frequency-dependent polarization of the individual (atomic or molecular) oscillators driven by the applied fields.

We can apply this idea to a higher class of inhomogeneous materials, such as metamaterials, where the inhomogeneities in a host background are much smaller than the wavelength of radiation, but yet much larger than the "atoms" or "molecules" that the material is composed of. Such a meso-structure would also not be resolved by the incident radiation, and the structure could be driven and polarized or magnetized by applied electromagnetic fields. Particularly near the resonance frequencies (if any), the structures can have a large polarizability. An array of such structural units can then be characterized by macroscopic parameters such as ε and μ that effectively define its macroscopic response to exciting electromagnetic fields, much like in a homogeneous material.

Metamaterials, in some sense, can be strictly distinguished from other structured photonic materials such as photonic crystals or photonic bandgap materials. In the photonic crystals or bandgap materials the stop-bands or bandgaps arise as a result of multiple Bragg scattering in a periodic array of dielectric scatterers. In fact, the periodicity of the structure in these cases is of the order of the wavelength, and hence homogenization in the classical sense cannot be performed. In metamaterials, the periodicity is by comparison far less important (Chen et al. 2006a), and all the properties mainly depend on the single scatterer resonances. Alternatively, one notes that the small periodicity and small size of the structural units imply that all the corresponding

Bragg scattered waves are evanescent and bound to the single scatterer. Consequently, the properties of a metamaterial are not resulting from interference between waves scattered off different points. Instead, the radiation probes the polarizability of the structural units as it moves through the medium, interacting with the polarizable objects in the same manner as in a homogeneous medium. Note that in the limit of long wavelengths, the phase shifts for the wave across a single structural unit are negligibly small and all units interact with the radiation in a similar manner.

1.5.2 Negative effective medium parameters

As discussed in Section 1.3.2, there is a large amount of dispersion in the material parameters at frequencies near the resonance. Below the resonance, the polarization is in phase with the applied driving field, whereas it is π out of phase above resonance. If the dissipation is sufficiently small, the resonance can be made very sharp so as to drive the real parts of ε and μ even toward negative values when the corresponding driving fields are the electric and the magnetic fields, respectively. Of course, the imaginary parts of ε and μ are also large at the resonance frequency and its immediate vicinity.

Thus, negative real parts of the material parameters should be regarded as a natural outcome of an *underdamped* and *overscreened* response of a resonant medium. Fundamentally, there is no objection to negative real parts of $\varepsilon(\omega)$ or $\mu(\omega)$ as long as other physical criteria are also satisfied such as causality. The latter implies for example that the frequency dispersive models for the permittivity and the permeability cannot be arbitrary, but should yield constitutive parameters that satisfy the Kramers-Kronig relationship of Eqs. (1.9).

In order to better understand the effect of negative material parameters, consider an isotropic medium where the $\text{Im}[\varepsilon(\omega)] \sim \text{Im}[\mu(\omega)] \simeq 0$, i.e., dissipation is assumed negligibly small at some frequencies (this would typically be a good approximation at frequencies somewhat away from the resonant frequency). We can conveniently characterize most electromagnetic materials by the quadrant where they lie in the complex $(\text{Re}(\varepsilon) - \text{Re}(\mu))$ plane as shown in Fig. 1.10.

Quadrant 1: This is the realm of usual optical materials with $\text{Re}(\varepsilon) > 0$ and $\text{Re}(\mu) > 0$. Electromagnetic radiation can propagate through these media and the vectors **E**, **H**, and **k** form a right-handed triad.

Quadrant 2: The usual form of matter that has $\text{Re}(\varepsilon) < 0$ and $\text{Re}(\mu) > 0$ is a plasma of electric charges. It is well known that a plasma screens the interior of a region from electromagnetic radiation. Indeed, all electromagnetic waves are evanescent inside a plasma and no propagating modes are allowed. This is directly expressed by the constitutive relation, which reduces to

$$\mathbf{k} \cdot \mathbf{k} = \varepsilon\mu\omega^2/c^2 < 0 \qquad (1.26)$$

for a plane wave $\exp[i(\mathbf{k} \cdot \mathbf{r} - \omega t)]$. Inside such a negative dielectric medium, no real solutions for the wave vector are possible. Note that dielectric materials can also exhibit Lorentz dispersion near an excitonic or optical phonon resonance and $\text{Re}(\varepsilon) < 0$ over a small frequency band above the resonance frequency.

Quadrant 4: This quadrant is the dual of quadrant 2, with $\text{Re}(\varepsilon) > 0$ and $\text{Re}(\mu) < 0$. Here, too, a wave incident on a medium of this family decays evanescently within the medium and no propagating modes are sustained. Due to the absence of magnetic monopoles, there can be no exact analogue of an electric plasma but there are natural examples of some antiferromagnetic and ferrimagnetic materials with a resonance at microwave frequencies that exhibit $\text{Re}(\mu) < 0$ in a frequency band above the resonance frequency.

Quadrant 3: This is the quadrant of primary interest in this book, directly related to the concept of metamaterials. The properties $\text{Re}(\varepsilon) < 0$ and $\text{Re}(\mu) < 0$ yield a dispersion condition that allows a real wave-vector in the medium, i.e., waves are propagating. Consider the Maxwell equations for a time-harmonic plane wave in the medium:

$$\mathbf{k} \times \mathbf{E} = \omega \mu_0 \mu \mathbf{H}, \tag{1.27a}$$

$$\mathbf{k} \times \mathbf{H} = -\omega \varepsilon_0 \varepsilon \mathbf{E}. \tag{1.27b}$$

Since $\varepsilon < 0$ and $\mu < 0$, it is clear that the vectors $\mathbf{E}$, $\mathbf{H}$, and $\mathbf{k}$ form a left-handed triad, which is the property that has given to these media their first very popular (and historical) name of *left-handed media*.[††] However, the triad of the vectors $\mathbf{E}$, $\mathbf{H}$, and the Poynting vector $\mathbf{S}$ still remains right-handed as $\mathbf{S} = \mathbf{E} \times \mathbf{H}$. This indicates that the Poynting vector and the wave-vector are antiparallel. This fundamental characteristic indicates that the left-handed media support backward waves, and provides a heuristic argument for declaring that such media have a negative refractive index. As a matter of fact, the refractive index may be defined by $\mathbf{k} = \hat{S} n \omega / c$, where $\hat{S}$ is the unit vector along the energy flow. Since $\mathbf{k}$ and $\hat{S}$ are in opposite directions, it can be inferred that $n < 0$. A rigorous explanation of this choice of the negative sign for the square root $n = -\sqrt{\varepsilon \mu}$ involves the consideration of causal boundary conditions and dissipative media, whose discussion is postponed to Chapter 5.

1.5.2.1 Terminology

We end this chapter by briefly touching upon the terminology used in the literature and in this book for media with $\text{Re}(\varepsilon) < 0$ and $\text{Re}(\mu) < 0$.

Historically, the first name was given in Veselago (1968): *left-handed media* (LHM). As mentioned before, this terminology refers to the left-handed triad formed by the vectors $\mathbf{E}$, $\mathbf{H}$, and $\mathbf{k}$. Some authors, however, criticized this name because of the confusion it may induce with the optical properties of

---

[††]Note that this terminology is not related to the polarization state of the wave or to the chirality of the medium, but only to the triad of the three vectors $\mathbf{E}$, $\mathbf{H}$, and $\mathbf{k}$.

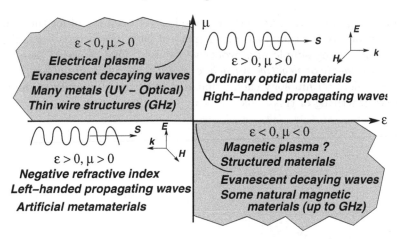

Figure 1.10 Schematic showing the classification of materials based on their dielectric and magnetic properties. The wavy lines represent materials that allow propagating waves, and the axes set in quadrants 1 and 3 show the right and left-handed nature of $\mathbf{E}, \mathbf{H}, \mathbf{k}$ vectors. The waves in quadrants 2 and 4 decay evanescently inside the materials and are schematically depicted. (Reproduced with permission from Ramakrishna (2005). © 2005, Institute of Physics Publishing, U.K.)

chiral materials, or even with the polarization state of an electromagnetic wave (right-handed or left-handed circularly polarized). In the frame of this book, however, chirality is only briefly mentioned when necessary, and therefore such confusion is very unlikely. We therefore use the historical terminology of left-handed media with consistency and without ambiguity to refer to media that have simultaneously a negative permittivity and a negative permeability.

A second very popular name for these media directly refers the important consequence of negative refraction: *negative refractive media* (NRM). Although the concept of negative refraction is more reductive than that of a simultaneously negative permittivity and permeability, this terminology has the appeal of referring to one of the most interesting physical properties of these new media so that we shall use it in this book as well.

A few other names have also been proposed in the literature. For example, following Ziolkowski and Heyman (2001), many authors prefer calling them *double negative* media, as opposed to *double positive* media when $\text{Re}(\varepsilon) > 0$ and $\text{Re}(\mu) > 0$, *single negative electric* media when $\text{Re}(\varepsilon) < 0$ and $\text{Re}(\mu) > 0$, or *single negative magnetic* media when $\text{Re}(\varepsilon) > 0$ and $\text{Re}(\mu) < 0$. Other authors have preferred to call them *backward media* (Lindell et al. 2001), which describes the negative direction of the phase vector. Another name is *negative phase velocity media* (NPVM) (McCall et al. 2002), also referring to one of the consequences of a simultaneously negative permittivity and permeability.

Our standpoint is that no perfect name has yet been found for these media: a name that would encompass all or most of their characteristics and properties, while being simple and evocative enough. Our choice of using either *left-handed media* or *negative refractive media* is therefore imperfect, but constitutes a mere terminology choice that had to be made in order to undertake the task of writing of this book.

2

Metamaterials and homogenization of composites

The description of a metamaterial as a homogeneous medium involves averaging over the fluctuations of the electromagnetic fields at two levels. As explained in Section 1.3, the macroscopic Maxwell equations are obtained by averaging the rapidly fluctuating electromagnetic fields at atomic or molecular lengthscales over volumes that contain enough number of polarizable or magnetizable atoms/molecules. Within this framework, susceptibilities for bulk materials can be defined.. In the case of metamaterials, the structural units of the metamaterial (see Chapter 3) are assumed to be sufficiently large on a molecular scale so that they can confidently be described by their bulk dielectric permittivity and magnetic permeability, and yet sufficiently small compared with the lengthscales over which the applied fields vary (typically a wavelength). Hence, only the fields due to the first few multipoles of the charge and current distributions induced in the structures contribute to the macroscopic polarization over lengthscales large compared to the metamaterial units. In other words, the fine structure of the charge and current distributions over the structural units is not discernible, but only a few averages such as the corresponding dipolar fields or (rarely) the quadrupolar fields can be resolved through the macroscopic polarization and magnetization. These average quantities determine the effective dielectric permittivity and the magnetic permeability tensors of the *bulk metamaterial*.

The metamaterials this book is devoted to are usually composed of metallic rods, sheets, and rings, usually but not necessarily organized in a periodic fashion. When it can be defined, a unit cell of such metamaterial is typically smaller than the exciting electromagnetic wavelength λ, usually on the order of $\lambda/5$ to $\lambda/10$, but sometimes as small as $\lambda/10000$ as in the case of Swiss roll metamaterial (see Chapter 3). The homogenization question thus arises naturally: inasmuch as dielectric materials composed of atoms can be characterized by an effective dielectric constant at microwave and optical frequencies, do the present metamaterials, discrete in their structure, behave as continuous media in the long wavelength regime? If yes, how large should *large* be before the homogenization approximation breaks down? How do we determine the values of the homogenized parameters? What is their range of validity?

More specifically to the realm of metamaterials, the question is posed in the following terms: given a medium of rings and rods occupying a certain region of space and given an incident electromagnetic radiation, can we find an effective medium within some spatial boundaries (to be determined) that presents the same scattered fields as the original medium? In addition, in order to be a real effective medium, we have to require that the homogenized parameters be independent of the size of the bulk of the metamaterial and independent on the direction of the incident wave, although the effective parameters can be anisotropic and spatially dispersive.

The homogenization analysis is most conveniently carried out on slabs of metamaterials, infinite along the transverse directions ($\hat{x}$ and $\hat{y}$) and finite along the normal ($\hat{z}$) direction, which we assume to be the direction of propagation of the wave. In addition, the criterion for homogenization can be translated into properties of the reflection and transmission coefficients, so that the homogenization procedure can also be rephrased as follows: can we find an equivalent homogeneous slab of effective parameters between two boundaries z_1 and z_2 such that the reflection and transmission coefficients measured from the metamaterial and the homogeneous medium are identical across all frequencies for various slab thicknesses and incident directions? From the microscopic point of view, the criterion is stated as follows: can we replace the metamaterial with a homogeneous medium of specified material parameters so that the fields in the structured metamaterial, when averaged over some reasonable volume, correspond to the fields in the homogeneous medium? These are the questions addressed in the present chapter, from both a physical and mathematical point of view.

2.1   The homogenization hypothesis

Materials composed of small elements are known to respond as continuous media when the operating wavelength is much larger than the individual constituents. A classical example of such materials are natural dielectrics that can be described by a single parameter, the electric permittivity ε. All the negative and positive charges in a dielectric medium are bound to their location by atomic forces and are therefore not free to move like in a conductor. Under the influence of an external electric field, however, these assemblies of negative and positive charges may slightly reorganize, which results in the creation of bound electric dipoles. From a macroscopic point of view, the orientation of these dipoles generates a polarization vector $\mathbf{P}$ that influences the electric flux density $\mathbf{D}$ such as $\mathbf{D} = \varepsilon_0 \mathbf{E} + \mathbf{P}$, where $\mathbf{E}$ is the external applied electric field. This allows one to define a general permittivity ε such that $\mathbf{D} = \varepsilon_0 \varepsilon \mathbf{E}$, where naturally ε is defined in terms of the polarization vector. Similarly, magnetic media are described by a magnetic permeability μ, and ε and μ represent the constitutive parameters essential to the macroscopic Maxwell equations. In our case, the metamaterials are composite structures designed to exhibit specific electromagnetic properties at some particular wavelengths that are much larger than the elementary constituents. It is therefore legitimate to look for homogeneous or effective medium parameters, typically an effective permittivity and an effective permeability.

Like in the case of more standard media, these constitutive parameters directly represent the properties of the medium: isotropic metamaterials should be described by scalar constitutive parameters while anisotropic ones should be described by second rank tensors $\bar{\bar{\varepsilon}}$ and $\bar{\bar{\mu}}$, losses induce an imaginary part to these parameters, frequency dispersion yields frequency-dependent parameters, non-locality makes them spatially dispersive (dependent on k), the passive nature of metamaterials forces the imaginary parts to be positive, reciprocity imposes conditions on the tensors, etc.

The homogenization procedure therefore involves two steps:

1. A hypothesis, more or less refined, on the characteristics of the medium (isotropic vs. anisotropic, lossless vs. lossy, etc). Paradoxically, the properties of the metamaterial are initially unknown, and yet a model has to be chosen to perform the homogenization procedure. Therefore, the model should also be tested *a posteriori*.

2. The determination (or retrieval) of the corresponding constitutive parameters, in their tensorial form in the most general case, using numerical or analytical algorithms.

The first aspect is as important as the second and should not be overlooked, since no matter how elaborate the retrieval algorithm, it is going to be bound by the initial assumptions. For example, various algorithms have been proposed to retrieve scalar ε and μ for metamaterials, but this does not imply that the metamaterial under study is actually isotropic. This only reflects the nature of the assumption: that the metamaterial is *supposed to be* isotropic and that one is interested in retrieving scalar constitutive relations. If, however, the metamaterial is in reality anisotropic, mapping the constitutive tensors onto scalar parameters can induce various artifacts that do not represent the physical reality. Hence, before drawing physical conclusions from retrieved parameters, one must be sure that the *a priori* model is valid. Various physical arguments can be used to that effect.

Once the overall model has been ascertained, the next step is to obviously determine the numerical values of the components of the permittivity and permeability tensors. The number of unknowns is directly related to the model chosen and dictates the number of equations that must be obtained from measurements (either experimental measurements or numerical simulations). For example, a simple lossy isotropic dielectric material exhibits two unknowns, the real and imaginary parts of the scalar permittivity, which can be obtained from one complex measurement such as the reflection and/or the transmission coefficients. The case presented by metamaterials is slightly more complex since these media are known to have a non-unity magnetic permeability, thus requiring more measurements. In addition, the permittivity and the permeability are frequency dispersive, and hence need to be determined at each frequency.

We should also note that there are more mathematically rigorous homogenization theories that analyze the fields as an asymptotic expansion in terms of the microscopic lengthscale associated with the inhomogeneities. Two scales characterize the variation of the fields inside a metamaterial: one, macroscopic, describes the fields over the bulk metamaterial, while a second one, microscopic, describes the fields over small lengthscales inside a unit. The electromagnetic fields can be written as $E(x, y)$ where the fields have a slow variation with the variable x representing the changing in the fields from unit to unit, and a fast variation with the variable y which represents the variation within the units. The fast variable can be taken to be $y = x/\xi$ where $\xi = a/\lambda$, the ratio of the unit cell size to the wavelength of radiation. The principle of the asymptotic expansion is to expand the field in a series involving powers of the fast variable and retain only the leading order terms in the limit of $\xi \to 0$. A rigorous demonstration of this method is out of the scope of this book and we refer the interested reader to Milton (2002) for an excellent treatment of this topic.

Two methods can usually be employed in order to retrieve the constitutive parameters. The first one is based on a set of measurements of the external emergent quantities, *viz.* the frequency-dependent reflection and transmission coefficients (the total number of measurements depends on how much information is needed to unambiguously determine all the unknowns in the problem). Note that these quantities are typically measured in experiments. This is the method that has been used historically for the retrieval of the permittivity of standard dielectrics, and needs to be slightly revisited in the case of metamaterials. We devote the next couple of sections to this topic. The second method for the retrieval of the constitutive parameters is based on the knowledge of the electromagnetic fields inside the metamaterial, i.e., the internal fields. This method is thus better suited for numerical simulations as the internal fields are not easily experimentally accessible, but some experimental approaches to measure these fields have been proposed as well. More details are given in Section 2.5.

2.2 Limitations and consistency conditions

Before detailing the homogenization procedures per se, it is necessary to say a few words about the limitations of homogenization. As a matter of fact, like we have previously outlined, executing an algorithm and obtaining numerical values for, say, a scalar permittivity and permeability does not represent a proof of homogenization (the algorithm is indeed *designed* to output numerical values, and would do so as soon as it is given some inputs, even if these inputs do not have a physical justification). Although we are interested in homogenizing metamaterials, the range of applicability of such process must be carefully examined. In particular, homogenization would not be valid under the following conditions:

1. When the constituents are not much smaller than the operating wavelength. This includes metamaterials with unit cells large or comparable to the wavelength of radiation, as well as photonic crystals.* In this case, the periodicity of the structure (if it exists) may become an important factor that affects the retrieved values of the permittivity and permeability.

2. When the wave propagation inside the material cannot be described by a single propagating mode. Such situation occurs for example close to the resonance of some constituents of the metamaterials, in photonic crystals when multiple Bragg diffraction modes need to be included, or simply in multi-mode waveguides.

Failing to comply with these conditions may produce unphysical artifacts in the frequency-dependent retrieved parameters, the most common of which are an anti-resonance of the permittivity (with an associated negative imaginary part) at the location of the resonance of the permeability, and possibly a truncated resonance of the index of refraction as pointed out in Koschny et al. (2005). These artifacts have been suggested to be due to the fact that the periodicity of the medium becomes visible at the corresponding frequencies, making the homogenization hypothesis less accurate. A conservative ratio of about 30 between the wavelength and the size of the unit cell has been proposed for the homogenization hypothesis to be very well justified, although most of the metamaterials realized to date exhibit a ratio of about 10 at best.

Additionally, an important test of homogenization is to verify that different methods yield comparable values for the effective medium parameters. This is

---

*More details on photonic crystals are provided in Chapter 4. For the purpose of the present discussion, it is enough to know that photonic crystals are dielectric structures exhibiting a periodicity close to half wavelength, and thus operate in a Bragg diffraction regime.

particularly true for comparison between those methods that utilize the external emergent quantities and those methods that use the internal fields. Note also that the retrieved effective medium parameters should be independent of the particular location of the volume over which the fields are averaged. Equivalently, in methods where the reflection and transmission coefficients are utilized, the retrieved effective medium parameters should be independent of the angle of incidence. In many metamaterials, when the structures are not substantially smaller than the wavelength of light, these conditions can be violated. One should note that even in these cases, it might be possible to define more restricted *equivalent medium parameters* that are specific to a set of angles of incidence and so on. However, the *equivalent medium parameters* cannot be consistently used for the calculations of other properties such as the rate of dissipation in the medium or the scattering property of another object embedded in the metamaterial. An example of such a restricted effective medium theory are the extensions of the Maxwell-Garnett theory based on the Mie scattering coefficients discussed in Section 2.5.2.

Finally, it has been argued (Belov et al. 2003) that spatial dispersion phenomena may occur in some lattice-based structures and should be incorporated in the effective medium model. This non-locality of the constitutive relations, usually apparent in the small wavelength limit, has been found in the large wavelength limit for a metamaterial composed of a series of parallel conducting wires when the incident wave-vector exhibits a component parallel to the axis of the wires. Although our subsequent in-plane incidences ensure that non-local effects are negligible, it is important to keep this effect in mind and accounted for in the treatment of general oblique incidences on metamaterial structures.

2.3   Forward problem

Once a model for the constitutive parameters has been physically justified, enough information needs to be gathered in order to determine the unknown values that populate the constitutive tensors. Traditionally, one method has been overwhelmingly used, which is based on the measurement of the complex reflection and transmission coefficients. This method is appealing because of its relative simplicity to set up in an experimental configuration, as well as its numerical efficiency since the electromagnetic fields need only to be computed over surfaces instead of volumes. In this procedure, the reflection and transmission coefficients (R and T) are used as the linking information between the incident electromagnetic field and the constitutive parameters, the latter relationship being essentially the purpose of inversion or parameter retrieval algorithms.

2.3.1 Relation between R and T and the electromagnetic fields

The first step is to determine the relationship between the incident, reflected, and transmitted fields and the reflection/transmission coefficients. In order to comply to the previous hypothesis, a medium is supposed to be of infinite extent in the transverse $\hat{x}$ and $\hat{y}$ directions and of finite extent in the propagation direction $\hat{z}$. The single propagating mode here is taken to be a plane wave, equivalent to the TEM mode in a parallel plate waveguide. The incident field propagating in the $-\hat{z}$ direction can therefore be written as

$$\mathbf{E}_{inc}(x, y, z) = \hat{e} E_0 \, e^{i\mathbf{k}_i \cdot \mathbf{r}}, \tag{2.1}$$

where $\hat{e}$ is the polarization of the incident field in the xy plane, $\mathbf{k}_i = \hat{x} k_x + \hat{y} k_y - \hat{z} k_z$ is the incident wave-vector of amplitude k, and $\mathbf{r} = \hat{x} x + \hat{y} y + \hat{z} z$ is the position vector. A homogeneous effective medium is sought between the boundaries $z = z_1$ and $z = z_2$, thus defining an incident and transmitted region. In the incident region, the total electric field at $z = z_1$ is given by

$$
\begin{aligned}
E_{tot}(x, y, z_1) &= E_0 \, e^{ik_x x + ik_y y} \left[e^{-ik_z z_1} + R \, E_0 \, e^{ik_z z_1} \right] \\
&= E_{inc}(x, y, z_1) + R \, E_0 \, e^{ik_x x + ik_y y} \, e^{ik_z z_1}, \tag{2.2}
\end{aligned}
$$

where R is the reflection coefficient at the first boundary. From this relation, R is obtained as

$$R = \frac{E_{tot}(x, y, z_1) - E_{inc}(x, y, z_1)}{E_0 \, e^{ik_x x + ik_y y} \, e^{ik_z z_1}} = \frac{E_{scat}(x, y, z_1)}{E_{inc}(x, y, z_1)} \, e^{-2ik_z z_1}. \tag{2.3}$$

In the transmitted region, the total field at $z = z_2$ is given by

$$E_{tot}(x, y, z_2) = T \, E_0 \, e^{ik_x x + ik_y y} \, e^{-ik_z z_2}, \tag{2.4}$$

where T is the transmission coefficient given by

$$T = \frac{E_{tot}(x, y, z_2)}{E_{inc}(x, y, z_2)}. \tag{2.5}$$

Consequently, once the incident and total electric fields are known at $z = z_1$ and $z = z_2$ (either from measurements or numerical simulations), the reflection and transmission coefficients are completely specified.

2.3.2 Determining the electromagnetic fields

A typical configuration is illustrated in Fig. 2.1 for the particular case of a periodic medium: Fig. 2.1(a) shows a material composed of a succession of unit cells that are much smaller than the wavelength (the elements are for the moment supposed to be non-resonant), while Fig. 2.1(b) shows the

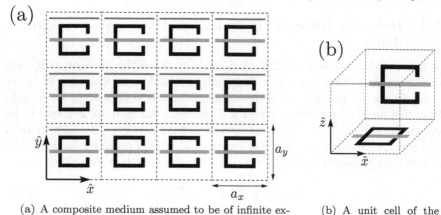

(a) A composite medium assumed to be of infinite extent along the $\hat{x}$ and $\hat{y}$ directions.

(b) A unit cell of the composite medium.

Figure 2.1 Schematic illustration of a periodic medium composed of split rings and rods.

details of a unit cell in three dimensions. In practice, the material should contain many unit cells (the overall size of the material should be of a few wavelengths), while in simulation, the assumption of an infinite medium in the lateral directions ($\hat{x}$ and $\hat{y}$ in our case) is often excellent and considerably reduces the computational requirements. Hence, Fig. 2.1(b) illustrates a unit cell that needs to be analyzed with the proper boundary conditions along the $\hat{x}$ and $\hat{y}$ directions. The material is of finite extent in the $\hat{z}$ direction, which is the principal direction of propagation of the electromagnetic radiation.

The computation of the reflection and transmission coefficients in Eqs. (2.3) −(2.5) requires the knowledge of the scattered electromagnetic fields. Various numerical algorithms are available toward this end, some of the most common ones being the Finite-Difference Time-Domain Method (FDTD), Finite Element Method (FEM), the Transfer Matrix Method, or an integral equation method such as the Method of Moments. The first three methods have the appeal of an extreme generality and mathematical simplicity: they have been applied for many years to solve a plethora of electromagnetic problems, and are well documented in various references. We shall therefore not contribute to an additional description of these methods but refer the reader to references such as Silvester and Ferrari (1996), Taflove and Hagness (2005). The integral equation approach, on the other hand, is usually less commented so that we provide subsequently some details on how to apply it to periodic metamaterials composed of metallic elements. The expert, the uninitiated reader or a reader not interested in the details of the method can therefore skip to the last paragraph of this section.

The integral equation approach is typically mathematically more involved than methods such as the FDTD or the FEM, but often results in reduced computational times because most of the structure's complexity can be embedded in the kernel of the method, the Green's function. Eqs. (2.3) and (2.5) indicate that it is necessary to know the electric field at various locations in order to compute the reflection and transmission coefficients. The electric field is obtained from the electric sources (magnetic sources can be added by duality) as

$$\mathbf{E}(\mathbf{r}) = i\omega\mu \iiint d\mathbf{r}' \, \bar{\bar{G}}(\mathbf{r}, \mathbf{r}') \cdot \mathbf{J}(\mathbf{r}'), \tag{2.6}$$

where $\mathbf{r}$ denotes the observation position while $\mathbf{r}'$ denotes the source position, $\bar{\bar{G}}$ is the dyadic Green's function and $\mathbf{J}$ represents the current sources that can be expressed in a similar manner as the electric field:

$$\mathbf{J}(\mathbf{r}) = \iiint d\mathbf{r}' \, \delta(\mathbf{r} - \mathbf{r}') \, \mathbf{J}(\mathbf{r}'). \tag{2.7}$$

The triple integral in Eq. (2.6) is defined over the primed coordinates, i.e., over the volume where the sources are defined. In the case of metamaterials for example, the sources are typically surface currents induced on metallic structures (the geometrical units composing typical metamaterials are presented in detail in Chapter 3), so that the three-fold integral reduces to a two-fold integral over the surface of the metallizations.

Combining these equations with the vectorial wave equation for the electric field yields

$$\nabla \times \nabla \times \mathbf{E}(\mathbf{r}) - k^2 \mathbf{E}(\mathbf{r}) = i\omega\mu \mathbf{J}(\mathbf{r}), \tag{2.8}$$

where k is the wavenumber. It can be directly shown that the Green's function satisfies the relation (Tai 1993)

$$\nabla \times \nabla \times \bar{\bar{G}}(\mathbf{r}, \mathbf{r}') - k^2 \bar{\bar{G}}(\mathbf{r}, \mathbf{r}') = \bar{\bar{I}}\delta(\mathbf{r} - \mathbf{r}'). \tag{2.9}$$

Finally, the dyadic Green's function itself can be expressed in terms of a scalar Green's function $g(\mathbf{r}, \mathbf{r}')$ as

$$\bar{\bar{G}}(\mathbf{r}, \mathbf{r}') = \left(\bar{\bar{I}} + \frac{1}{k^2} \nabla\nabla \right) g(\mathbf{r}, \mathbf{r}'). \tag{2.10}$$

The problem of computing the electric field is therefore reduced to the problem of finding the scalar Green's function g and walking the previous equations backward: from g determining $\bar{\bar{G}}$ using Eq. (2.10), and subsequently determining the electric field $\mathbf{E}$ using Eq. (2.6) knowing the sources in the problem. This last step is usually not straightforward and requires the use of a numerical method such as the Method of Moments (Harrington 1993). In the case of a free-space background, the scalar Green's function is simply given by

$$g(\mathbf{r}, \mathbf{r}') = \frac{e^{ik|\mathbf{r} - \mathbf{r}'|}}{4\pi|\mathbf{r} - \mathbf{r}'|}. \tag{2.11}$$

The appeal of an integral equation-based approach is to reduce the numerical burden (typically to solve Eq. (2.6) like already mentioned) by incorporating as much information analytically into the Green's function as possible. In the case of a periodic metamaterial for example, the periodicity can be incorporated into the expression of g, thus enabling the numerical method to be applied only to a unit cell rather than to the entire structure (note that a similar technique can be applied within the FDTD or FEM, upon applying the proper boundary conditions at the edges of the unit cell). Upon defining a_x and a_y to be the dimensions of the unit cell in the $\hat{x}$ and $\hat{y}$ directions, respectively, the periodic scalar Green's function is immediately obtained from the Bloch theorem as

$$g(\mathbf{r}'') = \sum_{n_1, n_2} \exp(i\mathbf{k} \cdot \mathbf{R}) \frac{e^{ik_0|\mathbf{r}'' - \mathbf{R}|}}{4\pi |\mathbf{r}'' - \mathbf{R}|}, \tag{2.12}$$

where $\mathbf{r}'' = \mathbf{r} - \mathbf{r}'$ and

$$\mathbf{R} = n_1 \mathbf{a}_x + n_2 \mathbf{a}_y \tag{2.13}$$

is a vector of the lattice, n_1 and n_2 being two integers. In the case of a square lattice for example, $\mathbf{a}_x = \hat{x} a_x$, $\mathbf{a}_y = \hat{y} a_y$, and the Green's function is written as

$$g(x'', y'', z'') \sum_{n_1, n_2} e^{i(k_x n_1 a_x + k_y n_2 a_y)} \frac{e^{ik_0 R_{n_1 n_2}}}{4\pi R_{n_1 n_2}}, \tag{2.14a}$$

where $x'' = x - x'$, $y'' = y - y'$, $z'' = z - z'$ and where

$$R_{n_1 n_2} = \sqrt{(x - x' - n_1 a_x)^2 + (y - y' - n_2 a_y)^2 + (z - z')^2}$$
$$= \sqrt{s(1)^2 + s(2)^2 + s(3)^2}. \tag{2.14b}$$

For numerical purposes, the sum in Eq. (2.14a) is truncated at an upper limit N_s which needs to be determined from convergence considerations. These considerations reveal that Eq. (2.14a) is slowly convergent, making it somewhat inefficient to calculate the Green's function (Tsang et al. 2000b). An alternative approach is to convert Eq. (2.14a) into the spectral domain. Upon doing so, the expression becomes

$$g(x'', y'', z'') = \frac{i}{2\Omega} \sum_{l_1, l_2} \frac{1}{k_{z l_1 l_2}} e^{i\left[(k_x + \frac{2\pi l_1}{a_x}) x'' + (k_y + \frac{2\pi l_2}{a_y}) y''\right]} e^{ik_{z l_1 l_2}|z''|}, \tag{2.15a}$$

where

$$k_{z l_1 l_2} = \sqrt{k_0^2 - \left(k_x + \frac{2\pi l_1}{a_x}\right)^2 - \left(k_y + \frac{2\pi l_2}{a_y}\right)^2} \tag{2.15b}$$

with $\mathrm{Im}\{k_{z l_1 l_2}\} > 0$ and $\Omega = a_x a_y$ is the surface of a periodic unit. The numerical evaluation of this expression shows already a much faster convergence rate than Eq. (2.14a). This rate of convergence can still be increased by using a method proposed by Ewald (Ewald 1921). The latter starts by splitting the scalar Green's function into two components:

$$g(x'', y'', z'') = g_1(x'', y'', z'') + g_2(x'', y'', z''), \qquad (2.16a)$$

where

$$g_1(x'', y'', z'') = \frac{i}{4\Omega} \sum_{l_1=-N_1}^{N_1} \sum_{l_2=-N_1}^{N_1} \frac{1}{k_{zl_1l_2}} e^{i\left[(k_x + \frac{2\pi l_1}{a_x})x'' + (k_y + \frac{2\pi l_2}{a_y})y''\right]}$$

$$\left\{ e^{ik_{zl_1l_2}z''} \operatorname{erfc}\left(-\frac{ik_{zl_1l_2}}{2E} - Ez''\right) + e^{-ik_{zl_1l_2}z''} \operatorname{erfc}\left(-\frac{ik_{zl_1l_2}}{2E} + Ez''\right) \right\},$$
$$(2.16b)$$

$$g_2(x'', y'', z'') = \sum_{n_1=-N_2}^{N_2} \sum_{n_2=-N_2}^{N_2} \frac{1}{8\pi R_{n_1n_2}} e^{i(k_x n_1 a_x + k_y n_2 a_y)},$$

$$\left\{ e^{ik_0 R_{n_1n_2}} \operatorname{erfc}\left(R_{n_1n_2}E + \frac{ik_0}{2E}\right) + e^{-ik_0 R_{n_1n_2}} \operatorname{erfc}\left(R_{n_1n_2}E - \frac{ik_0}{2E}\right) \right\},$$
$$(2.16c)$$

where E is Ewald's parameter chosen as $E = \sqrt{\pi/\Omega}$ (Tsang et al. 2000b).

The expression of both g_1 and g_2 can be simplified if the parameter $k_{zl_1l_2}$ is real (for g_1) and if k_0 is real (for g_2). The simplifications use the properties of the error and complementary error functions:

$$\operatorname{erf}(a+ib) \doteq A + iB \qquad \operatorname{erfc}(a+ib) = 1 - A - iB \doteq C + iD \quad (2.17a)$$
$$\operatorname{erf}(a-ib) = A - iB \qquad \operatorname{erfc}(a-ib) = 1 - A + iB = C - iD \quad (2.17b)$$
$$\operatorname{erf}(-a+ib) = -A + iB \quad \operatorname{erfc}(-a+ib) = 1 + A - iB = (2-C) + iD$$
$$(2.17c)$$

In this case:

- If $k_{zl_1l_2}$ is real:

$$e^{ik_{zl_1l_2}z''} \operatorname{erfc}\left(-\frac{ik_{zl_1l_2}}{2E} - Ez''\right) + e^{-ik_{zl_1l_2}z''} \operatorname{erfc}\left(-\frac{ik_{zl_1l_2}}{2E} + Ez''\right)$$

$$= e^{ik_{zl_1l_2}z''} + e^{-ik_{zl_1l_2}z''} + 2i\operatorname{Im} e^{ik_{zl_1l_2}z''} \operatorname{erf}\left(Ez'' + \frac{ik_{zl_1l_2}}{2E}\right)$$

$$= 2e^{ik_{zl_1l_2}z''} - 2i\operatorname{Im} e^{ik_{zl_1l_2}z''} \operatorname{erfc}\left(Ez'' + \frac{ik_{zl_1l_2}}{2E}\right). \qquad (2.18a)$$

- If k_0 is real (i.e., when the background is lossless):

$$e^{ik_0 R_{n_1n_2}} \operatorname{erfc}\left(R_{n_1n_2}E + \frac{ik_0}{2E}\right) + e^{-ik_0 R_{n_1n_2}} \operatorname{erfc}\left(R_{n_1n_2}E - \frac{ik_0}{2E}\right)$$

$$= 2\operatorname{Re} e^{ik_0 R_{n_1n_2}} \operatorname{erfc}\left(R_{n_1n_2}E + \frac{ik_0}{2E}\right). \qquad (2.19)$$

Once the scalar Green's function is determined, the tensorial form of the Green's function needs to be evaluated from Eq. (2.10). Subsequently, we detail this procedure for the Ewald method only since it is the one that is the computationally most effective.

We shall first define $\tilde{g}_1$ as

$$\tilde{g}_1 = e^{ik_x x'' + ik_y y''} \left[e^{ik_{zl_1l_2} z''} \operatorname{erfc}\left(-\frac{ik_{zl_1l_2}}{2E} - Ez'' \right) \right.$$
$$\left. + e^{-ik_{zl_1l_2} z''} \operatorname{erfc}\left(-\frac{ik_{zl_1l_2}}{2E} + Ez'' \right) \right], \qquad (2.20)$$

where $k_x = k_{ix} + 2\pi l_1/a_x$ and $k_y = k_{iy} + 2\pi l_2/a_y$. Then

$$\frac{\partial \tilde{g}_1}{\partial x} = ik_x \tilde{g}_1 \qquad (2.21)$$

$$\frac{\partial \tilde{g}_1}{\partial y} = ik_y \tilde{g}_1 \qquad (2.22)$$

$$\frac{\partial \tilde{g}_1}{\partial z} = ik_z e^{ik_x x + ik_y y} \left[e^{ik_{zl_1l_2} z} \operatorname{erfc}(-\frac{ik_{zl_1l_2}}{2E} - Ez'')\right.$$
$$\left. - e^{-ik_{zl_1l_2} z} \operatorname{erfc}(-\frac{ik_{zl_1l_2}}{2E} + Ez'') \right] \qquad (2.23)$$

$$\nabla \frac{\partial \tilde{g}_1}{\partial x} = -\hat{x}\, k_x^2\, \tilde{g}_1 \; -\hat{y}\, k_x k_y\, \tilde{g}_1 \; + i\hat{z}\, k_x\, \frac{\partial \tilde{g}_1}{\partial z} \qquad (2.24)$$

$$\nabla \frac{\partial \tilde{g}_1}{\partial y} = -\hat{x}\, k_x k_y\, \tilde{g}_1 \; -\hat{y}\, k_y^2\, \tilde{g}_1 \; + i\hat{z}\, k_y\, \frac{\partial \tilde{g}_1}{\partial z} \qquad (2.25)$$

$$\nabla \frac{\partial \tilde{g}_1}{\partial z} = i\hat{x}\, k_x\, \frac{\partial \tilde{g}_1}{\partial x} \; + i\hat{y}\, k_y\, \frac{\partial \tilde{g}_1}{\partial y} \; + \hat{z}\, \frac{\partial^2 \tilde{g}_1}{\partial z^2}\,, \qquad (2.26)$$

where

$$\frac{\partial^2 \tilde{g}_1}{\partial z^2} = -k_z^2\, \tilde{g}_1 + \frac{4ik_z E}{\sqrt{\pi}}\, e^{ik_x x + ik_y y} e^{\frac{k_z^2 - 4E^4 z^2}{4E^2}}. \qquad (2.27)$$

From these relations, we obtain

$$\nabla\nabla \tilde{g}_1 = \begin{pmatrix} -k_x^2 \tilde{g}_1 & -k_x k_y \tilde{g}_1 & ik_x \frac{\partial \tilde{g}_1}{\partial z} \\ -k_x k_y \tilde{g}_1 & -k_y^2 \tilde{g}_1 & ik_y \frac{\partial \tilde{g}_1}{\partial z} \\ ik_x \frac{\partial \tilde{g}_1}{\partial z} & ik_y \frac{\partial \tilde{g}_1}{\partial z} & \frac{\partial^2 \tilde{g}_1}{\partial z^2} \end{pmatrix} \qquad (2.28)$$

and finally

$$\left(\bar{\bar{I}} + \frac{1}{k^2}\nabla\nabla \right) g_1(\mathbf{r},\mathbf{r}') = \frac{i}{4\Omega} \sum_{l_1,l_2} \frac{1}{k_{zl_1l_2}} \left(\bar{\bar{I}} + \frac{1}{k_0^2}\nabla\nabla \right) \tilde{g}_1. \qquad (2.29)$$

Next, we define

$$\tilde{g}_2 = \frac{e^{ik_0 R_{n_1 n_2}}}{R_{n_1 n_2}} \operatorname{erfc}(R_{n_1 n_2} E + \frac{ik_0}{2E}) + \frac{e^{-ik_0 R_{n_1 n_2}}}{R_{n_1 n_2}} \operatorname{erfc}(R_{n_1 n_2} E - \frac{ik_0}{2E}), \quad (2.30)$$

where $R_{n_1 n_2}$ has been defined in Eq. (2.14b). Then

$$\frac{\partial \tilde{g}_2}{\partial s_i} = \frac{\partial R_{n_1 n_2}}{\partial s_i} \delta \tilde{g}_2 \,, \tag{2.31a}$$

$$\frac{\partial}{\partial s_j} \frac{\partial \tilde{g}_2}{\partial s_i} \frac{\partial^2 R_{n_1 n_2}}{\partial s_j \partial s_i} \delta \tilde{g}_2 + \frac{\partial R_{n_1 n_2}}{\partial s_j} \frac{\partial R_{n_1 n_2}}{\partial s_i} \delta^2 \tilde{g}_2 \,, \tag{2.31b}$$

where $s_i \in \{x, y, z\}$, $s_j \in \{x, y, z\}$, and

$$\delta \tilde{g}_2 - \left[\frac{4E}{\sqrt{\pi} R_{n_1 n_2}} e^{\frac{k_0^2 - 4R_{n_1 n_2}^2 E^4}{4E^2}} \right.$$
$$+ \frac{1 - ik_0 R_{n_1 n_2}}{R_{n_1 n_2}} \frac{e^{ik_0 R_{n_1 n_2}}}{R_{n_1 n_2}} \operatorname{erfc}(R_{n_1 n_2} E + \frac{ik_0}{2E})$$
$$+ \left. \frac{1 + ik_0 R_{n_1 n_2}}{R_{n_1 n_2}} \frac{e^{-ik_0 R_{n_1 n_2}}}{R_{n_1 n_2}} \operatorname{erfc}(R_{n_1 n_2} E - \frac{ik_0}{2E}) \right], \tag{2.32}$$

$$\delta^2 \tilde{g}_2 \frac{8E}{\sqrt{\pi} R_{n_1 n_2}^2} (1 + R_{n_1 n_2}^2 E^2) e^{\frac{k^2 - 4R_{n_1 n_2}^2 E^4}{4E^2}}$$
$$+ \frac{2(1 - ik_0 R_{n_1 n_2}) - k_0^2 R_{n_1 n_2}^2}{R_{n_1 n_2}^2} \frac{e^{ik_0 R_{n_1 n_2}}}{R_{n_1 n_2}} \operatorname{erfc}(R_{n_1 n_2} E + \frac{ik_0}{2E})$$
$$+ \frac{2(1 + ik_0 R_{n_1 n_2}) - k_0^2 R_{n_1 n_2}^2}{R_{n_1 n_2}^2} \frac{e^{-ik_0 R_{n_1 n_2}}}{R_{n_1 n_2}} \operatorname{erfc}(R_{n_1 n_2} E - \frac{ik_0}{2E}). \tag{2.33}$$

In addition

$$\frac{\partial R_{n_1 n_2}}{\partial s_i} = \frac{s(i)}{R_{n_1 n_2}} \,, \tag{2.34a}$$

$$\frac{\partial^2 R_{n_1 n_2}}{\partial s_i^2} = \frac{R_{n_1 n_2}^2 - s(i)^2}{R_{n_1 n_2}^3} \,, \tag{2.34b}$$

$$\frac{\partial^2 R_{n_1 n_2}}{\partial s_j \partial s_i} = -\frac{s(j)\, s(i)}{R_{n_1 n_2}^3} \,. \tag{2.34c}$$

The second contribution to the dyadic Green's function is finally expressed as

$$\left(\bar{\bar{I}} + \frac{1}{k^2} \nabla \nabla\right) g_2(\mathbf{r}, \mathbf{r}') = \sum_{n_1, n_2} \frac{1}{8\pi R_{n_1 n_2}} e^{i(k_x n_1 a_x + k_y n_2 a_y)}$$

$$\left[\bar{\bar{I}} \tilde{g}_2 + \frac{1}{k_0^2} \begin{pmatrix} \frac{\partial^2 R_{n_1 n_2}}{\partial x^2} & \frac{\partial^2 R_{n_1 n_2}}{\partial y \partial x} & \frac{\partial^2 R_{n_1 n_2}}{\partial z \partial x} \\ \frac{\partial^2 R_{n_1 n_2}}{\partial x \partial y} & \frac{\partial^2 R_{n_1 n_2}}{\partial y^2} & \frac{\partial^2 R_{n_1 n_2}}{\partial z \partial y} \\ \frac{\partial^2 R_{n_1 n_2}}{\partial x \partial z} & \frac{\partial^2 R_{n_1 n_2}}{\partial y \partial z} & \frac{\partial^2 R_{n_1 n_2}}{\partial z^2} \end{pmatrix} \delta \tilde{g}_2 \right.$$

$$\left. + \frac{1}{k_0^2} \begin{pmatrix} \frac{\partial R_{n_1 n_2}}{\partial x} \\ \frac{\partial R_{n_1 n_2}}{\partial y} \\ \frac{\partial R_{n_1 n_2}}{\partial z} \end{pmatrix} \begin{pmatrix} \frac{\partial R_{n_1 n_2}}{\partial x} & \frac{\partial R_{n_1 n_2}}{\partial y} & \frac{\partial R_{n_1 n_2}}{\partial z} \end{pmatrix} \delta^2 \tilde{g}_2 \right]. \tag{2.35}$$

The total Green's function is given by the sum of the two dyadic components, individually expressed in Eq. (2.29) and Eq. (2.35), and can directly be used in a Method of Moment algorithm. Such an algorithm, being more specific in its applicability, is usually faster than other numerical methods to obtain the electromagnetic fields, and can therefore be used more efficiently for example in the design and optimization of split-ring metamaterials.

The determination of the reflection and transmission coefficients directly follows Eqs. (2.3) and (2.5), where the fields computed using the Method of Moment are integrated over a unit cell surface. An example of transmission coefficient as function of frequency is shown in Fig. 2.2 for the standard square rings whose dimensions are detailed in the caption of the figure. It is seen that when the metamaterial contains only rings, the transmission exhibits a stop-band around 15 GHz, which corresponds to the center frequency of the negative permeability band. The medium of rods only, not shown here, exhibits a very low transmission in agreement with a Drude model with a plasma frequency of about 27 GHz. Noticeably, when both rings and rods constitute the metamaterial, a pass-band appears at about 15.2 GHz, within the region of negative permeability and negative permittivity where propagating waves are supported, which actually corresponds to a negative index of refraction as discussed in Chapter 3. These conclusions have been confirmed by the retrieval process detailed subsequently.

2.4 Inverse problems: retrieval and constitutive parameters

2.4.1 Standard media

The previous section was concerned with the determination of the electromagnetic fields scattered by a slab of metamaterial structure in order to obtain the reflection and transmission coefficients. This section is devoted to the other aspect of the problem: how to determine the constitutive parameters of an effective medium once the reflection and transmission coefficients are known.

In a numerical or experimental setup, the configuration often looks like the one depicted in Fig. 2.3: the sample of material whose constitutive parameters are unknown is placed inside a parallel-plate waveguide and is illuminated by a TEM incidence. The reflection and transmission coefficients, in form of S parameters, are measured with respect to some reference planes situated at a distance d_1 and d_2 from the sample. The electric field in the three regions

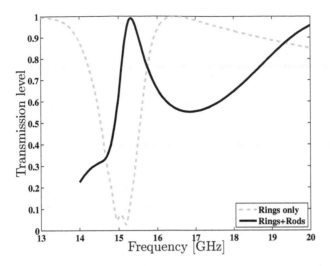

Figure 2.2 Typical transmission curves obtained using the periodic Green's function formalism. The ring-only medium exhibits a stop-band around 15 GHz, whereas the medium of rings+rods exhibits a pass-band at a similar frequency. The rod-only medium, not shown, presents a very low transmission at the frequencies shown (with a plasma frequency of about 27 GHz). The transmission levels have been obtained for a 2-layer metamaterial. The ring is the original square ring shown in Fig. 1.3 of size 3 mm × 3 mm in a lattice of 5 mm × 5 mm, gap and inter-ring spacing of 0.5 mm, and metallization thickness of 0.25 mm. The corresponding rod is centered on the ring, situated 1 mm away from the ring, and with a metallization thickness of 0.5 mm.

thus defined can be written as

$$\mathbf{E}_1 = \hat{y}(e^{ik_0 z} + C_1 e^{-ik_0 z}),$$ (2.36a)

$$\mathbf{E}_2 = \hat{y}(C_2 e^{ik_1 z} + C_3 e^{-ik_1 z}),$$ (2.36b)

$$\mathbf{E}_3 = \hat{y}(C_4 e^{ik_0 z} + C_5 e^{-ik_0 z}),$$ (2.36c)

and it is straightforward to see that $C_5 = 0$.

The coefficients $\{C_i\}$ ($i \in \{1, \ldots, 4\}$) are determined by ensuring the boundary conditions of the tangential electric and magnetic fields at $z = d_1$ and $z = d_1 + d$. The system obtained is

$$
\begin{cases}
C_1 e^{-ik_0 d_1} - C_2 e^{ik_1 d_1} - C_3 e^{-ik_1 d_1} & = -e^{ik_0 d_1} \\
\frac{1}{\eta_0} C_1 e^{-ik_0 d_1} + \frac{1}{\eta} C_2 e^{ik_1 d_1} - \frac{1}{\eta} C_3 e^{-ik_1 d_1} & = \frac{1}{\eta_0} e^{ik_0 d_1} \\
C_2 e^{ik_1(d_1+d)} + C_3 e^{-ik_1(d_1+d)} - C_4 e^{ik_0(d_1+d)} & = 0 \\
-\frac{1}{\eta_1} C_2 e^{k_1(d_1+d)} + \frac{1}{\eta_1} C_3 e^{-k_1(d_1+d)} + \frac{1}{\eta_0} C_4 e^{k_0(d_1+d)} & = 0
\end{cases}
$$ (2.37)

which contains four equations and four unknowns. The solution is straight-

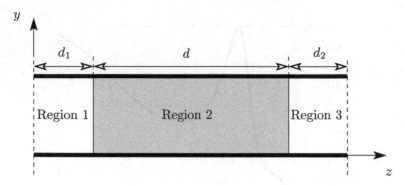

Figure 2.3 Measurement setup in a parallel-plate waveguide: the homogenized medium defines three regions and a reflection/transmission coefficient.

forward and can be written as

$$C_1 = e^{2k_0 d_1} \frac{r(1 - \Phi^2)}{1 - r^2 \Phi^2}, \tag{2.38a}$$

$$C_2 = e^{-id_1(k_1 - k_0)} \frac{2k_0 \mu_1}{(k_1 \mu_0 + k_0 \mu_1)(1 - r^2 \Phi^2)}, \tag{2.38b}$$

$$C_3 = e^{id_1(k_1 + k_0)} \frac{-2k_0 \mu_1 r \Phi^2}{(k_1 \mu_0 + k_0 \mu_1)(1 - r^2 \Phi^2)}, \tag{2.38c}$$

$$C_4 = e^{id(k_1 - k_0)} \frac{4k_0 \mu_0 k_1 \mu}{(k_1 \mu_0 + k_0 \mu_1)^2 (1 - r^2 \Phi^2)} = e^{id(k_1 - k_0)} \frac{1 - r^2}{1 - r^2 \Phi^2}, \tag{2.38d}$$

where

$$r = \frac{k_0/\mu_0 - k_1/\mu_1}{k_0/\mu_0 + k_1/\mu_1}, \tag{2.39a}$$

$$\Phi = \exp(ik_1 d), \tag{2.39b}$$

r being the reflection coefficient and Φ the phase factor of the transmission coefficient. The S parameters are subsequently defined from Eq. (2.3) and (2.5) as (Nicolson and Ross 1970)

$$S_{11} = \frac{\mathbf{E}_{\text{refl}}(z = 0)}{\mathbf{E}_{\text{inc}}(x = 0)} = C_1, \tag{2.40a}$$

$$S_{21} = \frac{\mathbf{E}_{\text{trans}}(z = d_1 + d + d_2)}{\mathbf{E}_{\text{inc}}(z = 0)} = C_4 \, e^{ik_0(d_1 + d + d_2)}. \tag{2.40b}$$

Also, by symmetry, we can compute S_{12} and S_{22} to obtain (Baker-Jarvis et al.

1990)

$$S_{11} = R_1^2 \frac{r(1 - \Phi^2)}{1 - r^2\Phi^2}, \tag{2.41a}$$

$$S_{12} = S_{21} = R_1 R_2 \frac{\Phi(1 - r^2)}{1 - r^2\Phi^2}, \tag{2.41b}$$

$$S_{22} = R_2^2 \frac{r(1 - \Phi^2)}{1 - r^2\Phi^2}, \tag{2.41c}$$

where we have defined

$$R_1 = e^{ik_0 d_1}, \tag{2.42a}$$

$$R_2 = e^{ik_0 d_2}. \tag{2.42b}$$

The problem is therefore reduced to finding a solution to Eqs. (2.41) for r and Φ from the measured S parameters. In these equations, the quantities d_1 and d_2 should also be treated as unknown since the exact distances to the reference planes are often unknown in experimental situations. As a consequence, either for stability or for the number of unknowns, it may appear that Eqs. (2.41) are not sufficient, which prompted Baker-Jarvis et al. (1990) to suggest additional measurements, using the equations depending on what quantities are known to a better precision. The simplest example is of course to invert Φ from its definition in Eq. (2.39b) to obtain

$$\Phi = e^{ik_1 d} \implies k_1 d = -i \ln(\Phi) + 2m\pi, \tag{2.43}$$

where m is a positive or negative integer. Once k_1 is known, the permeability is obtained from Eq. (2.39a) as

$$\mu_1 = \mu_0 \frac{k_1}{k_0} \frac{1 + r}{1 - r} \tag{2.44}$$

and the permittivity is obtained from the k_1 coefficient and the dispersion relation. Note that the choice of m is not always straightforward and deserves further attention, as discussed in the subsequent sections.

2.4.2 Left-handed media

The previous retrieval process has been used extensively on standard dielectric media. The same mathematical formalism can of course be applied to the retrieval of isotropic parameters of left-handed media, although some deeper considerations are necessary due to essentially two aspects:

1. The parameters of left-handed media can take negative values, which also requires them to be complex (to reflect the inherent lossy nature of the media). In the retrieval process, one has therefore to be careful that

no physical laws are violated by the retrieved parameters. In fact, it is judicious to use some physical requirements as inputs into the retrieval algorithm in order to converge to a unique and physical solution.

2. The parameters are frequency dispersive and may have very large values close to resonance. On the contrary, the frequency dispersion of standard media within the frequencies of interest is usually slowly varying. Mathematically, this is reflected by the fact that the parameter m in Eq. (2.43) is usually zero or at most one (it is usually possible to pick samples small compared to the effective wavelength), whereas in the case of left-handed media, it might vary much more. The determination of the proper m therefore becomes a key element of the inversion process.

Based on these considerations, we first rewrite the previous expressions of the S parameters as function of the index of refraction n and impedance z of the medium:

$$S_{11} = \frac{R_{01}(1 - e^{2ink_0d})}{1 - R_{01}^2\, e^{2ink_0d}}, \tag{2.45a}$$

$$S_{21} = \frac{(1 - R_{01}^2)e^{2ink_0d}}{1 - R_{01}^2\, e^{2ink_0d}}, \tag{2.45b}$$

where $R_{01} = (z - 1)/(z + 1)$. Note that for the moment, we have assumed that $d_1 = 0$ and that the fields are computed at the second boundary. Let us introduce an intermediary step and first invert for the index of refraction n and the impedance z from these equations as (Smith et al. 2002a)

$$z = \pm\sqrt{\frac{(1 + S_{11})^2 - S_{21}^2}{(1 - S_{11})^2 - S_{21}^2}}, \tag{2.46a}$$

$$e^{ink_0d} = X \pm i\sqrt{1 - X^2}, \tag{2.46b}$$

where $X = (1 - S_{11}^2 + S_{21}^2)/(2S_{21})$. The advantage of this intermediary step is that the index of refraction and the impedance are physical quantities upon which a series of requirements can be imposed. Typically, the fact that the medium is passive implies that

$$z' \geq 0, \tag{2.47a}$$

$$n'' \geq 0, \tag{2.47b}$$

where the prime denotes the real part and the double prime denotes the imaginary part, and the sign ambiguity in Eq. (2.46) can be lifted. The complexity due to the branch point of the logarithm function remains however, since the index of refraction is obtained from Eq. (2.46b) as

$$n = \frac{1}{k_0d}\left\{\left[(\ln(e^{ink_0d})'') + 2m\pi\right] - i\left[\ln(e^{ink_0d})\right]'\right\}. \tag{2.48}$$

The practical implementation of these equations for the retrieval of the index of refraction and the impedance is hindered by a few issues that need to be carefully examined:

1. The practical limitation of Eqs. (2.47).

2. The determination of the integer m.

3. The sensitivity of these parameters to noise.

4. The definition of the material boundaries.

1. The index of refraction n and the impedance z can *a priori* be determined from Eqs. (2.46) and the conditions of Eqs. (2.47). However, this method has an important practical limitation: both numerical simulations and experimental measurements introduce perturbations in the S parameters that translate into perturbations on n and z. These perturbations are typically small and can be filtered out unless the parameters themselves are close to zero, in which case a small perturbation can induce an unphysical sign change.

This issue can be resolved by introducing a threshold in the use of Eqs. (2.47): when the absolute value of z' is greater than the threshold, Eq. (2.47a) can be used. Otherwise, the sign of z is determined such that the corresponding index of refraction has a non-negative imaginary part. This condition is equivalent to $|e^{ink_0d}| \leq 1$, where

$$e^{ink_0d} = \frac{S_{21}}{1 - S_{11}\frac{z-1}{z+1}}. \tag{2.49}$$

2. The determination of the branch of n' is not a problem proper to left-handed media as can be seen from Eq. (2.43), but nonetheless it is usually not discussed when effective parameters of standard dielectric are retrieved. The reason is that the integer m in the exponential argument of Eq. (2.46b) can be related to the number of wavelengths that propagate inside the slab. By choosing short samples, it is therefore possible to ensure that the sample is smaller than one wavelength, thus automatically selecting m = 0 as solution. Such strategy works well with standard dielectric, where the constitutive parameters are usually reasonably small within the frequencies of interest and where their variations with respect to frequency are small. In the case of left-handed media, however, the parameters might take large absolute values, either at low frequencies for a Drude model or close to resonance for a Lorentz model. Therefore, the small thickness of the sample does not guarantee a sub-wavelength propagation distance due to the possibly large values of the permittivity and the permeability (and hence the effective wavelength inside the medium is small).

The process of determining the correct branch therefore needs to be carefully examined. Let us divide it into two steps. First, the branch at the initial frequency is determined. The imaginary parts of the permeability and permittivity can be expressed as

$$\mu'' = n'z'' + n''z', \tag{2.50a}$$

$$\varepsilon'' = (n''z' - n'z'')/|z|^2. \tag{2.50b}$$

The requirement of positive imaginary parts for both ε and μ leads to $|n'z''| \leq n''z'$. At low frequencies, n'' is usually close to zero so that $n''z'$ is small. Since z'' may not be small itself, n' should be small and m can be determined such that this condition is true. Upon doing so, either a single solution or multiple solutions can be obtained. In the latter case, each solution should be examined in order to make sure that the condition $|n'z''| \leq n''z'$ is satisfied at all subsequent frequencies as well, which usually only leaves one solution to the problem.

The second step in determining the correct branch is to obtain the parameters at all subsequent frequencies. This can be done by invoking the continuity of the permittivity and the permeability, and implementing an iterative scheme based on the parameters at the first frequency.[†] Assuming continuity with frequency, we can Taylor expand (Chen et al. 2004c)

$$e^{in(f_1)k_0(f_1)d} \simeq e^{in(f_0)k_0(f_0)d} \left(1 + \Delta + \frac{1}{2}\Delta^2\right), \tag{2.51}$$

where $\Delta = in(f_1)k_0(f_1)d - in(f_0)k_0(f_0)d$, k_0 is the wave-number in free-space, f_0 is the frequency at which all the parameters are known, and f_1 is the next frequency. Eq. (2.51) is a second-degree polynomial for the unknown $n(f_1)$ where all other terms are known (in particular, the left-hand side is directly obtained from Eq. (2.49)). The selection between the two solutions is performed by comparing $n''(f_1)$ with the value already obtained from Eqs. (2.48) and (2.46b). The closest root is selected as the correct one, whose real part can be used to determine the branch m.

3. The sensitivity to measurements is most prominent in two situations: when the transmission is either close to zero or close to unity. In the first case, which typically appears below the resonance band, $|S_{21}|$ is close to zero which yields large values of the X parameters in Eq. (2.46b) and hence, strong variations in the index of refraction (seen as unphysical spikes in the retrieved values). This situation is avoided by solving first for the impedance. The latter is indeed stable as can be seen by inspecting the first derivative

$$\frac{\partial z^2}{\partial S_{21}} = \frac{8S_{21}S_{11}}{[(1 - S_{11}]^2 - S_{21}^2]}. \tag{2.52}$$

[†]This method should be used cautiously when crossing a resonant region if losses are small. An alternative is to apply the method from low frequencies up to the resonance, and again in a backward fashion from very high frequencies down to the resonance.

The second situation is the mirror of this first one: the impedance is unstable, whereas the index of refraction is stable and should be used first in the retrieval process. Eqs. (2.45) can then be solved within a threshold.

4. The purpose of the homogenization procedure is to find the effective parameters of metamaterials, but also to find the boundaries of the effective medium. The metamaterials being usually composed of discrete metallic elements, it is not immediately clear where these boundaries should be.

The determination of the boundaries can be performed based on the requirement of homogenization, namely that the effective parameters of two slabs of different thicknesses should be identical at all frequencies. This principle, along with the fact that the impedance depends on the slab thickness, can be used to set up a minimization problem: the impedances of two different metamaterial slabs are computed, and their difference is optimized across the entire frequency spectrum of interest as function of the boundaries of the medium. The boundaries that reach the minimum in a pre-defined sense are chosen to be the boundaries of the effective medium.

In the case of periodic structures, this method yields the expected results that the effective boundaries coincide with the boundaries of the unit cell of the metamaterial. In case of non-periodic, or non-symmetric cells (Smith et al. 2005), the optimization might yield different results, which are, of course, dependent on the particular case under study.

A typical retrieval result is shown in Fig. 2.4 (Chen et al. 2004c) for the permittivity and the permeability of a wire-split ring composite metamaterial, for which the four issues aforementioned have been resolved as indicated. Note the good agreement of the permittivity with a Drude model and of the permeability with a Lorentz model, which justifies the effective models used to represent metamaterials. Note also that the region close to the resonance of the permeability, where the retrieval results are not accurate and where unphysical artifacts occur, has not been shown.

2.5 Homogenization from averaging the internal fields

The effective medium parameters such as ε and μ usually relate two fields:

$$\langle \mathbf{D} \rangle = \varepsilon_0 \varepsilon \cdot \langle \mathbf{E} \rangle, \qquad \langle \mathbf{B} \rangle = \mu_0 \mu \cdot \langle \mathbf{H} \rangle, \qquad (2.53)$$

where the fields have been averaged over some volume in the medium and where the angular brackets $\langle \ \rangle$ indicate averages taken over some relevant volume. Here $\mathbf{E}$ and $\mathbf{H}$ are the applied electric and magnetic fields. The bulk average polarization or magnetization in the medium is

$$\langle \mathbf{P} \rangle = N \langle \mathbf{p} \rangle \qquad \langle \mathbf{M} \rangle = N \langle \mathbf{m} \rangle, \qquad (2.54)$$

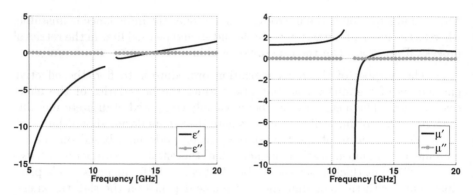

Figure 2.4 Typical retrieval results for the permittivity and the permeability of a split-ring medium (Chen et al. 2004c).

where N is the relevant density of the polarizable or magnetizable structural units, and **p** and **m** are the electric and magnetic dipole moment that develop in a structural unit. One principal task is therefore to determine the individual polarizabilities or magnetizabilities of the individual structural units, in which analytical and numerical modeling plays an important role. The dipole moments generated in an individual structural unit are related to the total fields at that given point and are not only determined by the applied electromagnetic fields, since the fields due to other induced dipole moments in the medium also affect the dipole moments. While these fields can be neglected if the medium is dilute enough (for small filling factor f), any homogenization procedure would need to also account for these fields. A somewhat elementary correction in this regard was discussed in Section 1.3.2 in calculating the bulk dielectric permittivity of a dense molecular medium (the Clausius-Mossotti relationship).

We subsequently discuss some rather elementary ideas by which one can determine effective medium parameters by carrying out appropriate averages of these fields inside the metamaterial. Again, the physical model of the metamaterial that allows us to make some assumptions about the nature of the electromagnetic fields in the metamaterial is as important as the algorithm that allows us to retrieve values for the effective medium parameters from the knowledge of the fields.

2.5.1 Maxwell-Garnett effective medium theory

The Maxwell-Garnett approach (Maxwell-Garnett 1904) has been a very successful theory in describing the effective dielectric properties of composite dielectric media. The composite medium is assumed to be composed of small particles of radius $r \ll \lambda$ and dielectric permittivity ε_i randomly embedded in another bulk medium with dielectric permittivity ε_h. The volume filling fraction of the included particles is taken to be f. The Maxwell-Garnett approach incorporates the distortions due to the dipole field on an average and has been very successful in describing the properties of dilute random inhomogeneous materials. If f is small, then the particles effectively do not feel the scattered fields of other particles, and one can write the dielectric permittivity for a dilute medium of small spherical particles simply as (Landau et al. 1984)

$$\varepsilon_{\text{eff}} = \varepsilon_h + 3f\varepsilon_h \frac{\varepsilon_i - \varepsilon_h}{\varepsilon_i + 2\varepsilon_h}. \tag{2.55}$$

The second term is immediately recognized to arise from the polarizability of the spherical particles,

$$\alpha = \frac{\varepsilon_i - \varepsilon_h}{\varepsilon_i + 2\varepsilon_h} 4\pi a^3. \tag{2.56}$$

The crucial assumption is that the size parameter of the particle $x = \sqrt{\varepsilon_i}ka \ll 1$ where $k = \omega/c$, which allows the assumption of a spatially uniform field over the entire region of the particle and consequently the use of the static polarizability of the spherical particle.

For a dense concentration of particles, one can carry out a correction to the applied fields to obtain the total fields on the lines of the Lorentz-Lorenz theory (see Section 1.3.2). The effective dielectric permittivity is determined by

$$\frac{\varepsilon_{\text{eff}} - \varepsilon_h}{\varepsilon_{\text{eff}} + 2\varepsilon_h} = \frac{N\alpha}{3}, \tag{2.57}$$

where α is the polarizability of individual particles given above. Hence we can obtain the Maxwell-Garnett effective medium parameter

$$\varepsilon_{\text{eff}} = \varepsilon_h \frac{\varepsilon_i(1 + 2f) + 2\varepsilon_h(1 - f)}{\varepsilon_i(1 - f) + \varepsilon_h(2 + f)}, \tag{2.58}$$

where the filling $f = 4\pi a^3 N/3$ and n is the number density of the spheres. The effective medium parameter is complex if the dielectric particles have a complex ε. This effective medium theory is an unrestricted theory in that the imaginary part arises from the actual dissipation of energy in the medium. Although the size of the particles does not appear in the final expression for ε, this is only valid when the size $a \ll \lambda$ or ideally for point dipoles.

The Maxwell-Garnett result can be generalized to include corrections for finite size of the particles. Three such extensions have been compared in Ruppin (2000a) and we briefly summarize here only the generalization of Doyle (1989) which was found to be the most successful of the three. This method consists of using the polarizability for the sphere which comes out of a Mie scattering calculation (Bohren and Huffman 1983). The Mie scattering coefficients are given in general by Bohren and Huffman (1983) as

$$a_m = \frac{n\psi_m(nx)\psi'_m(x) - \psi_m(x)\psi'_m(nx)}{n\psi_m(nx)\xi'_m(x) - \xi_m(x)\psi'_m(nx)}, \tag{2.59}$$

$$b_m = \frac{\psi_m(nx)\psi'_m(x) - n\psi_m(x)\psi'_m(nx)}{\psi_m(nx)\xi'_m(x) - n\xi_m(x)\psi'_m(nx)}, \tag{2.60}$$

where $x = ka$ and $n = \sqrt{\varepsilon_i/\varepsilon_h}$ is the relative refractive index of the spheres with respect to the background. Here ψ_m and ξ_m are the Riccati-Bessel functions of order m and related to the spherical Bessel and Hankel's functions as $\psi_m(x) = xj_m(x)$ and $\xi_m(x) = xh_m^{(1)}(x)$ and the primes imply differentiation with respect to the argument. The total extinction cross-section is given by

$$\sigma_{\text{ext}} = \frac{2}{x^2} \sum_{n=1}^{\infty} (2n+1) \ \text{Re}(a_n + b_n). \tag{2.61}$$

The electric dipole moment of the sphere is described by the term a_1, the quadrupole moment by a_2, and so on. The coefficients b_n denote the strengths of the magnetic multipoles. The electric polarizability of the sphere is given by

$$\alpha = i4\pi a^3 \frac{3a_1}{2x^3}. \tag{2.62}$$

Hence we obtain the effective medium dielectric permittivity as

$$\varepsilon_{\text{eff}} = \varepsilon_h \frac{x^3 + 3ifa_1}{x^3 - \frac{3}{2}ifa_1}. \tag{2.63}$$

A similar result can be obtained for the effective magnetic permeability as

$$\mu_{\text{eff}} = \mu_h \frac{x^3 + 3ifb_1}{x^3 - \frac{3}{2}ifb_1}. \tag{2.64}$$

It immediately becomes obvious that a system of dielectric particles can have a non-unit effective medium magnetic permeability if the magnetic Mie scattering resonance can contribute effectively as pointed out in Bohren (1986). In fact, this possibility has been effectively used to design magnetic metamaterials as discussed in Section 3.2.5. The extended Maxwell-Garnett theories always yield complex effective medium parameters. In other words, they predict a lossy medium even if the constituent particles or host medium are strictly non-dissipative. The imaginary parts arise due to the scattering phase shifts of the waves. The loss implied does not represent actual absorption of radiation in the medium leading to consequent heating of the medium, but rather represents scattering or radiation losses that are then absorbed by (assumed) absorbers at the boundaries of the medium (at the infinities). Thus, the extended Maxwell-Garnett theories are restricted effective medium theories where the parameters are valid for certain purposes (such as the extinction) while not valid for calculating other properties such as the amount of dissipation. This point has been extensively discussed by Bohren (1986).

2.5.2   Layered media as anisotropic effective media

As another example of a homogenizable medium, let us consider a periodically stratified medium consisting of alternating layers of two media. Let the dielectric permittivities of the two media be ε_1 and ε_2, and let d_1 and d_2 be the corresponding layer thicknesses. In the limit that the layer thicknesses are extremely small compared to the wavelength of light in the effective medium, one can assume reasonably uniform fields in the layered medium but subject to appropriate boundary conditions. If the electric field is applied parallel to the interfaces, then it is continuous across the layer. Hence we can write

$$E_1 = E_2 = \langle E_{\|} \rangle, \tag{2.65}$$

where E_1 and E_2 are the fields in the two layers. The average displacement field, averaged with respect to the volume fraction, is

$$\langle D_{\|} \rangle = \frac{D_1 d_1 + D_2 d_2}{d_1 + d_2} = \varepsilon_0 \frac{\varepsilon_1 d_1 + \varepsilon_1 d_2}{d_1 + d_2} \langle E_{\|} \rangle, \tag{2.66}$$

which allows us to to define the average relative permittivity in the parallel directions as

$$\varepsilon_{\|} = \frac{\varepsilon_1 + \varepsilon_2 \eta}{1 + \eta}, \tag{2.67}$$

where $\eta = d_2/d_1$. When the applied fields are normal to the interfaces, it is the normal component of the displacement field that is continuous across the layer interfaces. Hence we have

$$D_1 = D_2 = \langle D_{\perp} \rangle \tag{2.68}$$

and

$$\langle E_{\perp} \rangle = \frac{E_1 d_1 + E_2 d_2}{d_1 + d_2} = \frac{1}{\varepsilon_0} \frac{\frac{d_1}{\varepsilon_1} + \frac{d2}{\varepsilon_2}}{d_1 + d_2} \langle D_{\perp} \rangle, \tag{2.69}$$

and we can define the effective permittivity normal to the layers by

$$(\varepsilon_{\perp})^{-1} = \frac{\frac{1}{\varepsilon_1} + \frac{\eta}{\varepsilon_2}}{1 + \eta}. \tag{2.70}$$

The layering itself gives rise to an anisotropic response parallel and perpendicular to the layers. The possibility of mixing layers with negative dielectric permittivity (see Section 3.1.1) and positive dielectric permittivity produces an unusual anisotropic dielectric permittivity tensor. Depending on the relative values of ε_1, ε_2, and η, the parallel component can be negative while the perpendicular component can be positive or vice versa. Such anisotropic media are termed *indefinite media* (see Section 5.3.1 on page 202 for a more detailed analysis of indefinite media). As an example, in Fig. 2.5 we plot the effective permittivity components with frequency for a layered medium composed of fused silica and silver (layered along the z-direction). Silver has a negative permittivity at optical frequencies that can be modeled by the empirical formula $\varepsilon = 5.7 - 9^2/\omega^2 + i0.4$ (ω in eV units) and has negative dielectric permittivity, while silica has a positive dielectric permittivity. It can

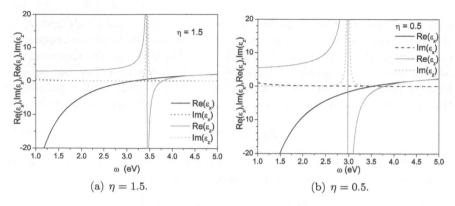

(a) $\eta = 1.5$.   (b) $\eta = 0.5$.

Figure 2.5 Anisotropic effective dielectric permittivities of a layered system consisting of alternating layers of silver and silica for two different values of η. The electric permittivity of silver is given by the empirical formula $\varepsilon = 5.7 - 9^2/\omega^2 + i0.4$, which fits the experimental data reasonably well. The dielectric properties of silica are modeled via the Sellmeier's three resonance formula (Buck 2004).

be clearly seen for the layered stack with $\eta = 1.5$ (more silica layer thickness) that ε_x and ε_z can have opposite signs in certain frequency bands. The effective medium permittivities of one such layered composite, where the dielectric permittivities and the thicknesses ratio of the layers are given in the figure caption, are shown in Fig. 2.5(a). At low frequencies ($\omega < 3.2eV$), $\varepsilon_x < 0$ and $\varepsilon_z > 0$. In the frequency bandwidth $3.5eV < \omega < 3.73$, $\varepsilon_x > 0$ while $\varepsilon_z < 0$. This anisotropy can be fruitfully used to obtain subwavelength image resolution as discussed in Chapter 8. Fine control over the behavior of the effective medium parameters is possible by changing the relative layer thicknesses. As shown in Fig. 2.5(b), for $\eta = 0.5$ (more silver layer thickness), both ε_x and ε_z have negative permittivity in the high frequency band. It should be stressed that even though the final expressions for ε involve only the ratio η and not the individual layer thicknesses, the expressions are valid only in the limit of extremely small layer thickness. For layers of finite thickness, the modulation of the fields inside the layers needs to be accounted for.

2.5.3 Averaging the internal fields in periodic media

Generally, in a given metamaterial, the structures are much more complex than spheres or ellipsoids. Furthermore, many metamaterials consist of units arranged on a periodic lattice. While the polarizabilities of the individual metamaterial units can be calculated or measured via the scattered fields, the average fields obtained via Maxwell-Garnett type theories do not often satisfy the conditions of continuity on the fields across the system in a rigorous man-

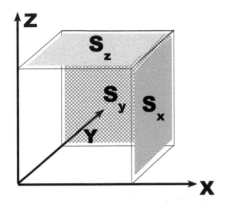

Figure 2.6 The procedure for averaging the internal microscopic fields consists of averaging the **E** and **H** fields over the edges of the unit cell of the simple cubic lattice, while the **D** and **B** fields are averaged over a face S_x for the x-component, and similarly for the other components.

ner. Consider subsequently a way of computing averages for electromagnetic fields in a lattice as discussed in Pendry et al. (1999). Let us start with the Maxwell equations in material media

$$\nabla \times \mathbf{E} = -\frac{\partial \mathbf{B}}{\partial t},$$

$$\nabla \times \mathbf{H} = \frac{\partial \mathbf{D}}{\partial t}.$$

These can be recast in integral form

$$\oint_C \mathbf{E} \cdot \mathrm{dl} = -\frac{\partial}{\partial t} \int_S \mathbf{B} \cdot \mathrm{d}\sigma, \qquad (2.71)$$

$$\oint_C \mathbf{H} \cdot \mathrm{dl} = \frac{\partial}{\partial t} \int_S \mathbf{D} \cdot \mathrm{d}\sigma, \qquad (2.72)$$

where S is the open surface bound by the closed curve C.

These equations themselves suggest a way of averaging the electromagnetic fields. Consider for simplicity that we have a simple cubic lattice for the metamaterial as shown in Fig. 2.6. The averaged fields $\mathbf{E}_{\mathrm{eff}}$ and $\mathbf{H}_{\mathrm{eff}}$ are defined by averaging the **E** and **H** along the sides of the unit cell:

$$E_{\text{eff}}^{(x)} = \frac{1}{a} \int_{(0,0,0)}^{(a,0,0)} E_x \mathrm{d}x, \quad E_{\text{eff}}^{(y)} = \frac{1}{a} \int_{(0,0,0)}^{(0,a,0)} E_y \mathrm{d}y, \quad E_{\text{eff}}^{(z)} = \frac{1}{a} \int_{(0,0,0)}^{(0,0,a)} E_z \mathrm{d}z,$$

$$(2.73)$$

$$H_{\text{eff}}^{(x)} = \frac{1}{a} \int_{(0,0,0)}^{(a,0,0)} H_x \mathrm{d}x, \quad H_{\text{eff}}^{(y)} = \frac{1}{a} \int_{(0,0,0)}^{(0,a,0)} H_y \mathrm{d}y, \quad H_{\text{eff}}^{(z)} = \frac{1}{a} \int_{(0,0,0)}^{(0,0,a)} H_z \mathrm{d}z.$$

$$(2.74)$$

Similarly, the $\mathbf{D}_{\text{eff}}$ and $\mathbf{B}_{\text{eff}}$ are defined by averaging them over the faces of the unit cell:

$$D_{\text{eff}}^{(x)} = \frac{1}{a^2} \int_{S_x} D_x \mathrm{d}\sigma_x, \quad D_{\text{eff}}^{(y)} = \frac{1}{a^2} \int_{S_y} D_y \mathrm{d}\sigma_y, \quad D_{\text{eff}}^{(z)} = \frac{1}{a^2} \int_{S_z} D_z \mathrm{d}\sigma_z,$$

$$(2.75)$$

$$B_{\text{eff}}^{(x)} = \frac{1}{a^2} \int_{S_x} B_x \mathrm{d}\sigma_x, \quad B_{\text{eff}}^{(y)} = \frac{1}{a^2} \int_{S_y} B_y \mathrm{d}\sigma_y, \quad B_{\text{eff}}^{(z)} = \frac{1}{a^2} \int_{S_z} B_z \mathrm{d}\sigma_z.$$

$$(2.76)$$

where S_x is the face in the yz plane, S_y is the face in the zx plane and S_z is the face in the xy plane. The averaged fields then relate the ε and μ component-wise. There is only one restriction in this procedure: that the unit cell boundaries should not intersect any of the structures contained within so that the continuity of the parallel components of $\mathbf{E}_{\text{eff}}$ and $\mathbf{H}_{\text{eff}}$ across the cell boundaries is maintained. In the case of periodically structured materials, the periodicity implies that the averaging need not be carried over orientational and density fluctuations in the metamaterial.

Consider the system of wires and rings shown in Fig. 2.1. It would be typical to average the H field along a line normal to the rings and along the edge of the cubic cell where the near-fields due to the rings are not very large. But the average for the B field would be across the area of the ring and would pick up the large fields due to the rings. Thus, to create a large effect on the effective medium parameters, we should design structures with large inhomogeneous fields. For sizes of the unit cell comparable to the wavelength, homogenization breaks down. This is automatically reflected in this procedure whereby the averages of different unit cells located at different points in the metamaterial yield different results. This procedure can also be easily extended to the retrieval of bianisotropic parameters (Smith and Pendry 2006).

2.6 Generalization to anisotropic and bianisotropic media

The information provided by the forward model as a set of reflection and transmission coefficients (for various incidences, polarizations, and frequencies if necessary) needs to be processed in an inversion algorithm. Typically, one needs to minimize the complex difference

$$\min \left(R_{\mathrm{mea}} - R_{\mathrm{simu}}(\bar{\bar{\varepsilon}}, \bar{\bar{\mu}}) \right), \tag{2.77}$$

where R_{mea} is the measured reflection coefficient (or obtained from the forward model) and R_{simu} is the calculated reflection coefficient for a given model as function of variable parameters. A similar minimization is performed on the transmission coefficient, and the constitutive dyads that achieve the minimum in a certain sense are chosen as the retrieved values.

The process of minimization therefore requires the possibility of computing R_{simu} and T_{simu} for the constitutive dyads chosen as the model of the metamaterial. The minimization can then be performed in essentially two ways:

1. Numerically, the minimization is transformed into an optimization problem that can be addressed with various methods such as genetic algorithms, differential evolution, and other deterministic or non-deterministic methods (the efficacy of the method depending on the complexity of the problem, the more complex the problem and the larger the number local minima to avoid).

2. Analytically, if R_{simu} and T_{simu} are known in closed-form (which is possible in some cases), inversion formulae can be obtained in order to solve for the constitutive parameters. The complexity of the formulae as well as of the mathematical derivations increases with the complexity of the model.

The necessity of computing R_{simu} and T_{simu} for many inputs (as the search algorithm progresses) essentially rules out the possibility of using a direct numerical approach. The latter are usually very flexible and powerful in terms of their simulation capabilities, but this flexibility also often comes at the expense of computational time: when hundreds or thousands of iterations are needed, this may become an important disadvantage. Whenever possible, it is therefore more desirable to develop a method that is less general but which addresses the specific problem under consideration in an efficient manner. In the next section, we present such a method for the computation of the fields in layered media where the constitutive parameters can be arbitrary bianisotropic tensors, as a generalization of the anisotropic case treated in Chew (1990).[‡]

2.6.1 Forward model

The following method is a generalization of the common Transfer Matrix Method for layered media (Born and Wolf 1999) for the general case when the materials can be anisotropic and bianisotropic. Let us consider a homogeneous layered medium in the $\hat{z}$ direction, where each layer is characterized by potentially fully populated bianisotropic tensors, as shown in Fig. 2.7.

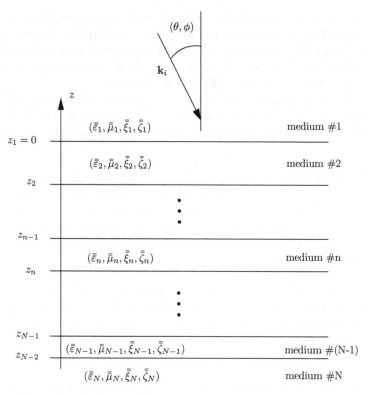

Figure 2.7 Configuration of the problem: a plane wave with wave vector $\mathbf{k}_i$ is incident with the polar angles (θ, ϕ) onto a multi-layered medium of arbitrary constitutive bianisotropic tensors.

[‡]It should be noted that to our knowledge, a numerical package does not exist that is able to compute the field in layered media when the bianisotropic parameters are all random tensors. However, some results in this direction have been presented in Ouchetto et al. (2006).

Assuming an $e^{-i\omega t}$ dependency, the source free Maxwell equations are written as

$$\nabla \times \mathbf{E}(\mathbf{r}) = i\omega \mathbf{B}(\mathbf{r}), \tag{2.78a}$$
$$\nabla \times \mathbf{H}(\mathbf{r}) = -i\omega \mathbf{D}(\mathbf{r}), \tag{2.78b}$$

with the constitutive relations in the Telegen representation (Weiglhofer 2003)

$$\mathbf{D}(\mathbf{r}) = \bar{\bar{\varepsilon}} \cdot \mathbf{E}(\mathbf{r}) + \bar{\bar{\xi}} \cdot \mathbf{H}(\mathbf{r}), \tag{2.79a}$$
$$\mathbf{B}(\mathbf{r}) = \bar{\bar{\mu}} \cdot \mathbf{H}(\mathbf{r}) + \bar{\bar{\zeta}} \cdot \mathbf{E}(\mathbf{r}). \tag{2.79b}$$

The constitutive parameters $\bar{\bar{\varepsilon}}$, $\bar{\bar{\mu}}$, $\bar{\bar{\xi}}$, and $\bar{\bar{\zeta}}$ are arbitrary random tensors, potentially fully populated, and homogeneous within each layer. In addition, within this section, it is understood that they are absolute values, related to the relative quantities, denoted by the subscript 'r', by

$$\bar{\bar{\varepsilon}} = \varepsilon_0 \bar{\bar{\varepsilon}}_r, \qquad \bar{\bar{\mu}} = \mu_0 \bar{\bar{\mu}}_r, \qquad \bar{\bar{\xi}} = \bar{\bar{\xi}}_r/c, \qquad \bar{\bar{\zeta}} = \bar{\bar{\zeta}}_r/c. \tag{2.80}$$

The dispersion relation within these general media can be obtained by eliminating all the vectors but $\mathbf{E}$ and setting the dispersion relation of the matrix operating on $\mathbf{E}$ equal to zero. One simply obtains

$$\left| \omega^2 \bar{\bar{\varepsilon}} + \left(\bar{\bar{k}} + \omega \bar{\bar{\xi}} \right) \cdot \bar{\bar{\mu}}^{-1} \cdot \left(\bar{\bar{k}} - \omega \bar{\bar{\zeta}} \right) \right| = 0, \tag{2.81}$$

where $\bar{\bar{k}} = \mathbf{k} \times \bar{\bar{I}}_3$. Using the preferential $\hat{z}$ direction to split the tensors into transverse and longitudinal parts, we write

$$\bar{\bar{\varepsilon}} = \begin{bmatrix} \bar{\bar{\varepsilon}}_{ss} & \bar{\bar{\varepsilon}}_{sz} \\ \bar{\bar{\varepsilon}}_{zs} & \varepsilon_{zz} \end{bmatrix} \tag{2.82}$$

and similarly for $\bar{\bar{\mu}}$, $\bar{\bar{\xi}}$, $\bar{\bar{\zeta}}$, which, upon inserting into the constitutive relations (2.79) and the Maxwell equations (2.78), yield

$$\nabla_s \times \hat{z} E_z(\mathbf{r}) + \hat{z} \times \frac{\partial}{\partial z} \mathbf{E}_s(\mathbf{r}) = i\omega \bar{\bar{\mu}}_{ss} \cdot \mathbf{H}_s(\mathbf{r}) + i\omega \bar{\bar{\mu}}_{sz} \cdot \hat{z} H_z(\mathbf{r})$$
$$+ i\omega \bar{\bar{\zeta}}_{ss} \cdot \mathbf{E}_s(\mathbf{r}) + i\omega \bar{\bar{\zeta}}_{sz} \cdot \hat{z} E_z(\mathbf{r}), \tag{2.83a}$$

$$\nabla_s \times \mathbf{E}_s(\mathbf{r}) = i\omega \bar{\bar{\mu}}_{zs} \cdot \mathbf{H}_s(\mathbf{r}) + i\omega \mu_{zz} \hat{z} H_z(\mathbf{r}) + i\omega \bar{\bar{\zeta}}_{zs} \cdot \mathbf{E}_s(\mathbf{r}) + i\omega \zeta_{zz} \hat{z} E_z(\mathbf{r}), \tag{2.83b}$$

$$\nabla_s \times \hat{z} H_z(\mathbf{r}) + \hat{z} \times \frac{\partial}{\partial z} \mathbf{H}_s(\mathbf{r}) = -i\omega \bar{\bar{\varepsilon}}_{ss} \cdot \mathbf{E}_s(\mathbf{r}) - i\omega \bar{\bar{\varepsilon}}_{sz} \cdot \hat{z} E_z(\mathbf{r})$$
$$- i\omega \bar{\bar{\xi}}_{ss} \cdot \mathbf{H}_s(\mathbf{r}) - i\omega \bar{\bar{\xi}}_{sz} \cdot \hat{z} H_z(\mathbf{r}), \tag{2.83c}$$

$$\nabla_s \times \mathbf{H}_s(\mathbf{r}) = -i\omega \bar{\bar{\varepsilon}}_{zs} \cdot \mathbf{E}_s(\mathbf{r}) - i\omega \varepsilon_{zz} \hat{z} E_z(\mathbf{r}) - i\omega \bar{\bar{\xi}}_{zs} \cdot \mathbf{H}_s(\mathbf{r}) - i\omega \xi_{zz} \hat{z} H_z(\mathbf{r}). \tag{2.83d}$$

The solutions in each homogeneous layer are plane waves of the form $e^{i\mathbf{k}_s \cdot \mathbf{r}_s}$ so that ∇_s is replaced by $i\mathbf{k}_s$ in Eqs. (2.83). Taking $\hat{z}$ as the propagation direction, the transverse components can then be separated from the longitudinal ones as

$$\mathbf{E}_z(\mathbf{r}) = \frac{1}{D}\left[-\frac{\xi_{zz}}{\omega}\mathbf{k}_s \times \bar{\bar{I}}_3 \cdot +\xi_{zz}\bar{\bar{\zeta}}_{zs} \cdot -\mu_{zz}\bar{\bar{e}}_{zs}\cdot\right]\mathbf{E}_s(\mathbf{r})$$

$$+ \frac{1}{D}\left[\xi_{zz}\bar{\bar{\mu}}_{zs}\cdot -\frac{\mu_{zz}}{\omega}\mathbf{k}_s \times \bar{\bar{I}}_3 \cdot -\mu_{zz}\bar{\bar{\xi}}_{zs}\cdot\right]\mathbf{H}_s(\mathbf{r}),$$
(2.84a)

$$\mathbf{H}_z(\mathbf{r}) = \frac{1}{D}\left[\frac{\varepsilon_{zz}}{\omega}\mathbf{k}_s \times \bar{\bar{I}}_3 \cdot -\varepsilon_{zz}\bar{\bar{\zeta}}_{zs} \cdot +\zeta_{zz}\bar{\bar{e}}_{zs}\cdot\right]\mathbf{E}_s(\mathbf{r})$$

$$+ \frac{1}{D}\left[-\varepsilon_{zz}\bar{\bar{\mu}}_{zs}\cdot +\frac{\zeta_{zz}}{\omega}\mathbf{k}_s \times \bar{\bar{I}}_3 \cdot +\zeta_{zz}\bar{\bar{\xi}}_{zs}\cdot\right]\mathbf{H}_s(\mathbf{r}),$$
(2.84b)

where

$$\mathbf{k}_x \times \bar{\bar{I}}_3 = \begin{pmatrix} 0 & 0 & k_y \\ 0 & 0 & -k_x \\ -k_y & k_x & 0 \end{pmatrix}$$
(2.85)

$$\text{and } D = (\varepsilon_{zz}\mu_{zz} - \xi_{zz}\zeta_{zz}).$$
(2.86)

Using Eqs. (2.84) into Eqs. (2.83), the longitudinal components H_z and E_z can be factored out, leaving only an expression in terms of the transverse components and their derivatives:

$$\frac{\partial}{\partial z}\mathbf{E}_s(\mathbf{r}) = \frac{1}{D}\left[-\frac{i\xi_{zz}}{\omega}\hat{z} \times \bar{\bar{I}}_3 \cdot \mathbf{k}_s \times \bar{\bar{I}}_3 \cdot \mathbf{k}_s \times \bar{\bar{I}}_3 \cdot +i\xi_{zz}\hat{z} \times \bar{\bar{I}}_3 \cdot \mathbf{k}_s \times \bar{\bar{I}}_3 \cdot \bar{\bar{\zeta}}_{zs}\cdot\right.$$

$$- i\mu_{zz}\hat{z} \times \bar{\bar{I}}_3 \cdot \mathbf{k}_s \times \bar{\bar{I}}_3 \cdot \bar{\bar{e}}_{zs} \cdot -i\omega D\hat{z} \times \bar{\bar{I}}_3 \cdot \bar{\bar{\zeta}}_{ss}\cdot$$

$$- i\varepsilon_{zz}\hat{z} \times \bar{\bar{I}}_3 \cdot \bar{\bar{\mu}}_{sz} \cdot \mathbf{k}_s \times \bar{\bar{I}}_3 \cdot +i\omega\varepsilon_{zz}\hat{z} \times \bar{\bar{I}}_3 \cdot \bar{\bar{\mu}}_{sz} \cdot \bar{\bar{\zeta}}_{zs}\cdot$$

$$- i\omega\zeta_{zz}\hat{z} \times \bar{\bar{I}}_3 \cdot \bar{\bar{\mu}}_{sz} \cdot \bar{\bar{e}}_{zs} \cdot +i\xi_{zz}\hat{z} \times \bar{\bar{I}}_3 \cdot \bar{\bar{\zeta}}_{sz} \cdot \mathbf{k}_s \times \bar{\bar{I}}_3 \cdot$$

$$\left. - i\omega\xi_{zz}\hat{z} \times \bar{\bar{I}}_3 \cdot \bar{\bar{\zeta}}_{sz} \cdot \bar{\bar{\zeta}}_{zs} \cdot +i\omega\mu_{zz}\hat{z} \times \bar{\bar{I}}_3 \cdot \bar{\bar{\zeta}}_{sz} \cdot \bar{\bar{e}}_{zs}\cdot\right]\mathbf{E}_s(\mathbf{r})$$

$$+ \frac{1}{D}\left[i\xi_{zz}\hat{z} \times \bar{\bar{I}}_3 \cdot \mathbf{k}_s \times \bar{\bar{I}}_3 \cdot \bar{\bar{\mu}}_{zs} \cdot -\frac{i\mu_{zz}}{\omega}\hat{z} \times \bar{\bar{I}}_3 \cdot \mathbf{k}_s \times \bar{\bar{I}}_3 \cdot \mathbf{k}_s \times \bar{\bar{I}}_3\cdot\right.$$

$$- i\mu_{zz}\hat{z} \times \bar{\bar{I}}_3 \cdot \mathbf{k}_s \times \bar{\bar{I}}_3 \cdot \bar{\bar{\xi}}_{zs} \cdot -i\omega D\hat{z} \times \bar{\bar{I}}_3 \cdot \bar{\bar{\mu}}_{ss}\cdot$$

$$+ i\omega\varepsilon_{zz}\hat{z} \times \bar{\bar{I}}_3 \cdot \bar{\bar{\mu}}_{sz} \cdot \bar{\bar{\mu}}_{zs} \cdot -i\zeta_{zz}\hat{z} \times \bar{\bar{I}}_3 \cdot \bar{\bar{\mu}}_{sz} \cdot \mathbf{k}_s \times \bar{\bar{I}}_3\cdot$$

$$- i\omega\zeta_{zz}\hat{z} \times \bar{\bar{I}}_3 \cdot \bar{\bar{\mu}}_{sz} \cdot \bar{\bar{e}}_{zs} \cdot -i\omega\xi_{zz}\hat{z} \times \bar{\bar{I}}_3 \cdot \bar{\bar{\zeta}}_{sz} \cdot \bar{\bar{\mu}}_{zs}\cdot$$

$$\left. + i\mu_{zz}\hat{z} \times \bar{\bar{I}}_3 \cdot \bar{\bar{\zeta}}_{sz} \cdot \mathbf{k}_s \times \bar{\bar{I}}_3 \cdot +i\omega\mu_{zz}\hat{z} \times \bar{\bar{I}}_3 \cdot \bar{\bar{\zeta}}_{sz} \cdot \bar{\bar{e}}_{zs}\cdot\right]\mathbf{H}_s(\mathbf{r}),$$
(2.87)

where

$$\hat{z} \times \bar{\bar{I}}_3 = \begin{pmatrix} 0 & -1 & 0 \\ 1 & 0 & 0 \\ 0 & 0 & 0 \end{pmatrix}. \tag{2.88}$$

Note that the second equation expressing $\partial \mathbf{H}_s(\mathbf{r})/\partial z$ is directly obtained from Eqs. (2.87) and from the duality condition so that we do not write it explicitly here. The two equations thus obtained can then be gathered in a matrix form as (Berreman 1972, Chew 1990, Norgen 1997):

$$\frac{\partial}{\partial z} \begin{pmatrix} \mathbf{E}_s(\mathbf{r}) \\ \mathbf{H}_s(\mathbf{r}) \end{pmatrix} \begin{pmatrix} \bar{\bar{F}}_{11} & \bar{\bar{F}}_{12} \\ \bar{\bar{F}}_{21} & \bar{\bar{F}}_{22} \end{pmatrix} \cdot \begin{pmatrix} \mathbf{E}_s(\mathbf{r}) \\ \mathbf{H}_s(\mathbf{r}) \end{pmatrix}, \tag{2.89}$$

where each $\bar{\bar{F}}_{ij}$ ($\{i,j\} \in \{1,2\}$) is the top-left 2×2 matrix obtained from Eqs. (2.87) and its dual (a 2×2 system that is obtained by discarding the zero components in Eqs. (2.87)). Written in vectorial form, the system (2.89) is a simple ordinary differential equation that admits exponential functions as solutions. Being a 4×4 system, it admits four eigenvalues, four eigenvectors, and four coefficients that need to be determined by applying the boundary conditions at the interfaces between all the layers. The eigenvectors correspond to the polarization states of the transverse electromagnetic field components, while the eigenvalues correspond to the wave-vectors supported by the medium. For example, if we consider an isotropic homogeneous medium, the four solutions correspond to the TE and TM polarizations (which happen to share similar wave-vectors) propagating in the upward and downward directions.

The eigenvectors and eigenvalues can thus be sorted so that the first two correspond to upward propagating waves while the last two correspond to downward propagating waves. For reasons that will be made clear in Chapter 5 in order to avoid all confusion in media with negative constitutive parameters, this ordering should be performed by examining the direction of power propagation (the Poynting vector) rather than just the sign of the wavenumbers.

Upon defining the vector $\mathbf{V}(z)$ as the 4×1 vector $[\mathbf{E}_s(\mathbf{r}); \mathbf{H}_s(\mathbf{r})]^T$ (where the subscript T denotes the transpose operator), we can rewrite Eq. (2.89) as $\frac{\partial}{\partial z}\mathbf{V}(z) = \bar{\bar{F}} \cdot \mathbf{V}(z)$, where the 4×4 tensor $\bar{\bar{F}}$ is straightforwardly defined from Eq. (2.89). Following the notation of Chew (1990), we denote by $\bar{\bar{a}}_n$ the tensor containing the four (sorted) eigenvectors of medium #n, and $\bar{\bar{\beta}}_n$ the diagonal 4×4 tensor containing the four (sorted) eigenvalues of medium #n. Thus, the transverse polarization states in medium n ($\mathbf{V}_n(z)$) are written as

$$\text{medium 1:} \quad \mathbf{V}_1(z) = \bar{\bar{a}}_1 \cdot e^{i\bar{\bar{\beta}}_1 z} \cdot \begin{bmatrix} \bar{\bar{R}} \\ \bar{\bar{I}}_2 \end{bmatrix} \cdot \begin{bmatrix} A_{31} \\ A_{41} \end{bmatrix}, \tag{2.90a}$$

$$\text{medium } n: \quad \mathbf{V}_n(z) = \bar{\bar{a}}_n \cdot e^{i\bar{\bar{\beta}}_n z} \cdot \mathbf{A}_n, \tag{2.90b}$$

$$\mathbf{V}_n(z) = \bar{\bar{P}}_n(z, z_{n-1}) \cdot \mathbf{V}_n(z_{n-1}). \tag{2.90c}$$

A_{31} and A_{41} are the amplitudes of the two polarizations of the down-going incident waves and are therefore known (if the first medium is free-space, A_{31} and A_{41} represent the amplitudes of the TE and TM incident waves). $\mathbf{A}_n$ is a 4×1 vector containing the coefficients of the four waves in medium n, and $\bar{\bar{P}}$ represents the propagator (or transfer) matrix (Chew 1990) defined as

$$\bar{\bar{P}}_n(z, z_{n-1}) = \bar{\bar{a}}_n \cdot e^{i\bar{\bar{\beta}}_n(z-z_{n-1})} \cdot \bar{\bar{a}}_n^{-1} . \tag{2.91}$$

These expressions can be directly used in the boundary conditions $\mathbf{V}_n(z_{n-1}) = \mathbf{V}_{n-1}(z_{n-1})$ expressing the continuity of the tangential electric and magnetic components and providing a recurrence relation in the vector $\mathbf{V}$:

$$\mathbf{V}_1(z_1) = \mathbf{V}_2(z_1) = \bar{\bar{P}}_2(z_1, z_2) \cdot \mathbf{V}_2(z_2) ,$$
$$\mathbf{V}_2(z_2) = \mathbf{V}_3(z_2) = \bar{\bar{P}}_3(z_2, z_3) \cdot \mathbf{V}_3(z_3) ,$$
$$\vdots$$
$$\mathbf{V}_{n-2}(z_{n-2}) = \mathbf{V}_{n-1}(z_{n-2}) = \bar{\bar{P}}_{n-1}(z_{n-2}, z_{n-1}) \cdot \mathbf{V}_{n-1}(z_{n-1}) ,$$
$$\mathbf{V}_{n-1}(z_{n-1}) = \mathbf{V}_n(z_{n-1}) . \tag{2.92}$$

Combining all the terms, we get

$$\mathbf{V}_1(z_1) = \bar{\bar{P}}_2(z_1, z_2) \cdot \bar{\bar{P}}_3(z_2, z_3) \cdot \ldots \cdot \bar{\bar{P}}_{n-1}(z_{n-2}, z_{n-1}) \cdot \mathbf{V}_n(z_{n-1})$$
$$= \bar{\bar{P}}_A(z_1, \ldots, z_{n-1}) \cdot \mathbf{V}_n(z_{n-1}) . \tag{2.93}$$

Upon using Eqs. (2.90a) and (2.90c), the state vector for the transverse components is written as

$$\mathbf{V}_n(z) = \bar{\bar{P}}_n(z, z_{n-1}) \cdot \bar{\bar{P}}_A^{-1}(z_1, \ldots, z_{n-1}) \cdot \bar{\bar{a}}_1 \cdot e^{i\bar{\bar{\beta}}_1 z_1} \cdot \begin{bmatrix} \bar{\bar{R}} \\ \bar{\bar{I}}_2 \end{bmatrix} \cdot \begin{bmatrix} A_{31} \\ A_{41} \end{bmatrix} . \tag{2.94}$$

The last medium is treated equally and yields

$$\mathbf{V}_N(z) = \bar{\bar{a}}_N \cdot e^{i\bar{\bar{\beta}}_N z} \cdot \begin{bmatrix} 0 \\ \bar{\bar{T}} \end{bmatrix} \cdot \begin{bmatrix} A_{31} \\ A_{41} \end{bmatrix} . \tag{2.95}$$

In these equations, $\bar{\bar{R}}$ and $\bar{\bar{T}}$ are 2×2 dyads that need to be solved for. Still assuming free-space as the surrounding medium, the polarization in the first and last medium can be decomposed into TE and TM so that the various components of the reflection and transmission dyads correspond to TE/TM reflection (transmission) due to a TE/TM incidence. Writing the recurrence formula from the first to the last medium,

$$\bar{\bar{a}}_1 \cdot e^{i\bar{\bar{\beta}}_1 z_1} \cdot \begin{bmatrix} \bar{\bar{R}} \\ \bar{\bar{I}}_2 \end{bmatrix} \bar{\bar{P}}_A(z_1, \ldots, z_{N-1}) \cdot \bar{\bar{a}}_N \cdot e^{i\bar{\bar{\beta}}_N z_{N-1}} \begin{bmatrix} 0 \\ \bar{\bar{T}} \end{bmatrix} \tag{2.96}$$

produces a matrix equation that can be solved for $\bar{\bar{R}}$ and $\bar{\bar{T}}$. Using Eqs. (2.90a) and (2.84), the fields in each medium can be entirely determined.

The method presented in this section is therefore able to produce not only the reflection and transmission dyads but also the field profile in each layer (which is not directly necessary in the retrieval process). The concatenation of 2×2 or 4×4 matrix multiplication makes this method very fast, and thus well suited in an inversion algorithm that requires multiple forward solutions for various inputs.

In addition, the method can be carried out analytically in some simple cases of constitutive tensors in order to obtain closed-form expressions for the reflection and transmission coefficients, necessary for any retrieval algorithm based on analytical solutions. For example, the following expressions can be directly obtained:

1. A half-space interfacing free-space and a biaxial medium, defined by the dyads

$$\bar{\bar{\varepsilon}}_1 = \varepsilon_0 \bar{\bar{I}}_3, \qquad\qquad \bar{\bar{\mu}}_1 = \mu_0 \bar{\bar{I}}_3 \qquad\qquad (2.97a)$$

$$\bar{\bar{\varepsilon}}_2 = \begin{bmatrix} \varepsilon_{2x} & 0 & 0 \\ 0 & \varepsilon_{2y} & 0 \\ 0 & 0 & \varepsilon_{2z} \end{bmatrix}, \qquad \bar{\bar{\mu}}_2 = \begin{bmatrix} \mu_{2x} & 0 & 0 \\ 0 & \mu_{2y} & 0 \\ 0 & 0 & \mu_{2z} \end{bmatrix}. \qquad (2.97b)$$

For $k_y = 0$, the TE coefficients (identified by the 'hs' subscript to denote the half-space case) are expressed as

$$R_{\mathrm{hs}} = \frac{k_{z1}\sqrt{\mu_{2x}\mu_{2z}} - k_{z2}\mu_1}{k_{z1}\sqrt{\mu_{2x}\mu_{2z}} + k_{z2}\mu_1}, \qquad (2.98a)$$

$$T_{\mathrm{hs}} = \frac{2k_{z1}\sqrt{\mu_{2x}\mu_{2z}}}{k_{z1}\sqrt{\mu_{2x}\mu_{2z}} + k_{z2}\mu_1}, \qquad (2.98b)$$

where

$$k_{z2}^2 = \omega^2 \varepsilon_{2y}\mu_{2z} - k_x^2. \qquad (2.99)$$

2. A slab of biaxial medium defined by similar constitutive parameters and surrounded by free-space produces a reflection and transmission coefficients that are expressed as

$$R_{\mathrm{slab}} = \frac{R_{\mathrm{hs}}(e^{2\Phi} - 1)}{R_{\mathrm{hs}}^2 e^{2\Phi} - 1}, \qquad T_{\mathrm{slab}} = \frac{R_{\mathrm{hs}}^2 - 1}{R_{\mathrm{hs}}^2 e^{2\Phi} - 1}, \qquad (2.100)$$

where $\Phi = ik_{z1}d$, d being the thickness of the slab.

3. A slab of bianisotropic medium defined by the following constitutive parameters:

$$\bar{\bar{\varepsilon}}_r = \begin{bmatrix} \varepsilon_{r_{xx}} & 0 & 0 \\ 0 & 1 & 0 \\ 0 & 0 & \varepsilon_{r_{zz}} \end{bmatrix}, \qquad \bar{\bar{\mu}}_r = \begin{bmatrix} 1 & 0 & 0 \\ 0 & \mu_{r_{yy}} & 0 \\ 0 & 0 & 1 \end{bmatrix}, \qquad (2.101\text{a})$$

$$\bar{\bar{\xi}}_r = \begin{bmatrix} 0 & 0 & 0 \\ 0 & 0 & 0 \\ 0 & -i\xi & 0 \end{bmatrix}, \qquad \bar{\bar{\zeta}}_r = \begin{bmatrix} 0 & 0 & 0 \\ 0 & 0 & i\xi \\ 0 & 0 & 0 \end{bmatrix}. \qquad (2.101\text{b})$$

These constitutive dyads correspond to the well-accepted model for the original split ring resonator (Pendry et al. 1999, Smith et al. 2000) (see Fig. 3.8) and have been studied in detail in Marques et al. (2002). The diagonal permittivity and permeability reflect the usual anisotropic property of the metamaterial due to the orientation of the metallic rods and rings. The bianisotropic terms instead are directly related to the shape of the metallic ring. In this case, the presence of oppositely oriented gaps in the two rings induces asymmetric currents and charge distributions in them due to the impinging magnetic field, generating symmetric dipole moments in one direction and asymmetric dipole moments in the orthogonal direction on either side of the ring. The asymmetric components do not cancel out and result in the generation of an electric polarization from an incident magnetic field, directly following the definition of bianisotropy in Eqs. (2.79). Conversely, an electric field propagating in the plane of the ring may impinge on either its symmetric side or its asymmetric side. If impinging on the asymmetric side, the induced current distribution is asymmetric, creating a current flow and therefore a magnetic field. We therefore have the creation of a magnetic field due to the interaction of the structure with the impinging electric field. Note, however, that the original ring is composed of two interlaced split-rings, whereas the explanation of the bianisotropic effect only requires one split-ring. The presence of the second split-ring would require a rigorous consideration of coupling and fringing fields between the two rings for a more accurate quantitative description. A methodology for including these effects is presented in Section 3.2.7.

Given the orientation of the rings in the metamaterial, it is more natural to look for a TM polarization (H fields normal to the plane of the rings) in this case. The reflection coefficients for both the half-space and slab cases are given by

$$R_{\text{hs}} = \frac{k_{z1}\varepsilon_{2z} - k_{z2}\varepsilon_1}{k_{z1}\varepsilon_{2z} + k_{z2}\varepsilon_1}, \tag{2.102a}$$

$$R_{\text{slab}} = \frac{R_{\text{hs}}\left(e^{2\Phi} - 1\right)}{\left(R_{\text{hs}}\right)^2 e^{2\Phi} - 1}, \tag{2.102b}$$

where

$$k_{z2}^2 = \frac{\varepsilon_{2x}}{\varepsilon_{2z}}\left(\omega^2\varepsilon_{2z}\mu_{2y} - k_x^2 - \omega^2\xi^2\right), \tag{2.103a}$$

$$\Phi = ik_{z2}d. \tag{2.103b}$$

2.6.2 Inversion algorithm

Two approaches can be considered for the inversion of the tensorial parameters: an analytical one where, like in the biaxial case, one looks for a formula yielding directly the unknown parameters, and a numerical one, typically based on an optimization scheme and a search algorithm. The analytical approach has the appeal of providing the exact solution without ambiguity and almost instantaneously. However, the situations where an inversion formula can be obtained is usually limited to simple tensors for which an analytical derivation is still manageable. The numerical approach, on the other hand, can always be applied but does not offer the guarantee of convergence and may find local minima as the solution. This can be somehow avoided by the use of both deterministic and stochastic algorithms, even though the final results still need to be checked for consistency. In addition, the numerical minimization of such highly nonlinear problems often requires very long computation time without providing the assurance of the existence of a unique solution.

In the case of the bianisotropy governed by Eqs. (2.101), the tensors are simple enough to allow for an analytical solution and, more interestingly, this model corresponds to the effective medium of a widely used split-ring resonator, as shown in Marques et al. (2002). The unknowns to be solved for $(\varepsilon_x, \varepsilon_y, \varepsilon_z, \mu_x, \mu_y, \mu_z, \xi)$, require multiple equations that can be obtained by varying the polarization and incidence angles with respect to the unit cell of the metamaterial. The number of unknowns and therefore the number of equations required depend on the initial assumption: for lossless media seven unknowns need to be determined whereas fourteen need to be determined in the case of a lossy medium (note that if ξ is a priori known to be real, we still treat it as complex and expect to retrieve a zero imaginary part). The incidences and polarizations that provide the necessary information are illustrated in Fig. 2.8, labeled as TE$_i$ and TM$_i$ ($i = 1, 2, 3$). The bianisotropic

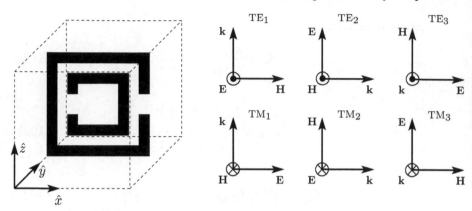

Figure 2.8 Illustration of multiple incidences and polarizations impinging on a unit cell of bianisotropic metamaterial governed by the model of Eqs. (2.101).

Table 2.1 Redefinition of the impedance and the index of refraction for the incidences defined in Fig. 2.8. Note that the TE$_2$ case is lossless.

Case	Dispersion relation	Impedance	Index of refraction
TE$_1$	$k_z^2=k_0^2\varepsilon_y\mu_x$	$\sqrt{\mu_x/\varepsilon_y}$	$\sqrt{\varepsilon_y\mu_x}$
TE$_2$	$k_x^2=k_0^2(\varepsilon_z\mu_y-\xi^2)$	$\mu_y/(\sqrt{\varepsilon_z\mu_y-\xi^2}+\mathrm{i}\xi)$	$\sqrt{\varepsilon_z\mu_y-\xi^2}$
TE$_3$	$k_y^2=k_0^2\varepsilon_x\mu_z$	$\sqrt{\mu_z/\varepsilon_x}$	$\sqrt{\varepsilon_x\mu_z}$
TM$_1$	$k_z^2=k_0^2(\varepsilon_x\mu_y-\varepsilon_x/\varepsilon_z\xi^2)$	$\varepsilon_x/\sqrt{\varepsilon_x\mu_y-\varepsilon_x/\varepsilon_z\xi^2}$	$\sqrt{\varepsilon_x\mu_y-\varepsilon_x/\varepsilon_z\xi^2}$
TM$_2$	$k_x^2=k_0^2\varepsilon_y\mu_z$	$\sqrt{\varepsilon_y/\mu_z}$	$\sqrt{\varepsilon_y\mu_z}$
TM$_3$	$k_y^2=k_0^2(\varepsilon_z\mu_x-\mu_x/\mu_y\xi^2)$	$\varepsilon_z/\sqrt{\varepsilon_z\mu_x-\mu_x/\mu_y\xi^2}$	$\sqrt{\varepsilon_z\mu_x-\mu_x/\mu_y\xi^2}$

property of this ring comes from the coupling between the y component of the magnetic field that produces an imbalance of charges via induced currents and hence, an electric field, and conversely, between the z component of the electric field that produces a magnetic field by driving a current via the polarization charges. Consequently, out of the six incidences considered, only three of them see the bianisotropy (TE$_2$, TM$_1$, TM$_3$) and only one (TE$_2$) sees it for both the electric and magnetic field (since it contains both E_z and H_y). The other incidences propagate through the medium as if it were isotropic, and therefore the retrieval of the associated constitutive parameters can be performed directly using the method presented in the previous section, providing that the index of refraction and the impedance are properly redefined (see Tab. 2.1 and also Appendix B).

Incidence TE$_2$ requires a particular attention since it contains both an H_y component that induces an electric dipole in the $\hat{z}$ direction, and an E_z component that induces a magnetic dipole in the $\hat{y}$ direction. Using the method presented in Section 2.6.1, the S_{11} and S_{21} parameters are found to be

$$S_{11} = \frac{\frac{\mu_{2y}-k_{2x}/k_0-i\xi}{\mu_{2y}+k_{2x}/k_0+i\xi}\left(1-e^{2ik_{2x}d}\right)}{1-\frac{(k_{2x}/k_0-\mu_{2y})^2+\xi^2}{(k_{2x}/k_0+\mu_{2y})^2+\xi^2}e^{2ik_{2x}d}}, \tag{2.104a}$$

$$S_{21} = \frac{\left(1-\frac{(k_{2x}/k_0-\mu_{2y})^2+\xi^2}{(k_{2x}/k_0+\mu_{2y})^2+\xi^2}\right)e^{ik_{2x}d}}{1-\frac{(k_{2x}/k_0-\mu_{2y})^2+\xi^2}{(k_{2x}/k_0+\mu_{2y})^2+\xi^2}e^{2ik_xd}}, \tag{2.104b}$$

where k_{2x} is the wavenumber in the propagation direction inside the medium, governed by the dispersion relation given in Tab. 2.1. In this general case, one cannot define a unique impedance and index of refraction so that the inversion of Eqs. (2.104) needs to be performed numerically as proposed in Chen et al. (2005d). In the special case of a lossless medium, however, Eqs. (2.104) can be simplified as follows:

$$S_{11} = \frac{R_{01}(1-e^{2ink_0d})}{1-R_{01}^2e^{2ink_0d}}, \tag{2.105a}$$

$$S_{21} = \frac{(1-R_{01}^2)e^{2ink_0d}}{1-R_{01}^2e^{2ink_0d}}, \tag{2.105b}$$

where $R_{01} = (z-1)/(z+1)$ is the half-space reflection coefficient, and z is the impedance redefined in Tab. 2.1. The inversion of Eqs. (2.105) can then be performed analytically and is straightforward.

The method presented above was first tested on analytical results: assuming a Lorentz model for the permittivity, the permeability, and the bianisotropic term, the reflection and transmission coefficients can be computed and subsequently used in the inversion algorithm. This procedure is therefore a self-check, where the output results are expected to be identical to the input ones, with a total control on all the parameters along the way. The results for both lossless and lossy media are shown in Fig. 2.9 and Fig. 2.10, respectively. In both cases, the method is seen to produce an excellent matching with the reference results, as is expected from an analytical approach.

The next step is to apply the inversion method to the ring of Fig. 2.8, with the reflection and transmission coefficients generated from the forward method shown in Section 2.3. Unlike the reflection and transmission data obtained from an analytical formula as above, those obtained from a numerical code are inherently imperfect and noisy to some extent, very much like those obtained from measurements.

In order to illustrate the necessity of using the bianisotropic retrieval when working with the ring shown in Fig. 2.8, which also highlights the importance of using a correct model when retrieving effective parameters of metamaterials, we first apply the isotropic retrieval of Section 2.4.2 along each direction. With six complex parameters to retrieve (the real and imaginary parts of ε_x, ε_y,

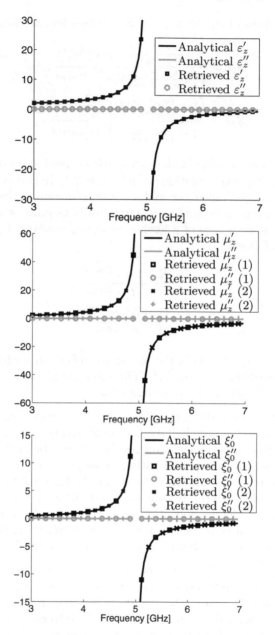

Figure 2.9 Retrieval results on analytical lossless Lorentz models for the permittivity, the permeability, and the bianisotropic term. The matching is seen to be excellent. The results are taken from Chen et al. (2005d; 2006e).

ε_z, μ_x, μ_y, μ_z) and twelve complex sets of data (six incidences that all give a complex reflection and transmission coefficient), we have twice as many equations as unknowns so that each parameter can be retrieved twice, as a consistency check. The results are shown in Fig. 2.11 where two important features are immediately seen: the two values retrieved for ε_x, ε_y, μ_x, and μ_z agree very well all across the frequency range, except for some numerical artefacts due to either noise or resonance of other parameters. However, the two retrieved results for ε_z and μ_y present significant discrepancies, especially close to resonance. These discrepancies prevent us from concluding the correct effective parameter, and illustrate how an incomplete model of a metamaterial may yield inconsistent results.

The bianisotropic study summarized in Tab. 2.1 reveals that the transmission and reflection coefficients involve ε_z, μ_y, and ξ in the incidences TE$_2$, TM$_1$, and TM$_3$ so that it is not surprising to obtain inconsistent results when the bianisotropy is not included. When it is included, however, the results are shown in Fig. 2.12 and reveal a significant improvement in the matching of the two results, giving us confidence that the results as well as the model this time are accurate.

It should now be clear that the inclusion of the bianisotropic term in the model of the split ring shown in Fig. 2.8 is essential. Failure to do so results in biaxial parameters (ε_x, ε_y, ε_z, μ_x, μ_y, μ_z) that are incidence dependent, which is clearly unsatisfactory for a homogeneous medium. This stresses the importance of first developing a correct physical model of the metamaterial under study and then of developing the corresponding inversion algorithm. However, such a model may not always be easy to predetermine, in which case one might try to resort to an all numerical solution without *a priori* assumptions. Needless to say that if no information is known in the tensorial constitutive relation, the problem contains 72 unknowns (nine complex unknowns in four tensors) and is highly nonlinear. The corresponding minimization problem becomes tremendously challenging to solve because of the presence of a very large number of local minima, in addition to requiring heavy computational resources. It is nonetheless interesting to take a closer look at such a method because of its generality.

The minimization problem can be written as follows: :

$$f(\bar{\mathbf{x}}) = \sum_{\phi,\theta} \sum_{i,j\in\{1,2\}} \left\{ |R_{ij}(\theta,\phi) - R_{ij}^m(\theta,\phi)|^2 + |T_{ij}(\theta,\phi) - T_{ij}^m(\theta,\phi)|^2 \right\},$$

(2.106)

where (R_{ij}, T_{ij}) are the reflection and transmission coefficients computed from the method presented in Section 2.3, while (R_{ij}^m, T_{ij}^m) are the measured (and therefore known) reflection and transmission coefficients. The sum over θ and ϕ indicates that multiple incident angles need to be incorporated in order to provide enough equations for the number of unknowns, whereas the sum of (i,j) indicates the two possible incident polarization states. All the parameters are included in the vector $\bar{\mathbf{x}}$ and the global minimum of f is searched, which corresponds to a relative error between the known parameters and the computed ones smaller than a pre-set threshold. Note that weighting factors are often included in the functional in order to improve convergence, as for example proposed in Chen et al. (2006d).

Essentially two factors influence the choice of the minimization algorithm: the search of the global minimum and the computational time. Deterministic search methods have the advantage of a relatively small computational time but they are very inefficient at searching for the global minimum when too many local minima are present. On the other hand, stochastic methods span the search space more thoroughly but have the disadvantage of being slower. One could therefore think of judiciously combining the two: starting first with a stochastic method (such as a Genetic Algorithm or differential evolution) in order to span the search space and select a few candidates as good initial points for the second stage, where a deterministic method (such as a simplex method) takes over and converges to the minima associated with the initial guesses.

Still, the highly nonlinear optimization problem with 72 parameters might not be solvable and some assumptions might have to be introduced. A reasonable one is that the bianisotropy terms $\bar{\bar{\xi}}$ and $\bar{\bar{\zeta}}$ are zero at low frequencies, thus reducing the number of unknowns by half. The solution obtained in this way can be used as a good initial guess for the retrieval of the first frequency in the recursive algorithm. A second good assumption, which is controllable, is that the material thickness is small, typically less than a wavelength, in order to avoid phase ambiguity. In a broad-band retrieval process, this might require working with two or more sample thicknesses depending on frequency.

With these assumptions, such a parameter retrieval algorithm has been shown to accurately retrieve the frequency dispersive parameters of rotated Omega media (consisting of metallic particles in the shape of Ω) and general bianisotropic media (Chen et al. 2006d). The first example, a rotated Omega medium, is characterized by the constitutive tensors of Eqs. (2.101), to which we add a random rotation along the three Euler angles (α, β, γ). In the $(\hat{x}, \hat{y}, \hat{z})$ coordinate system, the tensors are therefore potentially more complex than simply diagonal, yielding apparently more unknowns than really present in the problem. The various parameters of Eqs. (2.101) have been assumed to take the following analytical forms:

$$\nu_i = 1 - \frac{F_{\nu i} f^2}{f^2 - f_{\nu i}^2 + \mathrm{i}\gamma_{\nu i} f}, \qquad (2.107)$$

where f is the frequency, $\nu = \{\varepsilon, \mu\}$, $i = \{x, y, z\}$, and where a similar resonance model for ξ has been used with the parameters $(F_\xi, f_\xi, \gamma_\xi)$. The numerical values of all the parameters are given in Tabs. 2.2. Upon using the differential evolution algorithm at the first stage and two rounds of simplex method at the second stage, the optimized results for all the parameters are

2.6 Generalization to anisotropic and bianisotropic media

Table 2.2 Parameter definition of Eq. (2.107).

$f_{\nu i}$ [GHz]	i		
	x	y	z
ε	4	5	3.5
μ	5	4	3.5

$\gamma_{\nu i}$ [GHz]	i		
	x	y	z
ε	0.5	0.4	0.3
μ	0.4	0.3	0.2

$F_{\nu i}$	i		
	x	y	z
ε	0.5	0.3	0.4
μ	0.3	0.2	0.3

$$f_\xi = 4 \text{ GHz} \qquad \alpha = \pi/5$$
$$\gamma_\xi = 0.5 \text{ GHz} \qquad \beta = \pi/4$$
$$F_\xi = 0.4 \qquad \gamma = \pi/6$$

already in good agreement with the expected ones. Fig. 2.13 illustrates the agreement in the bianisotropic term, which is seen to be very good all across the frequency range.

As a second example, we shall mention the parameter retrieval for a general bianisotropic medium which is a chiroferrite medium, as studied in Dmitriev (2001):

$$\bar{\bar{\varepsilon}}_r = \begin{pmatrix} \varepsilon_{xx} & \varepsilon_{xy} & 0 \\ -\varepsilon_{xy} & \varepsilon_{xx} & 0 \\ 0 & 0 & \varepsilon_{zz} \end{pmatrix}, \qquad \bar{\bar{\mu}}_r = \begin{pmatrix} \mu_{xx} & \mu_{xy} & 0 \\ -\mu_{xy} & \mu_{xx} & 0 \\ 0 & 0 & \mu_{zz} \end{pmatrix},$$

$$\bar{\bar{\xi}}_r = \begin{pmatrix} \xi_{xx} & \xi_{xy} & 0 \\ -\xi_{xy} & \xi_{xx} & 0 \\ 0 & 0 & \xi_{zz} \end{pmatrix}, \qquad \bar{\bar{\zeta}}_r = \begin{pmatrix} -\xi_{xx} & -\xi_{xy} & 0 \\ \xi_{xy} & -\xi_{xx} & 0 \\ 0 & 0 & -\xi_{zz} \end{pmatrix}. \qquad (2.108)$$

Although some parameters are zero and others are identical to each other, it is important to realize that this information should not be used in a general retrieval, and should in fact come as a result of the method. Like in the previous case, an analytical model can be assumed for all parameters and the retrieval should be carried out in two steps: the first step is run with a stochastic method that explores the entire search space, whereas the second method is run with a deterministic method to converge to a series of solutions. As mentioned previously, successful retrieval results can be obtained in this case when two assumptions are implemented: that the bianisotropic tensors are negligible at small frequency (in fact, only at the first frequency in the frequency sweep algorithm) and that the medium is thin compared to the wavelength. Nevertheless, the encouraging results obtained in Chen et al. (2006d) should merely be viewed as a first step toward a more systematic characterization and retrieval of metamaterials for which a good physical model is yet unknown.

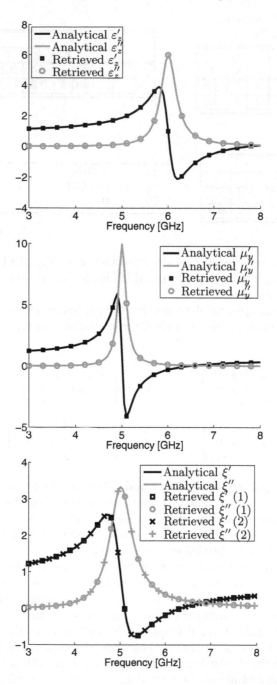

Figure 2.10 Retrieval results on analytical lossy Lorentz models for the permittivity, the permeability, and the bianisotropic term. The matching is seen to be excellent. The results are taken from Chen et al. (2005d; 2006e).

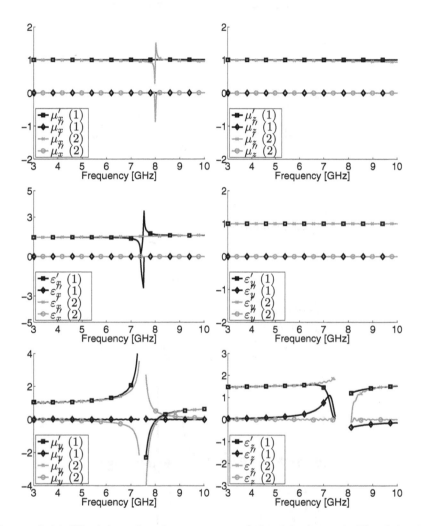

Figure 2.11 Biaxial retrieved parameters of the ring shown in Fig. 2.8 using the method of Section 2.4.2. The blanked regions correspond to frequencies where the mismatch between each corresponding two results is greater than a threshold. The results are taken from Chen et al. (2005d; 2006e).

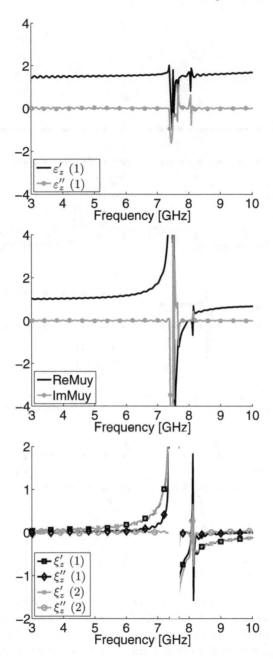

Figure 2.12 Bianisotropic retrieved parameters μ_y, ε_z, and ξ of the ring shown in Fig. 2.8 using the method of the present section. The blanked regions correspond to frequencies where the mismatch between each corresponding two results is greater than a threshold. The results are taken from Chen et al. (2005d; 2006e).

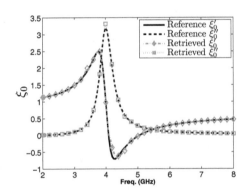

Figure 2.13 Comparison between the retrieved and the forward bianisotropic term of a rotated Omega medium. Results are from Chen et al. (2006d).

3

Designing metamaterials with negative material parameters

The brief historical description offered in Chapter 1 showed how the initial studies of materials with both $\varepsilon < 0$ and $\mu < 0$ (Veselago 1968) did not spark much interest at first because such materials were not available at the time. Their realization had to wait for more than three decades before the recipes for achieving $\varepsilon < 0$ and $\mu < 0$ (Pendry et al. 1996; 1999) paved the way for experimentalists to design these materials. It was the first experimental demonstrations of such media (Smith et al. 2000, Shelby et al. 2001b) at microwave frequencies that caused most of the scientific community to sit up and take notice. The subsequent fevered advances in this area are now history, but one has to remember that it was the new design principles within a physical paradigm that made this whole area relevant.

In this chapter, we discuss the design principles of metamaterials with various material parameters. General principles for the design of media with specified material parameters are laid down. In particular, metamaterials with negative constitutive parameters are of special interest. It is shown that in most cases, under-damped and over-screened resonances can lead to negative constitutive parameters (ε or μ) for the corresponding driving fields (the electric field **E** or the magnetic field **H**). Negative refractive index can be generated in many cases by putting together in a composite manner two structures that separately show $\varepsilon < 0$ and $\mu < 0$, although this presupposes that the two structures function independently without strongly interfering with each other, a condition that is not always automatically satisfied. We also show that the same structure can exhibit both dielectric and magnetic properties at optical frequencies through plasmonic resonances where the homogenization requirements put a premium on physically accessible designs and sizes. We also discuss chiral and bianisotropic materials, and the design principles for those media. The emphasis is on demonstrating how a variety of resonant phenomena can be utilized to generate various kinds of metamaterials through concrete examples. Finally, the possibility of actively tuning the resonance frequencies of the metamaterials by locally modifying the constituent materials in critically placed locations via non-linearities excited by externally applied fields is presented.

The implementation of metamaterials as a succession of split-ring resonators and rods yields a structure that is complex: the scattered electromagnetic field is very inhomogeneous, intense at some locations and weak at others, and influenced by neighboring elements in a way that is not straightforward to predict. Consequently, a rigorous electromagnetic analysis of metamaterials requires the use of advanced electromagnetic solvers, typically based on numerical methods. On the other hand, Chapter 2 has shown that metamaterials with small units can be homogenized, i.e., that effective constitutive parameters corresponding to homogeneous media can be assigned, based on which macroscopic scattering properties can be inferred. This suggests that the microscopic details of the electromagnetic field need not be totally resolved and only the knowledge of the average behavior is necessary. This principle has been used for example in Section 2.5 to infer the effective constitutive parameters based on the internal fields inside the metamaterial, the latter typically being obtained using one of the advanced electromagnetic solvers. This flexibility of not having to know the details of the scattered field can also be translated into a series of approximations of the fields at the microscopic level (i.e., inside the metallic constituents of the metamaterial), which is an approach that leads to the development of equivalent circuit models in which simple concepts such as the balance of the electromotive force (emf) can be applied. This approach is extensively used in this chapter since it is a simple yet intuitive tool to understand the physical working principle of split rings, to provide good estimates of their resonant frequencies, and to study the influence of various geometrical parameters on their performance.

Let us first consider Faraday's law in the integral representation form which, combined with the Stokes theorem, reads as

$$\oint_C d\ell \cdot \mathbf{E} = -\frac{\partial}{\partial t} \iint_S d\sigma \cdot \mathbf{B}. \tag{3.1}$$

Eq. (3.1) indicates that the surface integral of the magnetic induction $\mathbf{B}$ yields a magnetic flux Φ whose time variation is related to an electric field around the corresponding enclosing loop. Upon defining the emf as the line integral of the electric field, Faraday's law becomes

$$\text{emf} = -\frac{\partial}{\partial t}\Phi(t), \tag{3.2}$$

which indicates that the current in the loop, related to the emf, always flows in a direction such as to oppose the change in the magnetic flux $\Phi(t)$ that produced it. This result is known as Lenz's law. Upon resorting to some approximations on the regularity of the fields, Eq. (3.2) becomes the building block for the definition of equivalent circuits representing a metamaterial.

We should note that the equivalent circuits defined here are circuits indeed, but where the currents and voltages are defined based on a simplification of the full-wave Faraday's law of Eq. (3.1) and not based on Kirchhoff's approximation. In order to be convinced of this, consider the definition of the voltage between two points as the integral of the electric field between these two points (where the integration path is irrelevant as far as the starting and ending points are fixed). Next, let the two points merge, yielding a closed integration path. The voltage would then vanish whereas Eq. (3.1) indicates that the result is not zero but is equal to the time variation of the magnetic flux. Consequently, Kirchhoff's voltage law appears as an approximation of Faraday's equation in the electrostatic situation where $\nabla \times \mathbf{E} = 0$, whereas this is not the approximation the equivalent circuits are based upon.

3.1 Negative dielectric materials

We first address the problem of designing negative dielectric materials at any specified frequency, starting by considering natural materials (such as negative dielectric materials, *viz.*, plasmas of electrical charges) and continuing by considering engineered "designer" materials, which have become popular over the past few years.

3.1.1 Metals and plasmons at optical frequencies

Consider an electrically neutral plasma of separated charges. Similarly to the discussion on the Lorentz model for dielectric materials in Chapter 1, we assume that the mass of the negative charges (electrons) is much smaller than that of the positive charges (ion cores in a metal). Thus, it is only the motion of the electrons that contributes to the polarization in the medium. This model is most apt for a good metal as the conduction electrons are almost free and delocalized in a background of static positive ionic cores with overall charge neutrality. We subsequently show that such a plasma exhibits a negative dielectric permittivity for frequencies smaller than a particular frequency called the *plasma frequency*.

The equation of motion for an electron in a time harmonic electric field with angular frequency ω is

$$m\ddot{\mathbf{r}} + m\gamma\dot{\mathbf{r}} = -e\mathbf{E_0}\exp(-i\omega t), \tag{3.3}$$

where the magnetic contribution to the Lorentz force is neglected as usual.* This equation is identical to the one we began with in the discussion on the Lorentz model, only without the harmonic restoring force, as here the electrons are free. $m\gamma$ is the phenomenological damping force constant due to all inelastic processes. It is assumed that the wavelength of radiation is large compared to the distance traveled by the electron so that the electrons

---

*It is shown in Chapter 5 that such a neglect is not always possible and the magnetic contribution can lead to non-linear effects.

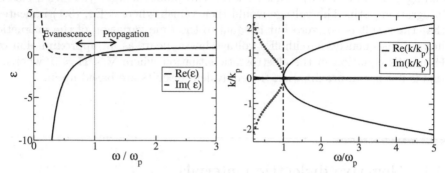

(a) Real and imaginary parts of the dielectric permittivity for a good plasma. The real part is negative for frequencies lesser than the plasma frequency.

(b) Dispersion of light in a good plasma.

Figure 3.1 Electromagnetic properties of plasmas. There is a dispersionless bulk mode at the plasma frequency (shown by the dashed line) and the dispersion for propagating modes with a real wave-vector above and near the plasma frequency is parabolic. Below the plasma frequency we only have evanescent waves with purely imaginary wave-vectors.

effectively see a spatially uniform field. These assumptions yield a polarization defined as the dipole moment per unit volume that can be expressed as

$$\mathbf{P} = (\varepsilon - 1)\varepsilon_0 \mathbf{E} = -Ne\mathbf{r} = -\frac{Ne^2/m\mathbf{E}}{w(\omega + i\gamma)}, \tag{3.4}$$

where N is the number density of the conduction electrons, and each electron is assumed to contribute independently to the polarization. Hence we obtain the relative dielectric permittivity as

$$\varepsilon(\omega) = 1 - \frac{Ne^2/(\varepsilon_0 m)}{w(\omega + i\gamma)} = 1 - \frac{\omega_p^2}{w(\omega + i\gamma)}, \tag{3.5}$$

where the plasma frequency is defined as $\omega_p \equiv \sqrt{(Ne^2)/(\varepsilon_0 m)}$.

It is immediately seen from the above that the real part of the dielectric permittivity is negative for $\omega < \omega_p$ and for negligible dissipation. The real and imaginary parts of the dielectric function are plotted in Fig. 3.1(a). For frequencies lower than the plasma frequency, the charges can move quickly enough to shield the interior of the medium from electromagnetic radiation. At frequencies larger than the plasma frequency, the medium behaves as an ordinary dielectric medium. Substituting the dispersive expression for ε into

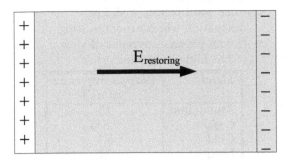

Figure 3.2 Schematic picture showing the bulk plasma oscillation of the cloud of electrons in a slab of metal. The schematic shows the excess positive and negative charge density that builds up on either side of the slab and gives rise to a restoring force on the oscillating electron cloud.

the Maxwell equations, we get

$$k^2 c^2 + \omega_p^2 = \omega^2. \tag{3.6}$$

Thus, waves with $\omega < \omega_p$ correspond to negative energy solutions and are evanescent with an imaginary wave-number. For $\omega > \omega_p$, the transverse modes of light appear to have become massive with a finite rest mass of $m_0 = \hbar\omega_p/c^2$.[†] The dispersion is parabolic at frequencies near ω_p and becomes asymptotic to the light cone at large frequencies as shown in Fig. 3.1(b).

The plasma frequency has a physical manifestation as the natural frequency of a collective excitation of the electron gas. Consider a small displacement (Δ) of the entire electron gas in a finite metal sample relative to the positive background (assumed homogeneous, too), as shown in Fig. 3.2 for a slab of metal. An accumulation of net negative electronic charges is created on one edge and of positive background charges on the other. The electric field due to these excess charges in the bulk can be calculated as $E = Ne\Delta/\varepsilon_0$ along the displacement direction, where N is the number density of the conduction electrons. The force on an electron is then given by $F = -Ne^2/\varepsilon_0 \, \Delta$, i.e. there appears a restoring force proportional to the displacement. The natural frequency for the oscillation of this entire electron gas comes out to be $\sqrt{(Ne^2)/(\varepsilon_0 m)}$, i.e. the plasma frequency. The bulk plasma oscillations are called plasmon modes and have a dispersionless character shown as the dashed line in Fig. 3.1(b). There are also other kinds of plasma oscillations that are confined to the interface of metal and a positive dielectric medium called *surface plasmons* to which we shall return in Chapter 7.

[†]Compare the dispersion equation with the relativistic equation of motion for a massive particle, $p^2 c^2 + m_0^2 c^4 = E^2$ with $p = \hbar k$ and $E = \hbar\omega$.

Table 3.1 Plasma frequencies of some good metals with low dissipation which are interesting from a plasmonics point of view. (* has high dissipation)

Metal	Plasma frequency (in eV)
Aluminium	15
Cesium	2.845
Gold	5.8
Lithium	6.6
Nickel	9.45
Palladium*	7.7
Potassium	3.84
Silver	3.735

Several "good" metals with low dissipation (small γ) can be reasonably well described as a Lorentz plasma. The plasma frequencies for many metals lie in the ultraviolet region of the spectrum. These include noble metals like gold and silver, alkali metals such as lithium, sodium and sotassium, and others such as aluminum. Tab. 3.1 gives the plasma frequency for some metals. Silver with the lowest value of γ and large enough value of ω_p is the best candidate for a negative dielectric material. The γ in this case actually does not correspond to the Drude conductivity of the metal, but the inter-subband transitions of the d-levels actually account for much of the dissipation in noble metals. In fact, the Drude Lorentz plasma model for the dielectric constant is only approximate for silver and gold.

We should note that some dielectric materials can also have interesting spectral (stop) bands with $\varepsilon < 0$. Going back to the Lorentz model discussed in Section 1.3.2, if the oscillator strength is large enough and the damping small enough in Eq. (1.15), then the dielectric constant becomes negative over the frequency band $\omega_0 < \omega < \omega_p$ where $\omega_p = \sqrt{(\omega_j^2 + \omega_0^2)}$ is the equivalent plasma frequency. In fact, there are optical phonon resonances in several dielectric and ionic materials that can become important. As an example, there is an optical phonon resonance in silicon carbide (SiC) at about 10 μm wavelength and a negative dielectric permittivity band above the resonance frequency. Given that the dissipation levels in dielectric materials can be much lesser than that in metals, such dielectric materials have great potential in the design of metamaterials for use as negative dielectric materials. This becomes particularly relevant at low frequencies (near-mid infrared and lower frequencies) when the dissipation in metals does not even allow for an easy description as a Lorentz plasma. The disadvantage of these dielectric materials is that the frequency range over which $\varepsilon < 0$ is often narrow unlike in metals.

3.1.2 Wire mesh structures as low frequency plasmas

While it is true that a good plasma should, in principle, have $\varepsilon < 0$ for frequencies lower than the plasma frequency all the way to zero frequency, the dissipation that is inherent to all negative materials spoils this property. As γ, the dissipation parameter, becomes comparable and even larger than the frequency (ω), the dielectric permittivity becomes more of a complex number and it is less accurate to describe the response using the negative real part of the permittivity alone. Further, at low frequencies the dielectric permittivity would have a very large magnitude implying that all radiation would decay within very short distances in the medium. Dielectric materials such as SiC have only a finite bandwidth with $\varepsilon < 0$. Hence, ideally one would seek a way of mimicking the action of a plasma at lower frequencies in a composite material.

It was proposed in Pendry et al. (1996; 1999) that dilute arrays of thin metallic wires behave as a low frequency plasma with a frequency stop-band from zero up to a cutoff frequency that can be attributed to the motion of the electrons in the wires. This metallic thin wire mesh structure is amenable to homogenization and description by a plasma-like dielectric permittivity function. The low frequency stop-band can then be attributed to negative dielectric permittivity and the cutoff frequency to the *plasma frequency*. The rationale behind this structure is to literally dope the vacuum with metal.

Consider an array of infinitely long, parallel, and very thin metallic wires of radius r placed periodically at a distance a in a square lattice with $a \gg r$ as shown in Fig. 3.3. The electric field is considered to be applied parallel to the wires (along the $\hat{z}$ axis) and the problem can be analyzed in the quasi-static limit since the wavelength of the radiation $\lambda \gg a \gg r$. The electrons are confined to move within the wires only, which has the first effect of reducing the effective electron density, as the radiation cannot sense the individual wire structure but only the average charge density. The effective electron density is immediately seen to be

$$N_{\text{eff}} = \frac{\pi r^2}{a^2} N, \tag{3.7}$$

where N is the actual density of conduction electrons in the metal.

There is a second equally important effect to be considered. The thin wires have a large self-inductance and it is not easy to change the currents flowing in them. Thus, it appears as if the charge carriers, namely, the electrons, had acquired a tremendously large mass. To see this, consider the magnetic field created at a distance ρ from a wire. On the average, we can assume a uniform **D** field within the unit cell. However, the current density is not uniform leading to a non-zero, non-uniform magnetic field that is large in regions close to the wires, which contributes to most of the flux. By symmetry there is a point of zero field in-between the wires and hence we can estimate the

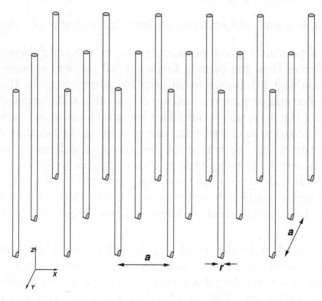

Figure 3.3 An array of infinitely long, thin metal wires of radius r on a lattice of period of a behaves as a low frequency plasma for an electric field oriented along the wires. (Reproduced with permission from Ramakrishna (2005). © 2005, Institute of Physics Publishing, U.K.)

magnetic field along a line between two wires at a distance ρ from the wire as

$$\mathbf{H}(\rho) = \hat{\phi}\,\frac{I}{2\pi}\left(\frac{1}{\rho} - \frac{1}{a-\rho}\right). \tag{3.8}$$

The vector potential associated with the field of a single infinitely long, current-carrying conductor is non-unique unless the boundary conditions are specified at definite points. In our case, we have a periodic medium that sets a critical length of $a/2$. We can assume that the vector potential associated with a single wire is

$$\mathbf{A}(\rho) = \begin{cases} \frac{\hat{z}I}{2\pi}\ln\left[\frac{a^2}{4\rho(a-\rho)}\right], & \forall\ \rho < a/2, \\ 0, & \forall\ \rho > a/2. \end{cases} \tag{3.9}$$

This choice avoids the vector potential of one wire overlapping with another wire and thus the mutual induction between the adjacent wires is addressed to some extent. Noting that $r \ll a$ by about three orders of magnitude in our model and that the current $I = \pi r^2 N(-e)v$ where v is the mean electron velocity, we can write the vector potential as

$$\mathbf{A}(\rho) = -\frac{\mu_0 \pi r^2 N e v}{2\pi}\ln(a/\rho)\hat{z}. \tag{3.10}$$

This is a very good approximation in the mean field limit. We have considered only two wires and the lattice actually has a four-fold symmetry. Hence the actual deviations from this expression are much smaller than in our case.

We note that the canonical momentum of an electron in an electromagnetic field is $\mathbf{p} - e\mathbf{A}$. Assuming that the electrons flow on the surface of the wire (for a perfect conductor), we can associate a momentum per unit length of the wire of

$$\mathbf{p} = -\pi r^2 n e \mathbf{A}(r) = \frac{\mu_0 \pi^2 r^4 N^2 e^2 v}{2\pi} \ln(a/r) = m_{\text{eff}} \pi r^2 N v, \qquad (3.11)$$

and thus an effective mass of

$$m_{\text{eff}} = \frac{\mu_0 \pi r^2 N e^2}{2\pi} \ln(a/r) \qquad (3.12)$$

for the electron.

Assuming a longitudinal plasmonic mode for the system, we have

$$\omega_p^2 = \frac{N_{\text{eff}} e^2}{\varepsilon_0 m_{\text{eff}}} = \frac{2\pi c^2}{a^2 \ln(a/r)} \qquad (3.13)$$

for the plasmon frequency. We note that a reduced effective electron density and a tremendously increased effective electronic mass would immediately reduce the plasmon frequency for this system. Typically one can choose $r = 10 \ \mu\text{m}$, $a = 10 \ \text{mm}$, and aluminium wires ($N \simeq 10^{29} \ \text{m}^{-3}$), which gives an effective mass of

$$m_{\text{eff}} = 1.12 \times 10^{-24} \ \text{kg}, \qquad (3.14)$$

which is more than six orders of magnitude larger than the bare electron mass, and a plasma frequency of about 28.6 GHz. Thus, we have succeeded in obtaining a negative dielectric permittivity material at microwave frequencies since the plasma frequency has been reduced by almost a factor of 10^6.

Note that the final expression for the plasma frequency in Eq. (3.13) is independent of the microscopic quantities such as the electron density and the mean drift velocity. It only depends on the radius of the wires and on the spacing suggesting that the entire problem can be recast in terms of the capacitances and inductances of the problem. This approach has been taken in Belov et al. (2002; 2003), and we present it below for the sake of completeness. Consider the current induced by the electric field along the wires related by the total inductance (self and mutual) per unit length (L),

$$E_z = -i\omega L I = i\omega L \pi r^2 n e v. \qquad (3.15)$$

Note that the dipole moment per unit volume in the homogenized medium is

$$P = -N_{\text{eff}} e r = \frac{N_{\text{eff}} e v}{i\omega} = -\frac{E_z}{\omega^2 a^2 L}, \qquad (3.16)$$

where $N_{\text{eff}} = \pi r^2 N/a^2$ as before. We can estimate the inductance L by calculating the magnetic flux per unit length passing through a plane between the wire and the point of symmetry between itself and the next wire where the field is approximately zero:

$$\phi = \mu_0 \int_r^{a/2} H(\rho)d\rho = \frac{\mu_0 I}{2\pi} \ln\left[\frac{a^2}{4r(a-r)}\right]. \qquad (3.17)$$

Noting $\Phi = LI$, and the polarization $P = (\varepsilon - 1)\varepsilon_0 E_z$, where ε is the effective permittivity, we obtain in the limit $r \ll a$,

$$\varepsilon(\omega) = 1 - \frac{2\pi c^2}{\omega^2 a^2 \ln(a/r)}, \qquad (3.18)$$

which is identical to the expression obtained using the plasmon picture. However, we lose here the physical interpretation of a low frequency plasmonic excitation.

It is simple to add the effects of finite conductivity in the wires, which has been neglected in the above discussion. The electric field and the current would then be related by

$$E_z = -i\omega LI + \frac{I}{\sigma\pi r^2}, \qquad (3.19)$$

where σ is the conductivity, which modifies the expression for the dielectric permittivity to

$$\varepsilon = 1 - \frac{\omega_p^2}{\omega\left(\omega + i\frac{\varepsilon_0 a^2 \omega_p^2}{\pi r^2 \sigma}\right)}, \qquad (3.20)$$

where ω_p is given by Eq. (3.13). Thus the finite conductivity of the wires contributes to the dissipation, appearing in the imaginary part of ε. For aluminium, the conductivity is $\sigma = 3.78 \times 10^7 \Omega^{-1}\text{m}^{-1}$ and yields a value of $\gamma \simeq 0.1\omega_p$, which is almost comparable to the values in real metals (Raether 1986). Thus the low frequency plasmon is stable enough against absorption to be observable.

In our above discussion, we considered only wires pointing in the $\hat{z}$ direction. This makes the medium anisotropic with a negative ε only for waves with an electric field along the $\hat{z}$ direction. The medium can be made to have a reasonably isotropic response by having a lattice of wires oriented along all the three orthogonal directions as shown in Fig. 3.4. In the limit of large wavelengths, the effective medium appears to be isotropic as the radiation fails to resolve the underlying cubic symmetry yielding a truly three-dimensional low frequency plasma. For very thin wires, the polarization in the direction orthogonal to the wires is small (Rayleigh-like) and can be neglected. Thus, the waves only sense the wires parallel to the electric field and correspondingly have a longitudinal mode. The effects of connectivity of the wires along different directions at the edges of the unit cell have also been examined (Pendry

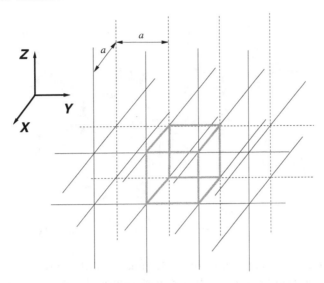

Figure 3.4 A three-dimensional lattice of thin conducting wires behaves as an isotropic low frequency plasma. (Reproduced with permission from Ramakrishna (2005). © 2005, Institute of Physics Publishing, U.K.)

et al. 1998). A wire mesh with non-intersecting wires was also shown to have a negative ε at low frequencies below the plasma frequency, but had strong spatial dispersion for both the longitudinal bulk plasmon mode and the transverse modes above the plasma frequency (Belov et al. 2003).

Next, we discuss the effects of having discontinuous conducting wire segments (or commonly called cut wires) instead of infinitely long continuous wires. The main difference is the presence of the capacitance between the ends of the adjacent wire segments. Consider the metamaterial structure shown in Fig. 3.5 which also allows for a tuning of the capacitance between the wire segments by adjusting the area or by embedding a dielectric in-between the edge plates. The capacitance between the cut wires is approximately $C_s = \varepsilon_0 \varepsilon_h A/d$ where A is the area of the capacitive edges, d is the distance between them, and ε_h is the relative dielectric permittivity of the medium in the capacitive gaps. The self capacitance of each wire segment is $C_i = \varepsilon_0 \varepsilon_i A/\ell$, where ℓ is the length of the wire segment and ε_i is the dielectric permittivity of the embedding medium. Note that the thickness of the wire has been neglected as well as the self capacitance since $\ell \gg d$. Incorporating the series capacitance per unit length in the expression for the electric field along the wire segments yields

$$E_z = -i\omega LI + \frac{I}{\pi r^2 \sigma} + \frac{1}{-i\omega C_s(\ell + d)}I. \tag{3.21}$$

Proceeding as before for the continuous wires, we can obtain the dielectric

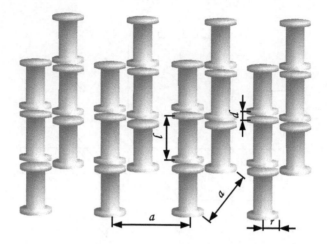

Figure 3.5 A metamaterial of discontinuous metallic wire segment behaves like a series of electric dipoles. In the metamaterial shown, the area of cross section (A) at the edges of the wire segments of length ℓ and the distance d between the wire segments are design parameters that can be used to tune the resonant frequency.

permittivity function for this composite to be of the Lorentz form:

$$\varepsilon(\omega) = 1 + \frac{\omega_p^2}{(\omega_0^2 - \omega^2 - i\Gamma\omega)},\qquad(3.22)$$

where ω_0 is the resonance frequency given by

$$\omega_0 = \sqrt{\frac{1}{LC_s(\ell+d)}} = \sqrt{\frac{2\pi c^2 d}{\varepsilon_h A(\ell+d)\ln(a/r)}}.\qquad(3.23)$$

ω_p is defined before by Eq. (3.13) and the dissipation parameter is

$$\Gamma = \frac{a^2}{\pi r^2}\frac{\varepsilon_0\omega_p^2}{\sigma}.\qquad(3.24)$$

This gives a frequency bandwidth of $[\omega_0, \sqrt{\omega_0^2 + \omega_p^2}]$ where the dielectric permittivity of the composite is negative.

   Dipole moments are resonantly formed across the capacitive gaps between the wire segments and the system exhibits a Lorentz-like dispersion. Thus, the response of the cut-wire system is more akin to a system of periodically placed dipoles. If the conductivity is large enough, the system exhibits a negative dielectric permittivity for a finite frequency bandwidth just above the resonance frequency ω_0. An advantage of this metamaterial is that the resonance frequency can be chosen by an appropriate choice of the capacitive gaps or the width of the capacitive gap in the metamaterial. Notice that interestingly while both the inductance and capacitance of the system determine the resonance frequency, the bandwidth for $\varepsilon < 0$ is mostly determined by the inductance. The dissipation is also only determined by the inductance and conductivity, unless a dissipative dielectric material is placed inside the capacitive gaps, in which case additional losses occur.

For example, consider 10 μm thick wire segments, 1 mm in length, spaced apart by 50 μm and placed on a sparse square periodic lattice of period 5 mm. If we assume a capacitive cross-section area of $\pi \times 10^{-10}$ m^2 and an embedding medium with a dielectric permittivity of $\varepsilon_h = 4$, we obtain $w_p \simeq$ 28 GHz, and $w_0 = 569$ GHz. For denser and much smaller scale wires of $r = 0.5$ μm, $a = 100$ μm, $\ell = 10$ μm, $d = 1$ μm embedded in a medium with dielectric permittivity $\varepsilon_h = 4$ and a lattice periodicity of $a = 50$ μm, we obtain $w_0 \simeq 13.9$ THz, i.e. in the mid-infrared region of the spectrum and $w_p = 7$ THz. Thus, we have a complete flexibility in the choice of the negative dielectric frequency range through the design of the metamaterial.

There is, however, a problem when we consider scaling up the structures to high frequencies ($\sim$10 THz and above). Metals as mentioned before do not behave as Ohmic conductors at high frequencies, but rather as plasmas with negative dielectric permittivity. We therefore need to analyze a system of cut wires at high frequencies when the bulk dielectric permittivity of the metal constituting the structure is modeled as a Drude plasma. It is assumed, however, that the frequencies are much lower than the bulk plasmon frequency so that the dielectric permittivity function can be approximated as

$$\varepsilon_m(\omega) = 1 - \frac{\omega_m^2}{\omega(\omega + i\gamma)} \simeq -\frac{\omega_m^2}{\omega(\omega + i\gamma)}. \tag{3.25}$$

The metal, therefore, has a negative dielectric permittivity whose magnitude is large. Let us consider the metamaterial of Fig. 3.5 with an incident electric field parallel to the wire segments. The cut-wire segments can be understood as primarily two capacitors in series per unit cell – one between the adjacent cut-wires and another one across the wire segment. The capacitance across the metallic wire segment is then essentially negative due to the large negative dielectric permittivity of the metal. A negative capacitance is equivalent to an inductance, which resonates with the positive capacitance across the gap between the wire edges. Treating the wire segment as a negative capacitance is justified by our assumption of a large magnitude for the bulk dielectric permittivity of the metal. This implies that all the electric field lines along the length of the wire segment between the two end-faces are concentrated inside the metal segment only with very few fringing field lines in the embedding medium. Hence the contribution to the self-capacitance of the wire segment arising from the free space outside can be neglected. The potential drop across a length ℓ_0 of the cut-wire is therefore

$$V = \frac{I}{-i\omega C} = E_z \ell_0, \tag{3.26}$$

where the series capacitance, C, is given by

$$\frac{1}{C} = \left(\frac{1}{C_s} + \frac{1}{C_m} \right) \frac{\ell_0}{\ell + d}. \tag{3.27}$$

Here $C_m \simeq \varepsilon_0 \varepsilon_m \pi r^2 / \ell$ approximately and $C_s = \varepsilon_0 \varepsilon_r A / d$ as before. Again relating the induced current to the induced polarization, we can calculate an effective dielectric function for the composite as

$$\varepsilon(\omega) = 1 + \frac{(\pi r^2 / a^2)(\ell + d)/\ell \omega_m^2}{(\pi r^2 / A)(d/\ell) \omega_m^2 - \omega(\omega + i\gamma)}. \tag{3.28}$$

Thus, we obtain again a Lorentz-like dispersion for the dielectric permittivity. There are some interesting aspects to the above expression. The resonance frequency occurs at a frequency lower by a ratio $(\pi r^2 / A)(d/\ell)$ than the bulk frequency of the metal. If we were to make the capacitive gaps and the wire segments of equal sizes we would not obtain this reduction. The bandwidth of the $\varepsilon < 0$ band is determined by the filling fraction $f = (\pi r^2 / a^2)$ primarily along with the bulk plasma frequency. Thus the system can be considered literally as a *diluted metal*. Note that the dissipation parameter in the system turns out to be the same as in the bulk metal. This is due to our assumption of all the fields being concentrated in the metal, and thus the dissipation is the same as if the metal were all pervading. The actual levels of dissipation should be expected to be slightly lower when one accounts for the positive capacitance coming from the embedding medium.

The above model is expected to accurately describe metamaterials with resonances at the near-infrared and the mid-infrared frequencies. In this case, our assumptions for the bulk dielectric function as a large and negative real number hold. At higher optical frequencies, however, the assumption that the electric field is entirely concentrated in the metal wire segments with very little fringing fields except in the capacitive gaps breaks down. Each wire segment can itself resonate as an electric dipole when the contribution to the self capacitance due to the fields in the surrounding medium is of the same order as the negative self capacitance arising from the fields inside the metal. In other words, we have a resonance of the wire segments: these are actually the localized plasmonic resonances of the particles. These plasmonic resonances can interact and hybridize between the wire segments and give rise to new bands and bandgaps. A comprehensive treatment of these collective plasmonic excitations is beyond the scope of this book and is a matter of active research. At far lower frequencies, the dissipation parameter γ dominates and the plasmonic nature of the particles is barely felt. Hence the previous treatment as Ohmic wires would model the system better at low frequencies.

The photonic response of superconducting cylinders made of high T_c superconducting materials has been considered (Takeda and Yoshino 2003) and it was shown that the system has a low frequency cutoff that is much smaller than the corresponding energy scale for the transition temperature of the bulk superconductor. The role of a plasmon in a superconductor as a massive *Higgs boson* has been stressed in Anderson (1963). It has been pointed out by Pendry et al. (1996; 1998) that in such thin-wire superconductor structures, the frequency of the plasmon could well be smaller than the superconducting gap making such media of great interest in a fundamental manner. Note, however, that the use of a superconducting material for low, if not zero, resistivity material is valid only for electromagnetic radiation with frequency smaller than the frequency of the superconducting gap for the Cooper pairs. Larger frequency radiation simply excites the charge carriers across the superconducting gap and the system exhibits resistive losses.

3.1.2.1 Other photonic metallic wire materials

Metal dielectric composites have been long studied for their rich electrodynamic response. For example, composites of randomly oriented long conducting fibers have been known to exhibit very large values of the permittivity (Lagarkov and Sarychev 1996, Sarychev and Shalaev 2000). Dense wire media and similar rod media also known as *fakir's beds* have been of interest to electrical engineers for a long time (Brown 1960, Rotman 1962, King et al. 1983). However, these structures were usually considered when the periodicity of the mesh as well as the diameters of the wires was of the order of the wavelength. Hence these systems are not easily amenable to homogenization and to the definition of effective medium parameters. In dense metallic structures the electromagnetic radiation generally penetrates only one or two unit cells, effectively interacting with only a two-dimensional surface layer of the structure. In large periodicity photonic structures, Bragg scattering becomes important when the structure lengthscale and the wavelength become comparable. Thus the structure cannot be effectively homogenized.

For example, the existence of a low frequency stop-band from zero frequency up to a cutoff frequency in a diamond lattice wire mesh as well as photonic bandgaps due to Bragg scattering at higher frequencies was demonstrated in Sievenpiper et al. (1996). The low frequency stop-band could be attributed to the flow of electrons in the interconnected metallic mesh, but at the plasma frequency the lattice size was barely half the free space wavelength. Thus the modes have a considerable spatial dispersion as in a photonic crystal. It is not completely clear if this structure could be homogenized in the classical sense. An interesting transmission line model (Brown 1960) gives

$$\varepsilon = \frac{\varepsilon_0}{(ka)^2} \left\{ \cos^{-1}\left[\cos(ka) + \frac{\pi}{ka\ln(a/(2\pi r))}\sin(ka) \right] \right\}^2, \qquad (3.29)$$

which can be shown to agree with our expression for a low frequency plasmonic medium in the limit of very thin wires (Belov et al. 2002). Sievenpiper et al. (1996) also considered capacitively loaded wires by cutting selected wires in the wire mesh medium and demonstrated the presence of resonant modes associated with the cut-wires. The photonic response of metallic cylinders embedded in a dielectric host has been considered in Pitarke et al. (1998) and agreed with the classical Maxwell-Garnett results when the wavelength was at least twice as large as the spacing between the cylinders. However, these cases do not ideally fall in the category of effective media.

3.2   Metamaterials with negative magnetic permeability

Although electric and magnetic fields are just two manifestations of the same fundamental phenomenon, the response of most materials is extremely one-sided and partial to the electric fields. One can find good dielectric materials at almost any frequencies, from radio to the ultraviolet frequencies. The magnetic response of most materials, on the other hand, tends to die out at high frequencies beyond the microwave frequencies. Magnetization usually results from unpaired electronic spins or electronic orbital currents. Shape-dependent collective excitations of these systems (Landau et al. 1984) can result in resonances at frequencies of even a few hundreds of Gigahertz in some ferromagnetic, anti-ferromagnetic or ferromagnetic materials. But these are rare and usually have very small bandwidths. Thus, even magnetic activity, let alone negative magnetic permeability, is special at frequencies beyond the microwave regime.

In fact, Landau et al. (1984) tried to give a very general argument that magnetic activity from real electronic currents (including orbital currents) should be negligible at optical frequencies and beyond. Their argument, however, crucially depends on neglecting the displacement currents arising from changes in polarization compared to real electronic currents. In many sub-micron sized metallic systems, the displacement currents can become very large, for instance at the surface plasmon resonances, and can dominate over the electronic currents which become small at high frequencies. Hence such systems can effectively develop magnetization even at high frequencies.

We subsequently analyze structured materials which give rise to magnetic activity and, in particular, to a negative magnetic permeability. The crucial issue is to design subwavelength sized structures whose resonances can be driven by the magnetic component of the electromagnetic field so that the corresponding effect can be reflected in the dispersion of the magnetic permeability. We first analyze how to obtain magnetic activity along any one direction in Sections 3.2.1 to 3.2.6. The designs presented in these sections are necessarily highly anisotropic, showing a magnetic activity only for magnetic

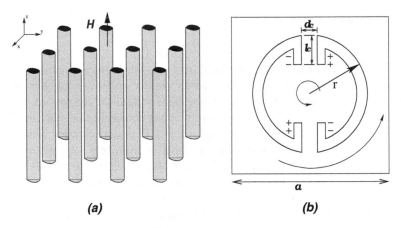

Figure 3.6 (a) An applied magnetic field along the axes of a stack of conducting cylinders induces circumferential currents that shield the interior. The resulting effective medium is diamagnetic. (b) The split-ring structure: the capacitance across the splits in the ring causes the structure to be resonant. The dimensions of the capacitive gaps l_c and d_c become design parameters to control the amount of capacitance in the loop.

fields oriented along one specific direction. We then generalize these ideas to structures that display magnetic activity along all three directions, with improved designs that have a more isotropic magnetic response.

Before pursuing with the analysis of artificial magnetic structures, it is important to note that we treat the radiation as a classical electromagnetic field and the structures as classical systems. Then, given that magnetic fields do not affect the free energy of a statistical system in thermal equilibrium, one can obtain magnetism only as a quantum phenomenon. Hence one could question our result of magnetism obtained in a classical treatment. This apparent paradox can be resolved by noting that the material structures that we consider and the radiation are at different temperatures and are not at equilibrium. Thus, the (AC) magnetism obtained here are, in fact, induced for systems driven away from thermal equilibrium.

3.2.1 Diamagnetism in a stack of metallic cylinders

Most materials have a natural tendency to be diamagnetic as a consequence of Lenz's law. Consider the response of a stack of metallic cylinders to an incident electromagnetic wave with the magnetic field along the axis of the cylinders as shown in Fig. 3.6(a). The cylinders have a radius r and are placed in a square lattice with period a. The oscillating magnetic field along the cylinders induces circumferential surface currents which tend to generate a magnetization opposing the applied field. The axial magnetic field inside the cylinders is

$$H = H_0 + j - \frac{\pi r^2}{a^2} j, \qquad (3.30)$$

where H_0 is the applied magnetic field and j is the induced current per unit length of the cylinder. A long magnetized cylinder appears as if there are two separated magnetic monopoles with opposite magnetic charges at its ends. Thus, there is a depolarizing field that is present inside the cylinder due to the depolarizing fields of the other cylinders. This depolarizing field can be assumed to be spatially uniform in the limit of infinitely long cylinders and uniform distribution of cylinders. The third term arises due to this depolarizing magnetic field. Thus the mutual inductance between the cylinders is taken into account to the first order here. In fact, if L is the self inductance of one cylinder and L' is the total inductance of all the other cylinders, one can show that the mutual inductance M of the coupling between them can be calculated to be

$$M = \frac{\Phi_L}{I} = \lim_{n \to \infty} \frac{(\pi r^2/na^2)\Phi_d}{I},$$
$$= \frac{\pi r^2}{a^2} L, \qquad (3.31)$$

where $\Phi_d = (n-1)LI$ is the total flux due to the depolarizing fields of all other cylinders, and Φ_L is the flux inside the cylinder due to the currents on the cylinder under consideration. The ratio $f = (\pi r^2/a^2)$ is called the filling fraction in the metamaterial.

The emf around the cylinder can be calculated from Lenz's law and is balanced by the Ohmic drop in potential:

$$i\omega\mu_0\pi r^2 \left(H_0 + j - \frac{\pi r^2}{a^2} j \right) = 2\pi r \rho j, \qquad (3.32)$$

where time harmonic fields are assumed and where ρ is the resistance per unit length of the cylinder surface. The frequencies are assumed to be low enough to have only a small skin effect.

The system of cylinders can be homogenized by adopting an averaging procedure (discussed in Chapter 2) that consists of averaging the magnetic induction B over the area of the unit cell while averaging the magnetic field H over a line along the edge of the unit cell. The averaged magnetic field is

$$B_{\text{eff}} = \mu_0 H_0, \qquad (3.33)$$

while the averaged H field outside the cylinders is

$$H_{\text{eff}} = H_0 - \frac{\pi r^2}{a^2} j. \qquad (3.34)$$

Using the above, we obtain the effective relative magnetic permeability

$$\mu_{\text{eff}} = \frac{B_{\text{eff}}}{\mu_0 H_{\text{eff}}} = 1 - \frac{\pi r^2/a^2}{1 + i2\rho/(\mu_0\omega r)}. \qquad (3.35)$$

Thus, the real part of the μ_{eff} is always less than one (diamagnetic) and greater than zero here. This diamagnetic screening effect has been known since some time for superconducting cylindrical shells (Kittel et al. 1988) and a diamagnetic effective medium is also obtained with percolation metallo-dielectric composites (Sarychev and Shalaev 2000).

Note here that

$$\lim_{\omega \to \infty} \mu_{\text{eff}}(\omega) = 1 - \frac{\pi r^2}{a^2}, \tag{3.36}$$

which violates the causal limit of $\mu_{\text{eff}} \to 1$ as $\omega \to \infty$. This is an artifact of our assumption of the metal as an Ohmic conductor at all frequencies. Obviously at very high frequencies, the inertia of the charge carriers prevents any appreciable currents from flowing and the assumption of an Ohmic conductor breaks down. In some sense this is also reflective of the quasi-static nature of our calculations. One should note that in spite of this improper high frequency limit, this model works quite well at frequencies where the quasi-static limit holds. Much before we go into the $\omega \to \infty$ limit, Bragg scattering and photonic band structure effects invalidate the assumptions inherent to the above calculations in any case. This is a feature of all the models for artificial magnetic materials that we present here.

3.2.2 Split-ring resonator media

In case of the cylinders, the induced currents made it appear as if magnetic monopoles were flowing up and down the cylinders, producing only an inductive response (the monopoles equivalently had no inertia). By introducing capacitive elements into the system, a rich resonant response can be induced. Consider an array of cylindrical metallic shells with gaps in them as shown in Fig. 3.6(b). For reasons that will become clear in Section 3.5, we introduce here two capacitive gaps placed symmetrically about the ring and not just a single gap. The capacitance per unit length along the cylinder can be tuned by either changing the length of the capacitive arms or by introducing a dielectric material into the gaps. Such loops with capacitive gaps have become well known subsequently as *split-ring resonators* – SRR for short. Some sort of split-ring resonator has been the basis of most of the metamaterials exhibiting negative magnetic permeability to date.

The SRR works on the principle that the magnetic field of the electromagnetic radiation drives a resonant LC circuit through the inductance, which results in a dispersive effective magnetic permeability. The induced currents flow in the directions indicated in Fig. 3.6 with charges accumulating at the gaps in the rings as shown. Balancing the emf generated around the circuit with the Ohmic drop due to the resistance yields

$$i\omega\mu_0\pi r^2 \left(H_0 + j - \frac{\pi r^2}{a^2}j \right) = 2\pi r\rho j - \frac{j}{i\omega C}, \tag{3.37}$$

where the effective capacitance per unit length along the leg of the cylinder is $C = \varepsilon_0 \varepsilon \pi \ell_c / 2 d_c$ and ε is the relative dielectric permittivity of the material in the capacitive gaps. The factor of 2 in the denominator of the capacitance accounts for the serial capacitance of the two capacitive gaps around the ring. We can associate a magnetic moment per unit length of each cylinder as

$$m = \pi r^2 j, \tag{3.38}$$

so that the magnetic dipole moment per unit volume is

$$M = \frac{m}{a^2} = \left(\frac{\pi r^2}{a^2} \right) j. \tag{3.39}$$

Noting as before that

$$H_{\text{eff}} = H_0 - \left(\frac{\pi r^2}{a^2} \right) j, \tag{3.40}$$

$M = \chi_m H_{\text{eff}}$, and $\mu_{\text{eff}} = (1 + \chi_m)$, we obtain the effective magnetic permeability as

$$\mu_{\text{eff}}(\omega) = \frac{B_{\text{eff}}}{\mu_0 H_{\text{eff}}}, \tag{3.41}$$

$$= 1 + \frac{(\pi r^2 / a^2) \omega^2}{\frac{1}{\mu_0 \pi r^2 C} - \omega^2 - i \frac{2\rho}{\mu_0 r} \omega}, \tag{3.42}$$

$$= 1 + \frac{f \omega^2}{\omega_0^2 - \omega^2 - i \Gamma \omega}. \tag{3.43}$$

Thus we have a resonant form of the permeability with a resonant frequency of

$$\omega_0 = \frac{1}{\sqrt{LC}} = c \left(\frac{2 d_c}{\varepsilon \ell_c \pi r^2} \right)^{1/2}, \tag{3.44}$$

that arises from the L-C resonance of the system where $c = 1/\sqrt{\varepsilon_0 \mu_0}$ is the speed of light in vacuum. The factor $f = \pi r^2 / a^2$ is as usual the filling fraction of the material. For frequencies larger than ω_0, the response is out of phase with the driving magnetic field and the μ_{eff} is negative up to the "magnetic plasma" frequency of

$$\omega_m = c \left(\frac{2 d_c}{(1 - f) \varepsilon \ell_c \pi r^2} \right)^{1/2}, \tag{3.45}$$

assuming that the resistivity of the material is negligible. It is seen that the filling fraction plays a fundamental role in determining the bandwidth over which the $\mu < 0$, whereas the dielectric permittivity of the embedding medium ε can obviously be used to tune the resonant frequency. The dissipation parameter can be expressed as $\Gamma = (2\rho)/(\mu_0 r)$ and directly depends on the resistivity of the rings: smaller rings have smaller resistive pathlengths, which

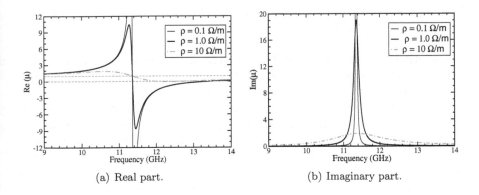

(a) Real part.   (b) Imaginary part.

Figure 3.7 Effective magnetic permeability for a system of split-cylinders with $r = 1.5$ mm, $\ell_c = 1$ mm, $d_c = 0.2$ mm, and the magnetic field along the axis for different values of ρ. The system has negative magnetic permeability for 11.358 GHz $< \omega <$ 13.41 GHz. As the resistance increases, the response of the system tends to tail off.

is reflected in the inverse dependence of Γ on the radius of the rings. A finite resistivity, in general, broadens the resonance peak, and in the case of very resistive materials the resonance is so highly damped that the region of negative μ can disappear altogether as shown in Fig. 3.7. Reducing the size of the ring while keeping the filling fraction large is obviously advantageous to increase the resonant effects, in addition to being desirable from the point of view of homogenization.

The resonance frequency and the negative μ band can be varied by changing the inductance (area) of the loop and the capacitance (the gap width d or the dielectric permittivity ε of the material in the capacitive gap) of the system. For typical sizes of $r = 2$ mm, $a = 5$ mm, $d_c = 0.1$ mm, and $\ell_c = 1$ mm, we have a resonance frequency of $f_0 = \omega_0/(2\pi) = 6.023$ GHz, and a "magnetic plasma frequency" of $f_m = \omega_m/(2\pi) = 8.54$ GHz. Note that at the resonance frequency, the free space wavelength is $\lambda_0 = 40$ mm implying a ratio of about $\lambda_0/a \simeq 8$ to 10, which is reasonably just large enough for the homogenization hypothesis to be satisfied. Obviously, if the capacitance can be increased while keeping the size of the SRR and the lattice period constant, the homogenized description becomes more accurate. The dispersion in the μ_{eff} for a slightly different SRR dimension and different values of the resistivity is shown in Fig. 3.7.

We note that the $\mu_{\text{eff}} \simeq 1 - \pi r^2/a^2$ asymptotically at large frequencies. This is due to the assumption of a perfect conductor in our analysis as pointed out earlier. Let us also note that μ_{eff} can attain very large values on the low-frequency side of the resonance. Thus, the effective medium of SRR is going to have a very large surface impedance $Z = \sqrt{\mu_{\text{eff}}/\varepsilon_{\text{eff}}}$. Such large impedances have also been obtained in capacitively loaded structured surfaces (Sieven-piper et al. 1999, Broas et al. 2001).

Essentially we have a design for a metamaterial whose geometrical sizes determine the frequency range where the medium exhibits magnetic activity. The size and dissipation levels determine the frequency range over which the magnetic permeability becomes negative. Our analysis indicates that by scaling down the size of the SRR while keeping other geometrical aspects constant, we should be able to obtain magnetic activity at higher frequencies. It is, however, the material properties of the metal constituting the SRR that do not permit this simple upscaling. The metals do not behave as Ohmic conductors at high frequencies and this actually presents a formidable obstacle to scaling up the frequencies of operation to the optical frequencies, as further discussed in Section 3.2.4.

3.2.2.1   Pendry's split rings

In the initial years of development of the field, a slightly different design for the split-ring resonators was fabricated and discussed. This was the design given in the original proposal for magnetic activity by Pendry et al. (1999). The original SRR is shown in Fig. 3.8 and has somewhat more capacitance; consequently a lower operating frequency and thus better satisfaction of the homogenization conditions are possible. The description of this SRR (which we call Pendry's SRR after its original proposer) is slightly more involved. We devote here a short discussion to the functioning of this particular structure primarily due to its historical importance in the evolution of this area.

The main difference that arises with the composite ring structure is that the large gap in each ring prevents the current from flowing around in a single ring and the circuit is completed across the small capacitive gap between the two rings. Thus, a large part of the capacitive loading arises from the capacitive interaction between the rings. The mutual capacitance is not very simple to estimate. In the most general case, there would be three resonances, one corresponding to each LC resonance of each ring and one resonance that arises from the mutual coupling.

If we assume that the gap (d) is very small compared to the radius (r) and that the capacitance due to the large gaps in any single ring is negligible, then the low frequency resonance is the one due to capacitive coupling across the rings. In this case, the flow of the currents in the two rings is in the same direction and the inner and outer loops have approximately the same inductance and self capacitance. Note that if $K_i(\phi)$ and $K_o(\phi)$ are the currents in the inner and outer loops that are functions of the azimuthal angle, and if j_r is the displacement current density that arises in-between the rings, we

Figure 3.8 Pictorial view of the cylindrical unit for the SRR proposed in Pendry et al. (1999). The inner conducting ring acts as a capacitive load on the outer ring. The charge buildup across the ends of the split rings, the direction of the currents, and the mutual capacitance are schematically depicted.

have

$$\frac{dK_i}{d\phi} = rj_r, \tag{3.46}$$

$$\frac{dK_o}{d\phi} = -rj_r, \tag{3.47}$$

so that the total circulating current in the two rings $K_i + K_o = K$ is uniform.

Under the assumption that the potential varies linearly with the azimuthal angle around the ring, one can write the equations for the currents (per unit length) in the two rings near the splits as

$$(Z_L + Z_C + Z_R)K_i + Z_{io}K_o = \text{emf}_H, \tag{3.48}$$

$$Z_{io}K_i + (Z_L + Z_C + Z_R)K_o = \text{emf}_H, \tag{3.49}$$

where the inductive impedance is $Z_L = -i\omega\mu_0\pi r^2$ for each ring (approximately as $d \ll r$), Z_C is the capacitive impedance, and Z_{io} is the mutual impedance.

We need to proceed with the calculation of the capacitance of the broken rings, which is quite a tedious process. The capacitance for each ring can be written as (Sauviac et al. 2004)

$$C = C_s + \frac{C_{\text{mut}}}{2}, \tag{3.50}$$

where C_s is the capacitance across the splits which can be neglected, and where the mutual capacitance can be approximated by $C_{\mathrm{mut}} = (\varepsilon_0 \varepsilon \pi r)/d$. Including the depolarization fields in the mutual inductance and adding the above two equations, we obtain for the total current K

$$K = \frac{-\mathrm{i}\omega\mu_0\pi r^2}{(1 - \frac{\pi r^2}{a^2})\mathrm{i}\omega\mu_0\pi r^2 - 2\pi r\rho + \frac{1}{\mathrm{i}\omega C}}. \tag{3.51}$$

Finally we obtain for the effective magnetic permeability,

$$\mu_{\mathrm{eff}} = 1 + \frac{f\omega^2}{\frac{2c^2 d}{\varepsilon\pi^2 r^3} - \omega^2 - \mathrm{i}\frac{2\rho\omega}{r\mu_0}}. \tag{3.52}$$

Thus, again we have a resonant form of the dispersion for the magnetic permeability. The main difference is the larger capacitance that can be attained here and a correspondingly lower operating frequency for a given size of the SRR than in the previous case. Also note that we have made several assumptions and approximations in our calculations that are usually violated in actual experimental implementations. For example, the sizes of the inner and outer rings can be considerably different and the capacitances across the splits can be comparable to, if not dominate over, the mutual capacitance. Experimentalists usually obviate this problem by including a few "fudge" factors that account for the extra inductances or capacitances. In most cases, the formula above can only be taken as a guideline and detailed numerical calculations become necessary to obtain better quantitative estimates. Furthermore, these SRRs also exhibit an electric polarizability for an electric field oriented in the plane of the SRR along the gap in the rings. We refer the reader to Sauviac et al. (2004) for more detailed calculations of such coupled SRRs. In fact, these SRR are bianisotropic as well: the magnetic field oriented along the axis of the SRR can induce an electric polarization in the plane of the SRR. This property arises due to the asymmetric charge densities that develop in the SRR and is discussed further in Section 3.5.

3.2.3   The Swiss Roll media for radio frequencies

Pendry and co-workers (Pendry et al. 1999) also proposed another resonant unit, the Swiss Roll, as the basis of magnetic metamaterials for low frequency operations. This structure conveniently avoids increasing the size of the units by having a large increased self inductance and enhances the homogeneous description as an effective medium since the ratio of the unit size to the free space wavelength is further decreased. Such Swiss Roll can be achieved by rolling up a metal sheet in the form of a cylinder with each coil separated by an insulator of thickness d as shown in Fig. 3.9. The current loop is completed here through the differential capacitance across the space between the metal sheets as shown. As before, the effective magnetic permeability for a system of such structures can be easily calculated.

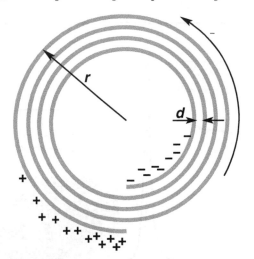

Figure 3.9 Cross-section of the cylindrical unit for the Swiss Roll meta-material. The Swiss Roll consists of a rolled-up sheet of a conductor with an insulating medium in-between that separates out the adjacent conducting sheets capacitively. The net inductance is significantly increased by the multiple loops.

If the number of coils N is large, one can assume a uniform current j in the sheet making up the Swiss roll in all the loops except the innermost and outermost loops where one can assume an average current of $j/2$. This difference can be neglected in the limit of large N. The total circular current in the ring is then $j_t = (N-1)j$. Thus, one can write for the total emf for each loop on the average, which is balanced by the resistive drop in the roll and the capacitance $C = (\varepsilon_0 \varepsilon 2\pi r)/d$ across adjacent loops as

$$i\omega\mu_0\pi r^2 \left[H_0 + (1 - \frac{\pi r^2}{a^2})j_t \right] = 2\pi r \rho j - \frac{j}{(i\omega C)}. \tag{3.53}$$

Note that the effective capacitance between the inner and outer loops of the Swiss roll can be approximated by $N-1$ capacitances in series and is given by $C/(N-1)$. Here N is the number of coils in the structure and it is assumed that total thickness of the wound layers $Nd \ll r$, the cylinder radius. Note that the effects of the depolarizing fields are included through the filling fraction in the above equation. Proceeding as before, one can write for the effective fields

$$B_{\text{eff}} = \mu_0 H_0, \tag{3.54}$$

$$H_{\text{eff}} = H_0 - \frac{\pi r^2}{a^2} j_t, \tag{3.55}$$

where the B fields are averaged over an area of the unit cell and the H fields

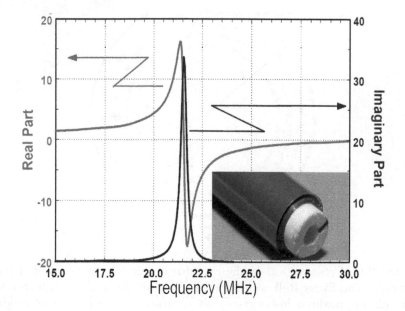

Figure 3.10 Dispersion of the real and the imaginary parts of the effective magnetic permeability measured for the Swiss Rolls. Inset: A photograph of a Swiss roll made by rolling up polyimide sheets with a thin layer of deposited copper in-between. The polyimide acts as the insulating capacitive gap between successive turns of the roll. The diameter of the Swiss Roll is about 1 cm. (Source: The photograph and the graph have been kindly provided by M. C. K. Wiltshire. Published in Wiltshire et al. (2003a) and reproduced with permission. © 2003, Optical Society of America.)

are averaged over a line lying entirely outside the Swiss roll. Hence we can obtain for the magnetic effective permeability

$$\mu_{\text{eff}} = 1 - \frac{\pi r^2/a^2}{1 - \frac{dc^2}{\pi^2 r^3 (N-1)\omega^2} + i\frac{2\rho}{\mu_0 \omega r(N-1)}}. \tag{3.56}$$

The relevant frequencies are the magnetic resonance frequency (ω_0) and the magnetic "plasma" frequency (ω_{mp}), respectively, given by

$$\omega_0 = c\sqrt{\frac{d}{\pi^2 r^3 N}}, \tag{3.57}$$

$$\omega_{mp} = c\sqrt{\frac{d}{(1-f)\pi^2 r^3 N}}. \tag{3.58}$$

This system also has the same generic resonance form of the permeability with frequency as the SRR structures, but the resonance frequency occurs at

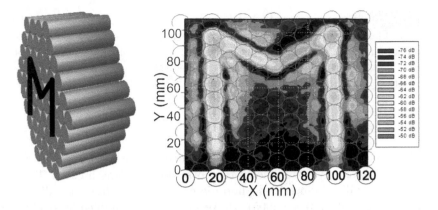

Figure 3.11 Left: Schematic picture showing the imaging of an M-shaped antenna emitting at 21.3 MHz by a stack of resonant Swiss rolls arranged in a hexagonal close packed structure. Right: The measured image of the M-shaped antenna, measured on the other end-face of the bundle of Swiss rolls. The image is transferred to the other end-face of the pack by virtue of the Swiss Rolls acting as magnetic flux tubes due to their high magnetic permeability. The locations of the Swiss Rolls are also marked out. (Source: The image has been kindly provided by M. C. K. Wiltshire. Published in Wiltshire et al. (2003a) and reproduced with permission. © 2003, Optical Society of America.)

a much lower frequency owing to the much larger inductance ($\sim N^2$) of the structure. Using the material EspanexB, which consists of a 12.5 μm poly-imide sheet with 18 μm of deposited copper, Wiltshire et al. (2003b) reported on sheets rolled-up into Swiss Rolls which had a resonant frequency near 21.3 MHz. A photograph of these Swiss Rolls along with the measured values of the magnetic permeability of these Swiss Rolls is shown in Fig. 3.10. The effective permeability of the Swiss Roll medium was determined by inserting a roll into a long solenoid, and measuring the changes in the complex impedance, where estimated corrections for the partial volume occupied and demagneti-zation have been made. We note that the relevant value of $a/\lambda \simeq 10^{-4}$ at these frequencies, which is almost the same as that for bulk material molecules and atoms at optical frequencies. Thus the effective medium approximation should be excellent in this context.

Note that at frequencies smaller than the resonance frequency, the real part of the magnetic permeability can take very large positive values, almost 15 in this case. Thus radio frequency fields can be strongly coupled into these Swiss rolls which can act as magnetic flux tubes. This has seen application in Magnetic Resonance Imaging (MRI) at radio frequencies (Wiltshire et al. 2001; 2003a). Fig. 3.11 shows an array of such Swiss rolls which were then used to channel down their length the image of an "M"-shaped RF antenna.

The image that was measured on the other side is seen to have preserved fine details. These Swiss rolls have also more recently been shown to work as a near-field lens with very high subwavelength image resolution (Wiltshire et al. 2006). The ideas of near-field imaging by materials with negative material parameters is discussed in greater detail in Chapter 8.

3.2.4   Scaling to high frequencies

The Maxwell equations indicate that one can scale the phenomena to higher frequencies by simply scaling down the corresponding lengthscales. This has been the overriding principle of photonic crystals made from non-dispersive dielectric materials (Sakoda 2005). However, the main problem in scaling the performance of metamaterials to higher infrared and optical frequencies is that the material parameters of the constituent materials disperse with frequency, which does not allow scaling in a straightforward manner. Primarily, metals no longer behave as perfect or Ohmic conductors, and the penetration depth of electromagnetic fields becomes considerable, while dissipation also increases. Consequently, the dispersive nature of the metals must be taken into account. Additionally, another cause of concern is the technological ability to accurately make large numbers and arrays of the resonant structures at the small length scales necessary.

The single ring SRR would be more suitable than more complex double rings from the point of ease of fabrication at micro and nanometric dimensions. Consider the split ring with two symmetric splits shown in Fig. 3.12(a). The dimensions of the SRR have been chosen to achieve operation at tens of terahertz and were originally proposed in O'Brien and Pendry (2002). The band structure and the transmission for infinitely long split-cylinders of silver obtained using a photonic band-structure calculation using the transfer matrix method (Pendry 1994) is shown in Fig. 3.12(b). The magnetic field is assumed to be along the axis of the split-cylinders and a plasma form for the dielectric function of silver, $\varepsilon(\omega) = \varepsilon_\infty - \omega_p^2/[\omega(\omega + i\gamma)]$, was used with the empirical values of $\hbar\omega_p = 9.013$ eV (Johnson and Christy 1972), $\gamma = 0.018$ eV, and $\varepsilon_\infty = 5.7$. Note the presence of a frequency gap that arises due to the negative effective μ of the structure at about 75 to 80 THz, while our simple circuit theory model in Section 3.2.1 predicts a resonance frequency of 104 THz for the L-C resonance. Consequently, the assumptions of a perfect conductor clearly cannot be made at these high frequencies. Using the parameter retrieval method presented in Chapter 2, the effective ε and μ were recovered from the emergent quantities, *viz.*, the complex transmission coefficient and reflection coefficients from a slab made of such SRR and are shown in Fig. 3.12 (c) and (d), clearly demonstrating a frequency band with $\mathrm{Re}(\mu) < 0$. The retrieval results also show $\mathrm{Im}(\varepsilon) < 0$ near the resonance and an anomalous dispersion of $\mathrm{Re}(\varepsilon)$ about the resonance frequency. This anti-resonant behavior arises because of the failure of this particular retrieval algorithm to

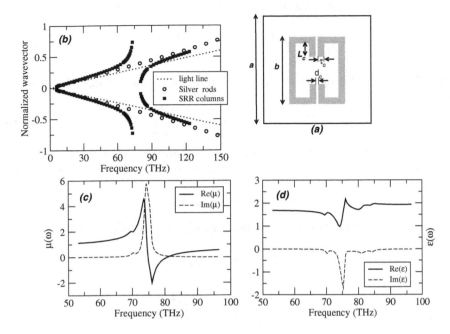

Figure 3.12 (a) Single ring with a double slit with $a = 600$ nm, $b = 312$ nm, $\tau_c = 24$ nm, $d_c = 24$ nm and $L_c = 144$ nm used for the calculations. (b) Photonic band structure of this system of columns for the S-polarisation (with **H** along the cylindrical axes). The light line (dotted) and the bands for silver columns with the same filling fraction are shown for comparison. The bandgap at about 75 THz arises due to negative μ. (c) Real and imaginary parts of μ and (d) real and imaginary parts of ε. (The figure is reproduced with permission from (Ramakrishna 2005) © 2005, Institute of Physics Publishing, U.K.)

properly account for the boundaries of the metamaterial slab.[‡] Thus, the retrieved effective medium parameters are restricted in this case. While they describe well the scattering properties of radiation from such metamaterials, they cannot be used to calculate, for example, other properties such as the heating rate due to absorption.

Let us build a simple model for this system to gain an insight into the high-frequency scaling properties. Assuming a strong skin effect, i.e. that the thickness τ_c of the metal shells is smaller than the skin depth ($\sim$20 nm at 100 THz), we can write the displacement current per unit length $j_\phi =$

[‡]When the size of the metamaterial units are extremely small compared to the wavelength, such spurious effects become minimal.

$-i\omega\varepsilon_0\varepsilon_m E_\phi$. The potential drop across each half of the ring is

$$V_r = \int_0^\pi E_\phi r\, d\phi = \frac{\pi r j_\phi}{-i\omega\varepsilon_0\varepsilon_m}, \tag{3.59}$$

and the potential drop across each capacitive gap (of arm length L_c) is given by

$$V_c = \frac{1}{C}\int I(t)\, dt = \frac{j_\phi \tau_c d_c}{-i\omega\varepsilon_0\varepsilon L_c}. \tag{3.60}$$

Using the fact that the total emf induced around the loop is $\oint E_\phi dl = i\omega\mu_0 \int H_{\text{int}} da$, we can equate the potential drop to the emf generated around the loop as

$$2V_r + 2V_c = i\omega\mu_0\pi r^2 H_{\text{int}}. \tag{3.61}$$

Using Ampere's law, the magnetic fields inside and outside the split cylinders can be related as

$$H_{\text{int}} - H_{\text{ext}} = j_\phi \tau_c, \tag{3.62}$$

yielding that

$$\frac{H_{\text{ext}}}{H_{\text{int}}} = 1 - \frac{\mu_0\varepsilon_0\omega^2\pi r^2\tau_c}{[2\pi r/\varepsilon_m + 2\tau_c d_c/(\varepsilon L_c)]}. \tag{3.63}$$

The averaged magnetic induction is $B_{\text{eff}} = (1-f)\mu_0 H_{\text{ext}} + f\mu_0 H_{\text{int}}$, where $f = \pi r^2/a^2$ is the filling factor. Averaging over a line lying entirely outside the cylinders for the magnetic field $H_{\text{eff}} = H_{\text{ext}}$, we obtain the effective magnetic permeability as

$$\mu_{\text{eff}} = \frac{B_{\text{eff}}}{\mu_0 H_{\text{eff}}} = 1 + \frac{f\varepsilon_0\mu_0\omega^2\pi r^2}{[(2\pi r/(\varepsilon_m\tau_c)) + 2d_c/(\varepsilon L_c)] - \varepsilon_0\mu_0\omega^2\pi r^2\tau_c}. \tag{3.64}$$

Assuming as before that $\varepsilon_m \simeq -\omega_p^2/[\omega(\omega+i\gamma)]$ for the infrared and optical frequencies, we obtain the generic form for the effective permeability:

$$\mu_{\text{eff}}(\omega) = 1 + \frac{f'\omega^2}{\omega_0^2 - \omega^2 - i\Gamma\omega}, \tag{3.65}$$

where the resonance frequency, effective damping coefficient, and the effective filling fraction are, respectively,

$$\omega_0^2 = \frac{1}{(L_i + L_g)C}, \qquad \Gamma = \frac{L_i}{L_g + L_i}\gamma, \qquad f' = \frac{L_g}{L_g + L_i}f. \tag{3.66}$$

Here $C = \varepsilon_0\varepsilon L_c/2d_c$ is the effective capacitance of the structure, $L_g = \mu_0\pi r^2$ is the geometrical inductance, and $L_i = 2\pi r/(\varepsilon_0\tau_c\omega_p^2)$ is an additional inductance that appears. Noting that the plasma frequency is $\omega_p^2 = Ne^2/(\varepsilon_0 m)$, we see that the additional inductance arises entirely due to the electronic mass so that L_i can be termed the *inertial inductance*.

The presence of this additional inductance can be explained by noting that at high frequencies, the currents are hardly diffusive and become almost ballistic because the distance through which the electrons move within a period of the wave becomes comparable to the mean free path in the metal. This means that if the frequencies are too high, the electrons can hardly be accelerated and the response falls. The mass of the electron contributes additionally to the effective self inductance.[§] The effective damping factor also becomes much larger as the size of the ring is reduced. This is due to the fact that the proportion of energy in the ballistic motion of the electrons increases compared to the energy in the electromagnetic field as the size is reduced and the resistive losses are then very effective indeed.

Thus even if the size of the ring were negligible, the inertial inductance would still be present preventing the scaling to higher frequencies. This effect has also been discussed in connection with using superconducting SRRs in the microwave region (Kumar 2002). The large increase in the damping as the dimensions are scaled down broadens the resonance and the permeability does not disperse violently, making the region of negative permeability vanish altogether. The increase in damping is a matter of great concern for operation at optical frequencies.

Numerical studies have confirmed the above model at high frequencies and the response of the SRR with two splits was shown to tail off in the infrared region ($\lambda \sim 5\mu m$) (O'Brien and Pendry 2002). By adding more capacitive gaps to lower the net capacitance and adjusting the dielectric constant of the embedding medium, it was numerically demonstrated that a medium of SRR with four splits can have negative μ at telecommunications wavelength of about 1.5 μm (O'Brien et al. 2004). A study with aluminum SRR having four splits also subsequently confirmed that the resonance frequency tended to saturate at few hundreds of terahertz (Zhou et al. 2005). An interesting use of pairs of parallel metal rods (with lengths ~100 nm) periodically embedded in a dielectric medium as an effective medium with negative magnetism at infrared frequencies was proposed in Panina et al. (2002). The current loop across the two rods is completed via the displacement currents in vacuum between the rods.

Suitably scaled down double ring SRRs have been demonstrated experimentally to have negative magnetic permeability at terahertz frequencies (Yen et al. 2004). SRRs with a size of 320 nm made of gold and a single capacitive gap were shown to have a magnetic resonance at about 100 THz (Linden et al. 2004) and the saturation of the resonance frequency with reducing size at about 300 THz was shown with these rings (Klein et al. 2006b). Novel loop structures on a metal surface which provide magnetic activity at about 75 THz was explored in Zhang et al. (2005b). The concept of using wire-

---

[§]The current $j = nev \sim ne(-i\omega eE/m)$, where E is the applied field. Then the potential drop over a distance ℓ is $V \sim m\ell/(ne^2)(\partial j/\partial t)$, implying an inductance that is linearly proportional to the electronic mass.

pairs as magnetic atoms was investigated in Dolling et al. (2005) where the transition of the behavior from a singly split SRR to a wire-pair was investigated with gold structures. The measurements, in general, have confirmed the results of the above model for scaling to high frequencies both qualitatively and quantitatively. There are also experimental implementations using gold nanopillar pairs (Grigorenko 2006) and arrays of paired strips of silver (Shalaev et al. 2005) that have tried to push the idea of magnetic activity and negative magnetic permeability into the visible spectrum ($\lambda \sim 725$ nm).

3.2.5 Magnetism from dielectric scatterers

It is clear from the discussion above that the essential key to strong magnetic activity is an under-damped resonance that results in large local enhancements of the magnetic field. In the designs presented above, the resonance is an L-C resonance that results from the geometric structure of the conducting materials. In addition, any other resonance that can be driven by the magnetic field can also be used. For example, magnetic activity via the Mie scattering resonances of magnetic nature (Bohren and Huffman 1983) in systems of small scattering particles has been realized as early as 1986 (Bohren 1986). The use of the magnetic Mie scattering resonances of a dielectric cylinder (O'Brien and Pendry 2002) and magneto-dielectric spheres (Holloway et al. 2003) for generating a negative magnetic medium have also been proposed. These Mie resonances are well known and give rise to the heavy photon bands in periodic photonic crystals of cylinders or spheres (Ohtaka and Tanabe 1996). Usually these bands arise at large frequencies when the corresponding wavelength is smaller than the unit cell size and the idea of an effective medium is not applicable. However, if the dielectric permittivity of the scatterers is large, these resonances can occur at much lower frequencies and these Mie resonances can be lowered well within the first Bragg band. In this case, the idea of an effective medium becomes useful and such polaritonic spherical or cylindrical inclusions along with metallic particles have been demonstrated to have a negative magnetic permeability and simultaneously negative dielectric permittivity, too (Yannopapas and Moroz 2005, Liang et al. 2007).

A dielectric particle can exhibit a variety of electromagnetic resonances depending on its geometry. One of the hallmarks of a resonance is the large magnitude of the electromagnetic fields that arise within the scatterer. We can aim to obtain scattering resonances whereby large and inhomogeneous magnetic fields within the scatterer contribute to a uniform magnetization in the medium. In our homogenization procedure of averaging the **B** fields over the area of the unit cell, the contribution of such inhomogeneous local fields are immediately picked up. Note that in the design of such materials, we are restricted to using mono-polar resonances only, because resonances with dipolar or higher multipolar terms tend to contribute nothing to the average **B** field over the unit cell.

In the above spirit, let us first analyze a stack of dielectric cylinders of radius R and consider Transverse Electric (TE) polarized light ($\mathbf{H}$ along the cylindrical axes) to be incident on the stack. The magnetic fields associated with scattering from a single cylinder in air are given by (Bohren and Huffman 1983)

$$H_z^{inc} = H_0 \sum_{m=0}^{\infty} \mathrm{i}^m J_m(kr) e^{\mathrm{i}m\phi}, \tag{3.67a}$$

$$H_z^{scatt} = -H_0 \sum_{m=0}^{\infty} \mathrm{i}^m a_m H_m^{(1)}(kr) e^{\mathrm{i}m\phi}, \tag{3.67b}$$

$$H_z^{int} = H_0 \sum_{m=0}^{\infty} \mathrm{i}^m c_m J_m(nkr) e^{\mathrm{i}m\phi}, \tag{3.67c}$$

where H_z^{inc} is the incident field of a plane wave, H_z^{scatt} are the scattered fields, H_z^{int} are the fields inside the cylinder, $k = 2\pi/\lambda$ and n is the refractive index of the cylinder material. J_m and $H_m^{(1)}$ are the Bessel function and Hankel's functions of the first kind and order m, respectively. The coefficients a_m and c_m are determined by matching the tangential components of the total electric and magnetic fields across the cylindrical boundaries and are obtained as

$$c_m = \frac{J_m(kR) - a_m H_m^{(1)}(kR)}{J_m(nkR)}, \tag{3.68a}$$

$$a_m = \frac{J_m'(nkR)J_m(kR) - nJ_m(nkR)J_m'(kR)}{J_m'(nkR)H_m^{(1)}(kR) - J_m(nkR)H_m^{(1)'}(kR)}, \tag{3.68b}$$

where the primes indicate a derivative with respect to the argument. The conditions for a resonance can be obtained from the poles of a_m. The characteristic equation obtained has complex solutions and illumination by radiation at a frequency close to the real part of a particular root causes the corresponding a_m and c_m to dominate over all the other scattering coefficients. Then the fields inside the cylinder have maximum contribution from this predominant eigenmode and essentially take the character of that function.

For example, the angularly symmetric ($m = 0$) resonance results in an enhanced magnetic field only in one direction, along the axis of the cylinder. In the case of an array of such cylinders, the system would appear as a conglomerate of magnetized rods with the displacement current in each rod being in the azimuthal direction. Substituting this mono-polar resonance contribution for the magnetic field in the definition for the effective magnetic permeability (O'Brien and Pendry 2002),

$$\mu_{\text{eff}} = \frac{B_{\text{eff}}}{\mu_0 H_{\text{eff}}}, \tag{3.69a}$$

$$= \frac{\frac{2\pi}{a^2} \int_0^R J_0(nkr) r \, dr + \int_R^{a/\sqrt{\pi}} [J_0(kr) - a_0 H_0^{(1)}(kr)] r \, dr}{J_0(ka/2) - a_0 H_0^{(1)}(ka/2)}, \tag{3.69b}$$

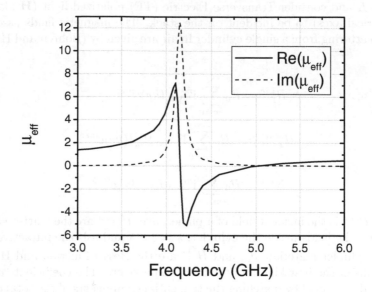

Figure 3.13 Effective magnetic permeability (μ_{eff}) obtained for a stack of long dielectric cylinders arranged on a square lattice with $\varepsilon = 200 + \text{i}5$ as predicted by Eq. (3.69a). The cylinders have a diameter of 4 mm and are placed on a square lattice of period 5 mm. The figure has been redrawn from the data of O'Brien and Pendry (2002). The data for the figure has been kindly provided by S. O'Brien.

where the integration is carried over a circle with the same area of the unit cell ($a^2 = \pi r^2$). The function is plotted in Fig. 3.13 for dielectric cylinders of of 4 mm diameter, placed apart on a square lattice of period 5 mm, and $\varepsilon = 200+\text{i}5$. The system possesses negative magnetic permeability if the dielectric permittivity of the corresponding cylinder is made large enough. This is due to the tremendous increase in the inhomogeneous local fields caused by confinement in the high dielectric rods. The magnetic Mie resonance can then occur at low enough frequencies such that it occurs well within the first Bragg band and the system is amenable to homogenization. This aspect was also confirmed with photonic band structure calculations and effective parameter retrieval from the reflection coefficients (O'Brien and Pendry 2002). Note that the ratio of the free space wavelength to the periodicities in the system can be about 15, which implies a good applicability of homogenization.

An important issue that needs to be overcome for the realization of magnetic metamaterials from dielectric scatterers is to obtain the constituent materials with a large enough dielectric permittivity. This seems well within reach as several ferro-electric materials such as barium strontium titanate

(Ba$_{0.6}$Sr$_{0.4}$TiO$_3$) have dielectric constants of a few hundreds with very small loss tangents of less than a percent (Sengupta 1997) at Gigahertz frequencies. At much higher frequencies, there are polaritonic crystals such as LiTaO$_3$, whose polaritonic resonance occurs at about 26.7 THz (Yannopapas and Moroz 2005), and the material has a large permittivity just below the resonance frequency.

Similarly, spheres of large dielectric permittivity can also be utilized to construct metamaterials of negative magnetic permeability via the magnetic Mie resonance. The effect in this case is intrinsically three-dimensional and this approach can easily be utilized to obtain an almost isotropic response. Consider a simple cubic crystal of periodically arranged spheres made of a polaritonic material. At the conditions for the magnetic Mie resonance, each sphere essentially behaves as a magnetic dipole. If there is not much loss of flux (depolarizing fields), one can imagine magnetic flux tubes running along the linear arrays of such magnetic dipoles formed. Then the description becomes almost analogous to the earlier situation of long cylinders with the magnetic flux along the cylindrical axis.

Even if the arrangement of the spheres did not satisfy these conditions, one can homogenize a system of such spheres accurately for small enough filling fractions (f). Consider spheres of radius R and dielectric permittivity of ε_s embedded in a homogeneous dielectric medium of dielectric permittivity ε_h. The Doyle extension of the Maxwell-Garnett effective medium theory discussed in Section 2.5.2 yields

$$\varepsilon_{\text{eff}} = \varepsilon_h \frac{x^3 + 3\mathrm{i}fa_1}{x^3 - \frac{3}{2}\mathrm{i}fa_1},$$

$$\mu_{\text{eff}} = \mu_h \frac{x^3 + 3\mathrm{i}fb_1}{x^3 - \frac{3}{2}\mathrm{i}fb_1},$$

where the size parameter $x = kR$, $\mu_h = 1$, $k = 2\pi\sqrt{\varepsilon_h}/\lambda$, and (a_1, b_1) are the Mie scattering coefficients representing the electric dipole and magnetic dipole terms, respectively. Using this, it has been shown that the homogenized effective permeability exhibits resonant features around the magnetic Mie resonance frequency and the effective magnetic permeability is negative for spheres with $\varepsilon_s = 100$ and of size $S = 0.5a_0$ where a_0 is the first nearest neighbor distance on an fcc lattice (Yannopapas and Moroz 2005). Photonic band structure calculations also confirmed the presence of a frequency band of negative magnetic permeability. The result for the effective medium parameters is shown in Fig. 3.14. In addition, Holloway et al. (2003) showed that a system similarly comprising of magneto-dielectric material has a similar negative magnetic permeability frequency band near the magnetic Mie resonance frequency. Note that the systems give the possibility of obtaining both negative effective dielectric permittivity as well as negative effective permeability, but at different frequencies as should be expected. By mixing different sized spheres, in principle, one could obtain both effective medium quantities to be

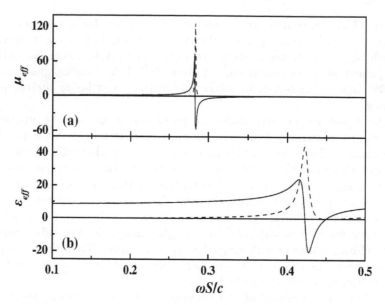

Figure 3.14 Effective magnetic permeability (μ_{eff}) and dielectric permittivity (ε_{eff}) obtained for a three-dimensional fcc lattice of spherical particles with $\varepsilon_s = 100$ and $S = 0.5a_0$ as predicted by Eqs. (2.63) and Eqs. (2.64). (Reproduced with permission from Yannopapas and Moroz (2005). © 2005, Institute of Physics Publishing, U.K.)

simultaneously negative at the same frequency. This is subsequently discussed in Section 3.3.

3.2.6 Arrangements of resonant plasmonic particles

The Mie resonances (plasmonic resonances) of very small metallic particles can be utilized to obtain magnetic media by virtue of a geometric loop-like arrangement of the particles (Alù et al. 2006a). The resonance frequency is, however, determined only by the plasmonic resonance frequency, which in turn is determined by the size and nature of the particle. The essence of this idea is to organize metallic particles in loop-like arrangements, whereby at frequencies near the plasmonic resonance, one can use the large displacement currents across each particle that arise at resonance to go around the loop and give rise to a magnetic dipole moment (Saadoun and Engheta 1992).

In fact, every plasmonic particle in itself can be considered as an LC circuit. Consider the charge accumulation on the surface of a spherical or cylindrical particle driven by a time harmonic field as shown in Fig. 3.15. The negative dielectric permittivity of the metal inside makes it a capacitor with negative capacitance (or equivalently an inductance), while the positive dielectric

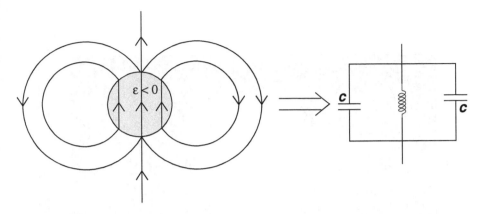

Figure 3.15 A plasmonic nanoparticle can be regarded as a resonant L-C circuit. The capacitance comes from the electric fields in the positive dielectric medium outside, while the negative dielectric permittivity inside the particle gives rise to an effective inductance.

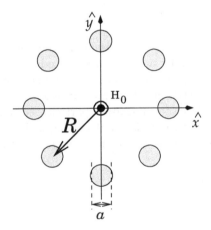

Figure 3.16 Loop-like arrangements of resonant plasmonic particles can give rise to a resonant excitation of the ring whereby the circulating displacement current around the ring makes it appear as a magnetic dipole.

permittivity of the medium (vacuum) for the fields outside enables a finite capacitance. Thus, all the necessary ingredients for an L-C resonance are gathered, creating circuits with light at nanoscale, or optical nanocircuits as proposed in Engheta (2007). The plasmonic resonance for very small nanoparticles, where the quasi-static approximation is valid, can just be thought of as due to a resonant L-C circuit. The dissipation in the metal brings in the resistive aspect to this circuit.

In a sense, putting together several resonant particles in the above manner

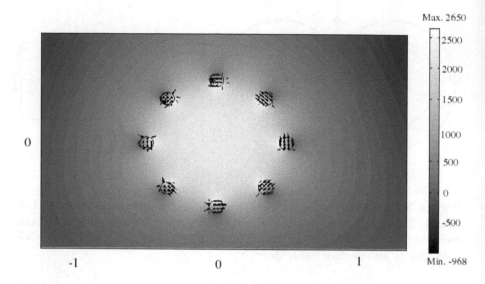

Figure 3.17 Response of small plasmonic nano-cylinders placed on a ring of larger radius to excitation by a line source placed at a distance of 1.2λ. The gray scale shows the magnetic field that is large and confined inside the ring. The arrows show the electric field that goes around in the azimuthal direction. The electric fields within the plasmonic particles is large and arises due to the plasmon resonance of the individual particles. The resonant "magnetic" excitation of the ring occurs at the same frequency as the plasmon resonance of the individual cylinders.

is an extension of the idea of the split ring resonator with N-splits. The resonance frequency in this system is, however, determined by the plasmonic resonance of the individual particles and not by the size of the ring. In the limit of small size of the particles, the resonant surface plasmon excitation of each particle dominates over the interaction of the particles and determines the resonance frequency.

Consider next a system of N spherical plasmonic particles of radii a, arranged symmetrically on a circle of radius R as shown in Fig. 3.16 (Engheta 2007). One can determine the induced magnetic polarization without exciting any electric polarization by considering the excitation by N-symmetrically incident plane waves with the magnetic field along the normal to the loop (Ishimaru et al. 2003, Alù et al. 2006a):

$$\mathbf{E} = \hat{\phi} \sum_{n=1}^{N} \frac{E_0}{N} e^{i\mathbf{k}_n \cdot \mathbf{r}}, \tag{3.70a}$$

$$\mathbf{H} = \hat{z} \sum_{n=1}^{N} \frac{H_0}{N} e^{i\mathbf{k}_n \cdot \mathbf{r}}, \tag{3.70b}$$

where $\mathbf{k}_n = \hat{r} k_b$, $E_0 = -\sqrt{\mu_b/\varepsilon_b} H_0$, and k_b is the wave-number in the embedding medium. This is a convenient way of isolating the magnetic response and is equivalent to considering a spatially constant magnetic field in the quasi-static limit. In the far-field limit, it has been shown that the scattered waves by such an arrangement are given by

$$\mathbf{E} = -\hat{\phi} \frac{iN k_b^3 R}{8\pi\varepsilon_b} \frac{\exp[ik_b r]}{r} p \sin\theta, \tag{3.71a}$$

$$\mathbf{H} = \hat{\theta} \sqrt{\frac{\varepsilon_b}{\mu_b}} \frac{iN k_b^3 R}{8\pi\varepsilon_b} \frac{\exp[ik_b r]}{r} p \sin\theta, \tag{3.71b}$$

where p is the induced electric dipole moment per particle in the loop. It can immediately be deduced that the above fields correspond to those radiated by a magnetic dipole of strength

$$\mathbf{m} = -\hat{z} \frac{i\omega}{2} pNR. \tag{3.72}$$

Each loop is therefore equivalent to a resonant magnetic dipole and we have an array of resonant magnetic dipoles.

The resonant response in a system of similar cylindrical plasmonic particles excited by a line source placed at a point about a wavelength away from the loop is shown in Fig. 3.17. The calculations were carried out by the Finite Element Method. An enhanced magnetic field in the interior of the loop and the polarization of the individual nanoparticles in the azimuthal direction can be clearly seen. In fact, the resonant field enhancement is so strong inside the loop that the fields of the exciting source are barely discernible in the figure. In the effective medium limit of small loop radius compared to the wavelength, one obtains a Lorentz-like dispersion of the magnetic permeability. Note that these loops also contribute to the effective dielectric permittivity of the medium. For electric fields in the plane of the loops, the net dipole moment can be taken to be the sum of the individual dipole moments induced on each nanoparticle. The effective magnetic permeability and dielectric permittivity calculated for a system of plasmonic nanoparticles (silver nanospheres of 16 nm radius) are shown in Fig. 3.18. For nanoparticles made of good metals such as gold, silver or potassium, one can obtain a negative magnetic permeability in a frequency band above the plasmon resonance frequency.

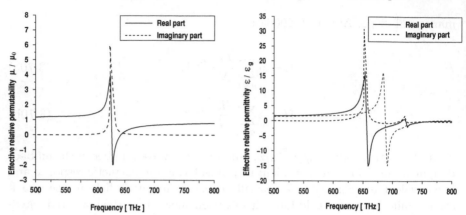

Figure 3.18 Dispersion in the effective medium parameters (μ, ε) for a meta-material consisting of loop-like units of plasmonic nanoparticles. For the figure shown, the metamaterial unit consists of six silver plasmonic nanospheres of radius 16 nm which are arranged on a circle of radius 40 nm. The number density of such loops is $(108 nm)^{-3}$. Reproduced from (Alù et al. 2006a) with permission © 2005, Optical Society of America, 2006.

3.2.7   Isotropic magnetic metamaterials

Most of the designs previously discussed for the realization of a magnetic activity are anisotropic: the effective medium was magnetically active only for electromagnetic waves that were polarized so that the magnetic field was normal to the plane of the SRR (along the cylindrical axis) for example. In that regard, the medium is characterized as being uniaxial. In fact, the structure based on the two-dimensional cylinders appears as a plasma medium to waves with the electric field polarized along the cylindrical axes due to the ability of the metallic cylinders to conduct currents in the axial direction. Thus, stacking cylindrical structures along the three axes would not result in a more isotropic magnetic response, but the medium would appear to be more like a plasma. To design more isotropic structures it is imperative to break the continuity along the axial directions. One simple solution would be to have planar SRR structures deposited on a substrate in arrays as shown in Fig. 3.19 and the arrays can be stacked up in the vertical direction, with the SRRs arranged one above the other. This would break the continuity along the vertical direction and the discrete rings arrayed in the vertical direction would mimic the action of a cylinder confining the magnetic flux inside them. Upon arranging the planar SRR arrays along the three orthogonal directions, one would create interleaving orthogonal planes of SRR which would render a more isotropic response. The respective SRRs would respond to the components of the exciting magnetic field along all three orthogonal directions.

In order to work, the stacking distance should be small enough such that

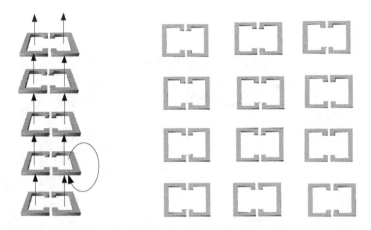

Figure 3.19 Left: Planar SRR stacked in the axial direction to mimic a continuous cylindrical SRR. The flux lines are confined in a quasi-solenoidal manner for small stacking distances. The figure also shows the possibility of fringing fields escaping the "flux tube" and contributing to the depolarizing field. Right: Planar SRRs arrayed in the plane. The planes can be stacked along the three orthogonal directions with care taken to ensure stacking of the SRRs in the axial direction (as in the left) and a three-dimensional SRR medium that displays a magnetic response can be made.

one can assume that the flux lines are confined within the column of SRRs along the axial (magnetic field) direction. However, this stacking distance in the axial direction is limited by the need to accommodate SRRs in the orthogonal directions as shown in Fig. 3.20, and hence the stacking distance can be taken to be the periodicity a in the plane of the SRR medium. Subsequently, there are fringing magnetic fields that escape out of the "flux tube" and the fringing effects of the magnetic field lines are not negligible, especially when the periodicity a is larger than the radius of the ring r.

A magnetic scatterer which consists of two orthogonal intersecting SRRs was introduced in Gay-Balmaz and Martin (2002). This structure exhibits an isotropic response for any wave incident on it normal to the plane of the rings. Balmaz and Martin have discussed the various orientations of the rings and concluded that they should exclusively intersect along symmetric points in order to produce an isotropic response. It should be mentioned that three orthogonal non-intersecting SRRs would be a possible candidate for a completely isotropic scatterer in three dimensions, but unfortunately the SRRs then tend to have non-degenerate resonance frequencies or different Q factors. An array of two-dimensional isotropic scatterers with random orientations can be ex-

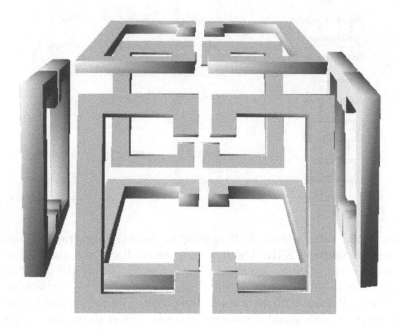

Figure 3.20   The unit of a three-dimensionally isotropic magnetic medium. The figure shows six planar SRRs of finite thickness oriented along all three orthogonal directions. By repeating the structure periodically, one generates a three-dimensional SRR medium.

pected to behave reasonably as an effective medium with an isotropic response. A similar crossed split ring structure for two-dimensionally isotropic media was proposed in Chen et al. (2003). A three-dimensional magnetic medium could, of course, be generated by randomly orienting such one-dimensional or two-dimensional SRR units, but this would be at the cost of reducing the filling fraction of the system. Consequently, the dispersion in the resulting effective isotropic medium might not be large enough to generate a negative magnetic permeability. A novel method to produce three-dimensionally oriented SRR arrays has been demonstrated recently (Islam and Logeeswaran 2006). The method consists of lithographically patterning the SRR on a bi-layer film and the residual stresses in the bilayer can then be used to curl up the patterned SRR out of the plane along a patterned weak hinge. The angle at which the SRR sticks out of the plane is determined by the thickness and the materials of the bilayer.

Finally, note that while the proposal of O'Brien and Pendry (2002) to use cylinders of high dielectric permittivity materials results in an anisotropic magnetic medium, the use of magneto-dielectric spheres (Holloway et al. 2003) or dielectric spheres of large permittivity (Yannopapas and Moroz 2005) results in a medium which is intrinsically three-dimensional and reasonably isotropic in the effective medium limit. This idea has been recently pushed to its second iteration by proposing a meta-metamaterial, a superstructure of metamaterials (Rockstuhl et al. 2007). In the first iteration, metal nanoparticles are densely packed in order to create a Lorentz resonance response in the permittivity. The metamaterial thus created is then shaped into spherical units which exhibit Mie resonances due to the excited magnetic mode. Therefore, in the second iteration, the metamaterial spheres are organized in a cubic lattice in order to create a meta-metamaterial, whose permeability exhibits a strong resonance in the visible spectrum.

3.3 Metamaterials with negative refractive index

Section 5.1 provides a rigorous justification to the fact that a medium with simultaneously $\mathrm{Re}(\varepsilon) < 0$ and $\mathrm{Re}(\mu) < 0$ can be characterized by a negative index of refraction. These considerations can be extended to anisotropic structures as well when $\mathrm{Re}(\varepsilon) < 0$ and $\mathrm{Re}(\mu) < 0$ apply to the corresponding directions of the electric and magnetic fields, respectively, of the electromagnetic radiation (see Chapter 5 for more details). Although it is easiest to conceptually understand the negative refractive index of a medium as $\mathrm{Re}(\varepsilon) < 0$ and $\mathrm{Re}(\mu) < 0$, it is perhaps not as fruitful to literally implement media in the same manner. For example, if one tries to embed magnetizable SRR inside a good uniform metal with $\varepsilon < 0$, one would not obtain a negative refractive index medium for the simple reason that the electromagnetic radiation just cannot penetrate into the metal and get the SRR to respond. Furthermore, a uniform metal everywhere would not allow the independent existence of the SRR or would just short out the SRR. Restated very simply, one requires a structure into which the radiation penetrates sufficiently and the structure would need to have resonances that can be independently driven by the electric field and the magnetic field. In addition, the presence of the resonant dielectric units should not interfere with the functioning of the resonant magnetic units themselves. Only by penetrating inside the structure does the radiation excite the resonances that give the medium its effective properties. Consequently, the medium would need to have closely spaced dielectric and magnetic resonances such that the frequency bands of negative dielectric permittivity and negative magnetic permeability overlap, in principle producing a negative refractive index.

One way of understanding how the separated electric and magnetic resonant elements interact with the wave is to imagine a body-centered cubic lattice with an electric dipole placed at one corner of the cubic and a magnetic dipole placed at the body center. In the presence of only one of these elements, the polarization in the medium screens out the incident radiation in the negative parameter frequency band. This happens due to the anti-phased response for either the electric field or the magnetic field. However, in the presence of both the electric and magnetic dipoles, the anti-phased response to both the electric and magnetic fields results in a propagating mode, although one with negative phase velocity. Combining the dispersions for the plasma-like dielectric medium and a resonant magnetic medium, we get

$$k^2 = \varepsilon\mu\omega^2/c^2 = \frac{(\omega^2 - \omega_p^2)(\omega^2 - \omega_m^2)}{c^2(\omega^2 - \omega_0^2)}, \qquad (3.73)$$

where ω_p and ω_m are the electric and magnetic plasma frequencies, respectively, and ω_0 is the magnetic resonance frequency ($\omega_m = \omega_0/\sqrt{1-f}$). An illustration of the frequency dependence of the permittivity, the permeability, and the index of refraction (directly related to the wave-number), is shown in Fig. 3.21. If ω_p is the largest of them, then we obtain a pass band in the region $\omega_0 < \omega < \omega_m$, which is referred to as the negative index band. Thus, in order to obtain a negative refractive index the presence of vacuum in which the resonant structures are embedded is essential. The negative medium parameters should all be considered to be effective medium parameters only. One cannot assume a uniform ε and μ due to one of the structures and calculate the response of the other structure. It is essential that each structure functions independently as if the other structure were absent. This places restrictions on how to build the interleaving isotropic metamaterials as one has to find essentially "null" points for one structure where the other structure can be placed so that the fields of one interfere with the other only minimally.

3.3.1 Combining the "electric" and "magnetic" atoms

The first medium with an effective $n < 0$ was reported in Smith et al. (2000), obtained by combining in a composite metamaterial the thin wire medium (which gives $\varepsilon < 0$) and the SRR medium (which gives $\mu < 0$) in which the normal of the SRRs was orthogonal to the wire direction. Admittedly it is not obvious that the composite metamaterial thus created would have a negative index of refraction n as the wires might interfere with the functioning of the SRR and vice versa, but the calculations and experimental measurements (Smith et al. 2000, Shelby et al. 2001a) were very suggestive of a real negative refractive index. The uniaxial composite obtained, shown in Fig. 3.22(a), consisted of wires of 0.8 mm thickness and SRRs with $\omega_0 = 4.845$ GHz. The numerical calculations showed that if thin wires are introduced into an SRR medium, a passband appears within the bandgap of

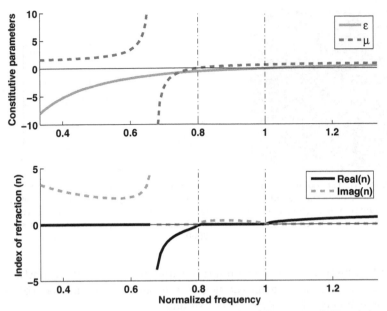

Figure 3.21 Evolution of the permittivity (ε), permeability (μ), and index of refraction (n) for a typical lossless Drude and Lorentz medium. The frequency axis has been normalized to the dielectric plasma frequency. Note how the sign of Real(n) is negative where $\varepsilon < 0$ and $\mu < 0$, and the presence of bandgaps where only one parameter is negative.

negative μ for radiation with the electric field along the wires and the magnetic field normal to the plane of the SRR. The computations suggested that the thin wires and the SRRs functioned independently and the composite metamaterial exhibited a passband with a negative refractive index. Typical experimental results on the transmission through waveguides filled with a medium of thin wires, a medium of SRRs only, and a composite medium with both thin wires and SRR are shown in Fig. 3.22(b): at the frequencies corresponding to the stopband due to a negative μ of the SRRs, an enhanced transmission appears when thin wires are introduced in addition to the SRR. These experimental results have often been cited as proof that the medium indeed exhibits a negative refractive index, although in reality, the only appearance of the transmission band could be due to a positive index of refraction as well, produced by the strong interaction between the wires and the SRRs. In order to clarify the remaining uncertainties, it was shown (Shelby et al. 2001b) that a prism made of such a composite metamaterial indeed refracted microwaves in the opposite direction of the normal compared to a prism of

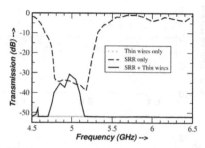

(a) Photograph of the composite thin wire arrays and SRRs that exhibits a negative refractive index (Smith et al. 2000). (Courtesy of D. R. Smith.)

(b) Transmission across a sample of SRRs (dotted line) and across the composite of thin wire arrays and SRRs (solid line). Data taken from Smith et al. (2000).

Figure 3.22 Illustration of a metamaterial and its passband properties. The enhanced transmission within the negative μ band visible in case (b) is interpreted as evidence of negative refractive index.

positive refractive material such as teflon.¶

Similar experiments were conducted with large samples in free space (Parazzoli et al. 2003, Greegor et al. 2003) where both the transmission and reflection were measured, and the absorption was found to be small enough to unambiguously demonstrate the negative refraction effect, with results consistent with calculations. These experiments have unequivocally demonstrated the existence of negative refractive index in the SRR-thin wire composites.

In order to understand the functioning of the SRR-thin wire composite, one notes that as long as the thin wires are not placed in regions where the highly inhomogeneous magnetic fields associated with the SRRs are present (along the axis of the SRR) and the SRR planes are placed such that they are at the points of symmetry between the wires (where the magnetic fields associated with the wires are minimal), the interference can be much reduced. Also note that the magnetic fields due to the wires fall off rather rapidly with distance from the wires and do not affect the SRRs significantly. As a result, the quasi-static responses derived previously for ε and μ remain valid in the negative refractive index band, and the SRRs along with the thin wires function independently as if located in vacuum. Nonetheless, the relative placement of the various components is crucial and might account for the differences reported in numerical and experimental data.

¶This is due to the fact that radiation refracts to the other side of the normal at the interface between positive and negative media, as explained in details in Section 5.2.1 and Section 5.3.2.

Using alternating layers of polaritonic spheres and plasmonic spheres as the basis of a negative refractive index medium has been proposed by Yannopapas and Moroz (2005). The large permittivity dielectric spheres enable negative magnetic permeability as discussed in Section 3.2.5 and plasmonic spheres give rise to a resonant Lorentz dispersion for the dielectric permittivity. The negative refractive index material thus realized is potentially a truly sub-wavelength structure with a wavelength-to-structure ratio as high as 14:1 at frequencies where polaritonic materials are available such as at middle infrared frequencies. It should also be noted that the loop-like arrangements of plasmonic nanoparticles that give rise to the negative magnetic permeability can themselves give rise to a negative refractive index if properly combined with a dielectric resonance (Alù et al. 2006a).

Note again that it is easier to fabricate anisotropic metamaterials with the magnetic permeability or the dielectric permittivity negative only for fields applied along certain directions. More isotropic designs necessarily involve interleaving orthogonal planes of SRR and thin wires (cut-wires). Designing and implementing a truly isotropic metamaterial that has a negative refractive index as well is a formidable design challenge. It is therefore very advantageous when the same structural unit can provide both the negative permittivity and the negative permeability as in the case of Alù et al. (2006a). However, these units have the same problems of anisotropy as those based on the more standard SRR. The proposals of Yannopapas and Moroz (2005) to use polaritonic and plasmonic spheres and that of Holloway et al. (2003) to use magneto-dielectric spheres also yield an intrinsically three-dimensional negative index medium. Although these proposals for isotropic negative refractive index materials are very attractive, for the moment they remain limited to microwave to mid-infrared frequencies due to the difficulty of finding dielectric media with very large dielectric permittivities at optical frequencies.

3.3.2 Negative refractive index at optical frequencies

The refractive index is a quantity that is usually and intimately associated with optical phenomena. Shortly after the demonstration of a negative refractive index phenomenon at microwave frequencies, the research trend quickly focused on obtaining negative refractive index metamaterials at optical frequencies, where many of the novel phenomena can actually be *"seen."* Since 2005, there have been claims of successful implementations of such metamaterials (Grigorenko et al. 2005, Dolling et al. 2006b, Shalaev 2007) and we devote the coming section to their description and to the study of their limitations. In addition, we also note that many of the metallic metamaterials at optical frequencies exhibit plasmonic resonances and negative phase velocity bands.

As explained in Section 3.2.4, it is not straightforward to scale up the performance of split ring resonators or cut wire media to optical frequencies. The plasma-like aspect of the metal begins to dominate over the Ohmic nature and the inertial inductance due to the finite electronic mass prevents a straightforward scaling to optical frequencies by simply reducing the size of the metamaterial units while keeping all other aspects fixed. A more fruitful approach seems to be to reduce the series capacitance by including more capacitive gaps which would then increase the resonant behavior to higher frequencies.

However, simply reducing the capacitance by adding more series capacitors while keeping the overall geometric size and the periodicity as well as other parameters fixed brings in another problem. As the resonant frequency increases, the ratio of the geometric size of the metamaterial units (or the periodicity) to the wavelength in the medium becomes larger ($a/\lambda \to 1$) and homogenization becomes increasingly questionable. As a/λ increases, the electromagnetic wave begins to discern the underlying structures of the metamaterial and, as a first effect, spatial dispersion appears. At even larger ratios of a/λ, the homogenization hypothesis itself breaks down and the average effective medium parameters lose their significance. Bragg scattering becomes more important and the system is better described in terms of the photonic band structure as introduced in the next chapter.

One could, however, wonder what prevents us from simply scaling down the sizes of the metamaterial resonant units as well while decreasing the overall capacitance of the system? For example, an operating frequency of about 200 THz would be obtained with the SRR considered in Section 3.2.4 with a unit cell size of $a = 300$ nm, an overall SRR dimension of about 180 nm, metallic sheets of uniform thickness $D = 24$ nm, and a separation $d_c = 24$ nm defining the capacitance in the structure, as shown in Fig. 3.12. While very small and difficult to fabricate, these dimensions are just within reach of experimental fabrication abilities by present-day high resolution electron lithography, focused ion beam etching or colloidal lithography. Since significant reductions of the dimensions from these numbers would render the design impractical for physical implementation, we only consider dimensions of this order for the metamaterial designs for high frequency operation. This implies a ratio $a/\lambda \sim 0.5$ for a wavelength of $\lambda = 600$ nm, indicating that the homogenization hypothesis is not reliable. This order of magnitude is typical of all the experimental implementations at optical frequencies thus far. While there are slightly simpler designs for metamaterials such as the plasmonic nanoparticles placed on nano-sized loops of Alù et al. (2006a),[‖] we consider those also as extremely difficult to implement except by some self-assembly techniques. In that case, the inherent disorder that occurs in the fabrication also reduces the effective filling fraction of the system and weakens the resonant nature of the system.

Let us consider the SRR made of silver with four splits for operation at

---

[‖] Typical sizes for optical frequency would have a radius of the ring of 38 nm, a nanoparticle size of 16 nm, with four particles symmetrically arranged on the ring.

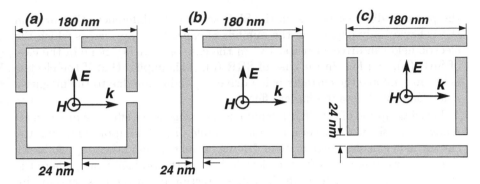

Figure 3.23 Three possible variations of the SRR with four capacitive gaps. SRR (b) and (c) have the same dimensions but are only rotated by 90° with respect to the incident electric field.

telecommunications frequencies discussed in Section 3.2.4. The operating frequency was deliberately kept low by embedding the SRR in a dielectric with $\varepsilon = 4$ and enhancing the capacitance of the SRR. This way, a respectable ratio of $a/\lambda \sim 1/6$ was maintained and one could just about accept the validity of homogenization and effective medium parameters. Subsequently, let us consider the different ways of placing the splits on the rings which are embedded in a dielectric as shown in Fig. 3.23 (a-c). In each case, the inductance and capacitances (except for the self capacitance) are similar and the net induced electric dipole moment on the rings should be zero in the quasi-static limit. The three systems are expected to behave similarly except for, perhaps, small changes in the resonance frequencies. In fact, designs (b) and (c) just differ in the relative orientations with respect to the electric field of the incident radiation. In Fig. 3.24 we show the effective magnetic permeability calculated from the reflection and transmission coefficients obtained by transfer matrix calculations for slabs of four layers thickness made out of the three structures. The values obtained experimentally for the bulk dielectric permittivity of silver have been used in the computations. We see that there are significant differences in the effective medium parameters obtained for the three structures. For example, compared to the original SRR (a), we see that the SRR (b) has a smaller bandwidth of negative μ and a smaller modulation in the values of the permeability ($\mathrm{Re}(\mu_{\min}) \sim -2$). In the case of SRR (c), however, a larger bandwidth of negative μ is obtained along with a deeper modulation of the band ($\mathrm{Re}(\mu_{\min}) \sim -3.5$). The relative orientations of the SRR capacitive gaps to the electric field therefore appear to induce significant differences in the behavior of the metamaterial. This rather unexpected phenomenon arises because of the significant phase shifts of the radiation across a unit cell that can arise in these metamaterials where homogenization is barely viable. The electric field can also drive currents around the loop due to the differences in the phase at different points on the SRR so that the differences that arise in the behavior only depend on the number of capacitive gaps across which the electric field can drive currents: two in the case of SRR (a), none in the case of SRR (b), and four in the case of SRR (c). This implies that if the electric field can reinforce the currents being driven by induction due to the magnetic fields, one could obtain a larger effect.

This description of the SRR medium as a homogeneous effective medium already shows signs of inaccuracy in our example. To a first approximation, the electric field can also participate in driving currents around the loop. Actually there are two resonances for the electric field in the problem: one, where the electric fields drive currents in the same direction down the legs on opposite sides of the SRR (a symmetric resonance), and another, where the electric fields drive oppositely oriented currents down the legs on opposite sides of the SRR (an anti-symmetric resonance). The antisymmetric resonance reinforces the inductive action of the magnetic field and conversely the possibility of having charge distributions on the capacitive gaps allows the electric field to couple strongly to the anti-symmetric resonance.

Next we take the logical step of removing the dielectric in which the SRR are embedded and consider the SRR to be in vacuum: the resonance should then occur at visible frequencies (far-red). Fig. 3.25 shows the band-structure calculated for two-dimensional SRR for P-polarized light (TM-modes). It is clear at first sight that the behavior in the three cases are extremely different. Most importantly we should note that in the case of SRR (a) and SRR (c), when the electromagnetic radiation can interact with the charge distributions across the capacitive gaps, propagating bands appear where the real part of the wave-number and the imaginary part of the wave-number have opposite signs – the real and imaginary parts of one wave-number are plotted in the same color (black or gray). In other words we have negative phase velocity bands[**] due to closely spaced electric resonance and magnetic resonance. For SRR (b), when the electric field cannot directly drive the charges across the capacitive gaps, we obtain a bandgap instead. It is seen that the SRR in this case is just acting as a system of resonant electric dipoles with a negative dielectric permittivity. If one alters the capacitance of the SRR, this bandgap moves up or down in frequency depending on whether the capacitance is smaller or larger. For example, removing the middle (smaller) legs of the SRR (b) causes the bandgap to move up to 1.8 eV energies due to the reduced capacitive area of interaction between adjacent SRR cells. The photonic bandgap due to Braggscattering is in the range of 1.9 to 2 eV (see Fig. 3.25, bottom left). Note that the larger number of capacitive gaps with which the electric field can interact in SRR (c) causes a larger bandwidth for the negative phase velocity. Another interesting aspect it that the negative phase velocity bands

---

[**]In these solutions, the Poynting vector decays along the positive direction and hence the imaginary part of the wave-vector should be positive.

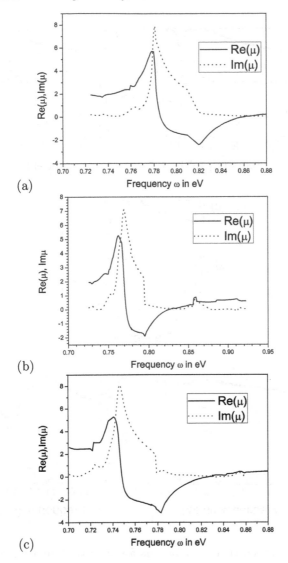

(a)

(b)

(c)

Figure 3.24 Effective magnetic permeability calculated from the reflection and transmission coefficients of layers made up of the three SRR structures (a, b and c) and the incident electromagnetic radiation shown in Fig. 3.23. Although the three should be equivalent in the quasi-static approximation, the orientation of the electric field relative to the capacitive gaps affects the response considerably.

of SRR (a) and (c) could have a positive or negative group velocity. These calculations confirm the experimental observations by Dolling et al. (2006b)

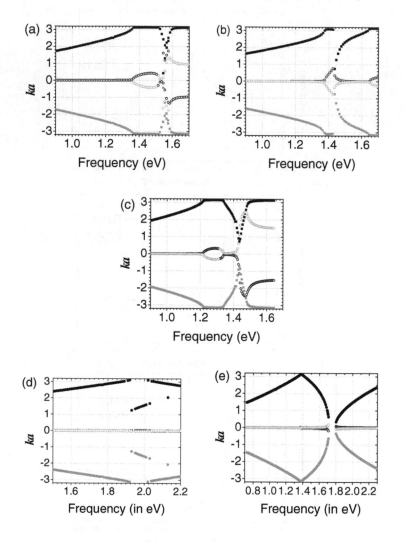

Figure 3.25 Band structure diagrams for the two-dimensional lattice of the three SRR structures made of silver and in vacuum with the incident electromagnetic radiation shown in Fig. 3.23. SRR(a) and SRR(c) have negative phase velocity bands while SRR(b) has only a bandgap corresponding to $\varepsilon < 0$ at similar frequencies. The bottom panel (left) shows the band structure for the two-dimensional lattice of SRR(b), but with the middle shorter stubs removed. The bandgap at about 1.95 eV is due to Bragg scattering. The bottom panel (right) shows the band structure for the two-dimensional lattice of SRR(c), but with the middle shorter stubs removed. The two plate-pairs have remarkably different band structures.

of all combinations of positive and negative phase and group velocities at different frequencies in similar fish-net structures described later in this section. In fact, the shorter middle stubs can be removed in SRR (c) to produce a negative phase velocity over a much larger bandwidth (see Fig. 3.25, bottom right) – the origin of the electric resonance is the plasmonic resonance of the remaining plates of the SRR. This system is, in fact, essentially a plate-pair system which is the two-dimensional analogue of the wire-pairs in three dimensions proposed and demonstrated by Shalaev et al. (2005) which had an effective negative refractive index of about -0.2 at a wavelength of 1.5 μm.

Tempting as it is to attribute the above behavior of negative phase velocity to negative refractive index, note that at these photon energies of about 1.4 eV ($\sim$870 nm) the ratio of $a/\lambda \sim 3$ and homogenization is hardly valid. We should refrain from proposing effective media parameters for such a system and calling it a negative refractive index medium, which is the reason why we do not present equivalent medium parameters for this system. However, the negative phase velocity band properties depend only on the localized resonances of the structure and not on the periodicity. To demonstrate this, we present a calculation based on the Finite Element Method in which a slab of such SRR is used as a flat Veselago lens to image sources as explained in Chapter 8. The imaging of two point sources is shown in Fig. 3.26 through a periodic slab of SRR and a disordered slab of SRR with even a ring missing. We can see that the imaging action is quite robust against the disorder and two images are clearly formed. This last example makes it clear that the origin of negative refraction in this system depends crucially on the localized resonances of the SRR system and is not due to band structure effects (as discussed in Chapter 4).

The behavior of the SRR medium is illuminative and enables us to clearly understand the nature of the negative phase velocity bands at optical frequencies. It clearly shows the increasing role of the electric field in the excitation of the plasmonic resonances of the SRR at high frequencies, and how the single structure can have both closely spaced electric and magnetic resonances. This is the origin of the negative phase velocity bands. However, the two-dimensional cylindrical structures discussed above are not easily fabricated by conventional deposition and etching techniques. The excitation of the planar SRR structures, which are amenable to fabrication by such conventional techniques, involves radiation incident along the plane with the magnetic field polarized normal to the SRR structures. This is rather difficult experimentally. To avoid this, a double-layered fish-net structure with thin wires running along one direction and thick plate-like structures in the orthogonal direction has been proposed in Zhang et al. (2005a). Here the incident radiation would be normal to the plane of the fish-net with the electric polarized along the thin wires. The magnetic field induces a resonance across the plates in the two layers and thus the structure can show an effective negative refractive index. This fish-net structure has been implemented at both telecommunication frequencies (Dolling et al. 2006a;b) and at optical frequencies, and shows

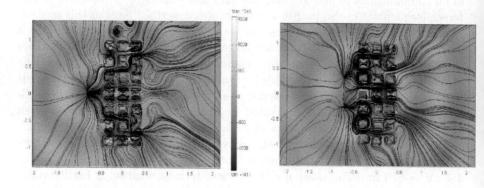

Figure 3.26 Left: Imaging of two point sources by a finite slab of period-ically placed SRR(c), which acts as a Veselago lens. Right: Imaging of the two sources by a slab of disordered SRR(c) demonstrating that the effect does not depend on the periodicity of the system. The magnitudes of the magnetic field along the SRR axis is shown in gray scale while the streamlines show the Poynting vector. The focusing action is clear despite the very small transverse width of the slab – the energy concentrates transversally at the location of the images. The images are present very close to those locations predicted for the Veselago lens.

a rich variety of phenomena including those described above for the SRR structures. However, these structures also have a ratio of $a/\lambda \sim 3$ and suffer from the same limitations set by the criteria of homogenization. It would not be rigorously correct to term these as effective negative refractive index media. Only the loop-like structures of Alù et al. (2006a) have a small enough ratio of $a/\lambda \sim 10$ and are amenable to homogenization when the magnetic permeability and the dielectric permittivity are both negative in a common frequency band. Most of the other metamaterials proposed for optical oper-ation so far either are not homogenizable or depend on inter-particle (unit) interactions for their operation. In principle, a metamaterial that depends on inter-particle interactions is not likely to be robust against disorder and cannot be recommended at nano-metric sizes where considerable disorder is almost unavoidable.

3.4 Chiral metamaterials

Optical effects with molecular chiral media, particularly the rotation of the plane of polarization for linearly polarized light upon reflection or transmission from a chiral medium, are well known since the times of Pasteur and Faraday. The idea of having chiral inclusions to affect wave properties is not very new either. [††] Chiral media, however, are interesting by themselves because they break time reversal symmetry for an electromagnetic wave propagating in them and have a sense of handedness. Usually the chirality of natural media is not very strong and the corresponding effects on radiation are weak. Metamaterials, however, yield an opportunity to strongly and resonantly enhance the chiral properties via chiral scatterers. Chiral metamaterials are not necessarily built up from intrinsically chiral materials but rather derive their chiral properties from the chiral geometric structure of the metamaterial units. For example, the breaking of time reversal symmetry for light has been demonstrated using a planar metamaterial consisting of a single layer of metallic particles with chiral shapes (Schwanecke et al. 2003).

An impetus to the development of chiral metamaterials came with the realization that one of the circularly polarized states of light could experience a negative refractive index under suitable conditions (Tretyakov et al. 2003, Pendry 2004a, Monzon and Forester 2005). Consider the following bi-isotropic constitutive relations in a chiral medium:

$$\mathbf{D} = \varepsilon_0 \varepsilon \mathbf{E} - \mathrm{i}(\xi/c)\mathbf{H}, \tag{3.74a}$$

$$\mathbf{B} = \mathrm{i}(\xi/c)\mathbf{E} + \mu_0 \mu \mathbf{H}. \tag{3.74b}$$

The above constitutive relations are reciprocal and ξ in general is a complex function of the frequency that also satisfies the Kramers-Kronig relations (see Eqs. (1.9)). The dispersion for a plane wave in this medium is given by

$$k_\pm = (\sqrt{\varepsilon\mu} \pm \xi)\omega/c, \tag{3.75}$$

and the two different dispersions correspond to the two circular polarized states of light. The different polarizations feel different material parameters:

$$\varepsilon_\pm = \varepsilon\left(1 \pm \frac{\xi}{\sqrt{\varepsilon\mu}}\right), \qquad \mu_\pm = \mu\left(1 \pm \frac{\xi}{\sqrt{\varepsilon\mu}}\right). \tag{3.76}$$

Consequently, if ε or μ becomes small or zero (as it happens near the plasma frequency), it is clear that the phase vector would be negative. The energy flow on the other hand is determined by the impedance for the wave and is

---

[††]We refer the reader to Lindell et al. (1994) for the basic ideas and an account of earlier work on bi-isotropic and chiral media.

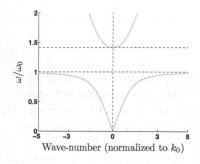 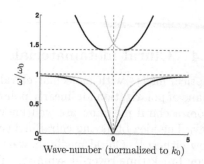

Figure 3.27 Left: band structure for light in a medium of resonant dipoles. There is a bandgap just above the resonance frequency. Right: band-structure for light in a chiral medium containing resonant dipole particles. The bands split for the two circular polarizations shown as black and gray lines. Just above the bandgap, one obtains a narrow band within which the phase vector and the group velocity have opposite directions for one of the helicities.

given by $Z = \sqrt{\mu/\varepsilon}$ in the case of reciprocal media, and is positive regardless of the choice of the branch. Such anti-parallel directions of the wave-vector and the energy flow are a typical sign of negative refraction. Note that this would also imply that the system would be able to support surface plasmon waves of a chiral nature and that these surfaces states should enable the perfect lens effect via amplification of the evanescent waves[‡‡] for one circular polarized state of light (Jin and He 2005).

Such a scenario can easily be realized by embedding electric dipoles in a chiral medium (assumed non-dispersive for the time being) (Pendry 2004a), in which case the dielectric permittivity is given by a Lorentz form (see Eq. (1.15)). The resulting dispersion for the two branches is plotted in Fig. 3.27. As ε goes through a zero, one clearly sees that there is a small frequency band where the phase vector is negative. This negative refractive index band in chiral media gives another possibility to have negative refractive index without developing a metamaterial that has both the ε and μ negative. Interestingly, there is a cross-over point ($k = 0$) in the bands where the group velocity ($\partial\omega/\partial k$) is non-zero while the phase velocity is zero. Actually this reversed phase vector and the energy flow were noted much earlier by Engheta et al. (1992) in the context of a dispersive chiro-plasma.

In the previous example, an underlying chiral medium where the electric dipoles were embedded was utilized. However, it is not easy to find atomic or molecular media with large chiral coefficients and the band with negative refraction has a very small bandwidth. Hence it would be better to have a metamaterial that contributes to the chirality as well. A variant of the

---

[‡‡]See Chapter 8.

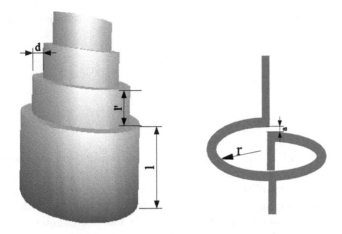

Figure 3.28 Left: chiral particle made by rolling up a strip of metallic sheet. Right: simple chiral loop-like particle. These particles have the property of bi-isotropic chirality along the axis.

Swiss roll where a strip of metal rolled up at an inclined angle to make a chiral structure (shown in Fig. 3.28, left) was considered in Pendry (2004a). It was shown that one could enhance the magneto-electric coupling due to the enhanced inductance in the system (the factor of N rolls contributes to this). A three-dimensional array of such rolled-up metallic strips was shown to have a band of negative refraction for one of the circular polarizations. To understand this effect, consider for simplicity the simpler chiral loop made of a metallic wire shown in Fig. 3.28, which has been well studied in the context of chiral antennas (Tretyakov et al. 1996; 2005). This chiral particle has a dielectric polarizability for electric fields applied parallel to its axis, but simultaneously generates a magnetization by virtue of the circular current in the loop. Similarly, the particle has a magnetic polarizability for a magnetic field applied along its axis which induces an electromotive force and currents around the loop, but simultaneously develops an electric polarization along the axis, as the currents cannot complete the circuit around the loop due to the gap. Thus, in these systems of chiral particles, there is a simultaneous excitation of the electric, magnetic, and chiral resonances. Usually, there is a hierarchy (Tretyakov et al. 2005) of the dielectric (α_{ee}), chiral (α_{em}), and magnetic (α_{mm}) polarizabilities of such particles with $\alpha_{ee} > \alpha_{em} > \alpha_{mm}$.

Typical resonant forms for these polarizabilities are (Tretyakov et al. 1996):

$$\alpha_{ee} = \frac{A_e}{\omega_0^2 - \omega^2 - i\gamma\omega}, \tag{3.77a}$$

$$\alpha_{em} = \frac{A_c\omega}{\omega_0^2 - \omega^2 - i\gamma\omega}, \tag{3.77b}$$

$$\alpha_{mm} = \frac{A_m\omega^2}{\omega_0^2 - \omega^2 - i\gamma\omega}, \tag{3.77c}$$

where A_e, A_m, and A_c are some constants depending on the geometry of the particle and ω_0 is the resonance frequency of the particle. Hence one can use these polarizabilities for a randomly oriented ensemble of such chiral particles and calculate the effective medium parameters within, say, the Maxwell-Garnett homogenization approach. It has been shown in Tretyakov et al. (2005) that such an ensemble can have a negative refraction band for one circularly polarized state in a narrow region of frequencies just above the bandgap.

3.5   Bianisotropic metamaterials

The previous sections have presented some geometries of split-ring resonators that achieve a negative permeability at some frequencies. The literature contains many more designs, creating too long a list to be discussed exhaustively here. Nonetheless, all the designs present one important similarity: they contain a ring-like structure around which a current can flow, but they also contain an interruption (a gap) in the metallization of the ring, so that the conduction current flow is interrupted (the current loop can be closed by displacement currents, which introduce a capacitive coupling). These structures therefore present some analogies with three-dimensional helical structures, which are known to exhibit chiral properties already discussed. An important difference, however, is that the split rings within metamaterials are usually two-dimensional [a three-dimensional ring has been nonetheless proposed and analyzed in Gay-Balmaz and Martin (2002)], so that the isotropy of the chirality is lost, and is reduced to a bianisotropy (Saadoun and Engheta 1992). Mathematically, this implies that the magneto-electric coupling terms in the constitutive relations of Eq. (3.74b) change from being scalar to being tensors in which only specific components are non-zero. An example of bianisotropic medium is provided in Eqs. (2.101). Physically, the loss of isotropy requires all the constituents to be arranged in an ordered manner so as to exhibit a collective macroscopic effect, unlike chiral constituents which can be embedded randomly in a host medium (Sihvola 2000). Metamaterials, which exhibit helical microstructures regularly organized in space (which does not

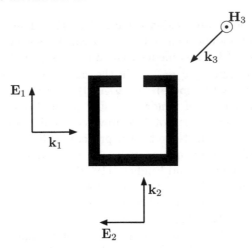

Figure 3.29 Illustration of various incidences of electromagnetic radiation on a split-ring resonator that do or do not induce a bianisotropic response.

imply *periodically* organized, however), are therefore very likely to exhibit bianisotropic properties, and should be studied in detail.

Bianisotropy can be induced in split-ring resonators by a geometrical asymmetry that creates a charge and current imbalance. Let us consider the simplified single ring of Fig. 3.29 and the three incidences for the radiation shown. The first incidence presents an electric field parallel to the two symmetric sides of the split-ring, whereas in the approximation of thin metallizations, the current in the perpendicular directions can be neglected. In addition, the small size of the ring compared to the incident wavelength ensures that the variations of $\mathbf{E}_1$ between the two sides are also negligible to a first approximation. Consequently, the charge distribution resulting from this incidence is symmetric and does not generate a circulating current. Under the second incidence $\mathbf{k}_2$, the situation is different because of the break in symmetry introduced by the gap. Within the same approximations, the charge distribution induced by the impinging $\mathbf{E}_2$ is asymmetric and generates a circulating current sustained by the capacitive coupling within the gap, which in turn generates a magnetic field. Consequently, a non-zero magnetic moment is created by the impinging electric field. This electro-magnetic coupling is reflected by the constitutive parameter $\bar{\bar{\zeta}}$ in Eq. (2.79). Conversely, the incidence $\mathbf{k}_3$ presents a magnetic field perpendicular to the axis of the ring which, in virtue of Ampère's law, creates a circulating current via the charge accumulation at the gap. Again due to the presence of the gap, the resulting charge distribution is asymmetric, results in charge accumulation about the capacitive gaps and therefore induces an electric dipole moment. This magneto-electric coupling is reflected by the constitutive parameter $\bar{\bar{\xi}}$ in Eq. (2.79). Note that these two effects are not independent: the incidence $\mathbf{k}_2$ yields a magnetic field parallel to $\mathbf{H}_3$, whereas the incidence k_3 yields an incidence parallel to $\mathbf{E}_2$. The parameters $\bar{\bar{\xi}}$ and $\bar{\bar{\zeta}}$ are therefore not independent either, and can be shown to be related by $\bar{\bar{\xi}} = \bar{\bar{\zeta}}^\dagger$ in the case of lossless media (where the † sign indicates a transpose complex conjugate) and by $\bar{\bar{\xi}} = -\bar{\bar{\zeta}}^T$ in the case of reciprocal media (where the T sign indicates a transpose) (Kong 2000). Note that the bianisotropic property of the split rings of Pendry was first pointed out by Marques et al. (2002), where an explicit form of $\bar{\bar{\xi}}$ is provided. This model is at the basis of the bianisotropic parameter retrieval procedure discussed in Section 2.6. Other examples of bianisotropic particles are the "Omega"-shaped particles first introduced in Saadoun and Engheta (1992): the current around the loop which makes the particle have a magnetic moment normal to the plane of the particle and gives rise to separated charges at the ends due to which an electric dipole moment in the plane is generated. Thus, ordered arrays of Omega particles as shown in Fig. 3.30 have a bianisotropic response. The reader is referred to Aydin et al. (2007) for a detailed study of metamaterials composed of small Omega particles.

The presence of bianisotropic effective medium parameters renders the propagation of the electromagnetic wave inside the medium more complex. However, this does not necessarily imply that bianisotropy should be avoided. For example, an electromagnetic wave propagating through a chiral medium (and thus through a bianisotropic medium under the proper incidence) experiences a reciprocal rotation of polarization, which can be used for example in the design of phase shifters (Saadoun and Engheta 1992). In addition, bianisotropy can also induce a negative refraction as shown in Section 5.3.4 and, since it is easier to achieve at optical frequencies than a negative permeability, it may effectively be used to generate interesting phenomena at the higher end of the spectrum. Initial work in this direction has been proposed in the microwave regime in Tretyakov et al. (2007) by obtaining a backward wave material perfectly matched to free-space at normal incidence.

In some other situations, however, bianisotropy is undesirable and should be avoided. When this is the case, the ring design should be modified accordingly so as to not induce a bianisotropic response or to cancel it by inducing two opposite responses. For example, a modified design has been proposed in Marques et al. (2002), whereby a second ring is placed behind the first one across the substrate after being rotated by 180 degrees. The two gaps being in symmetric positions with respect to the overall design of the ring, bianisotropy effects are drastically reduced. Alternative designs include the inclusion of more symmetrically placed gaps (as has been used in this book) or the total redesign of the ring, an example of which can be found in Chen et al. (2004a), Bulu et al. (2005). When there are two symmetrical splits present, for example, the dipole moments across opposite ends cancel each other and one only gets a weak electric quadrupole moment. The symmetry of the single ring with two symmetrically placed capacitive gaps renders the design less bianisotropic and electrically less active. This was in fact the un-

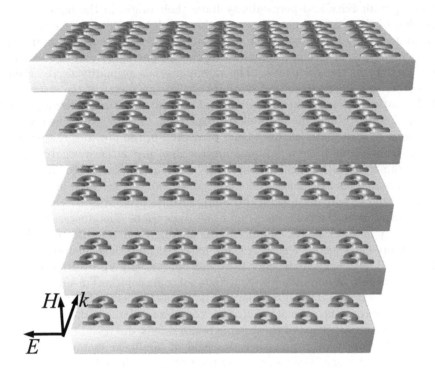

Figure 3.30 Schematic portrayal of stacked arrays of planar "Omega"-shaped metallic particles deposited on a dielectric substrate to form a strongly bianisotropic medium. The suggested incident directions to obtain a negative refractive index medium are also shown.

derlying reason for which we have only considered SRR with symmetrically placed gaps in the ring: it avoided a bianisotropic effect and allowed us to concentrate on the magnetic effects. The bianisotropy can also be effectively resolved, of course, by rotating adjacent SRRs with single gaps in the plane by 180° and the corresponding electric dipole moments would cancel.

3.6 Active and non-linear metamaterials

So far, the metamaterials discussed in this chapter have been linear passive systems. They consist of resonant structures that show large dispersion at frequencies near the resonance frequency. All the effects including negative material permittivity and permeability have their origin in the fast-varying dispersion of the material parameters in the vicinity of the resonances. Dissipation invariably accompanies the dispersion in these systems and many of the metamaterials absorb large amounts of energy, particularly at frequencies in the vicinity of the resonance. The metamaterials primarily derive their resonant nature due to structural resonances. The geometry of the structure and the nature of the constituent materials of the structure determine the resonant frequency and the dispersion properties. A variety of linear wave phenomena in such materials have been presented in the previous chapters.

However, one would like to do much better than just achieve negative material parameters. Although unachievable until a few years ago, negative material parameters are almost becoming *passé* today. Currently, one would like to be able to actively control the metamaterial properties by externally applied fields, tune the resonances, and the dispersion. Metamaterials have exceptionally large local field enhancements making them a fertile ground for nonlinear optical effects. One can even have reconfigurable metamaterials with external feedback. We wonder if we can compensate for dissipation and dispersion by gain or nonlinearities and have solitonic solutions in such media. What new effects would the dispersion in the magnetic permeability bring in? Eventually there is the ultimate desire for complete control of the properties of the metamaterial. Some of these aspects are briefly discussed hereafter.

It should be mentioned first that the field of nonlinear metamaterials is still in a nascent stage. Only a small number of effects have been discovered and discussed theoretically and very few among them have been implemented experimentally. Hence, it actually becomes possible to give a short account of the development in this area. In the first design for the Split Ring Resonators in Pendry et al. (1999), it was pointed out that the intense electric fields in the capacitive gaps would enable a large nonlinear response by embedding nonlinear materials. This suggestion was later followed up in O'Brien et al. (2004) where nonlinear bistable switching between positive and negative permeability of the SRR at telecommunication frequencies was numerically demonstrated using Kerr nonlinear materials. Nonlinear intensity switching of the effective negative refractive index was proposed in Zharov et al. (2003). Lapine et al. (2003) theoretically studied the nonlinearity of a metamaterial arising out of diode insertion into resonant circuit configurations. The resulting nonlinearity in the magnetization was shown to lead to three-wave coupling in the metamaterial at microwave frequencies (Lapine and Gorkunov 2004). An active and controllable metamaterial switch at terahertz frequencies based on a electrically resonant unit has been experimentally demonstrated by injecting charges into the capacitive regions of the unit, which changes the resonance conditions (Chen et al. 2006b). An interesting novel nonlinearity arising from the Lorentz magnetic force on electrons in motion ($\mathbf{v} \times \mathbf{B} \sim -i\omega\mathbf{E} \times \mathbf{B}$) and the consequent nonlinear second harmonic generation was demonstrated in Klein et al. (2006a). It is important to emphasize that many nonlinear effects, particularly those that require long interaction length or time in the metamaterials, are marginal in the presence of dissipation. The effects that involve smaller samples or distances, on the other hand, could be remarkably stable against dissipation.

Dissipation seems to be the ultimate limit on the demonstrability and utility of any nice effect possible with negative refractive index media. Thus, one of the important needs of the hour is to reduce the levels of absorption in the metamaterial. One intuitively feels that it can be achieved by embedding some media or device that contributes to amplification in the metamaterial design. While recognizing that the level of losses in metals is very large, particularly at optical frequencies, one hopes to bypass this problem either by including high gain laser media such as semiconductors or nonlinear media (through Raman or wave mixing processes) or by improving the design of the metamaterial constituents so as to concentrate the radiation in the embedding medium and outside the metallic regions where maximum dissipation occurs. Note that such strategy has already been suggested for microwave metamaterials by inserting lumped element amplifiers (Tretyakov 2001). Nonlinear wave mixing and optical parametric amplification have been proposed in nonlinear metamaterials (Popov and Shalaev 2006) and many researchers are considering the implementation of media with gain in metamaterial designs. Moreover, the additional tunability offered is particularly interesting for sustaining the surface waves (see Chapter 7), since introduction of gain outside the metamaterials offers an effective manner of extending the lifetime of the surface modes. It was also shown that amplification in neighboring regions could compensate for dissipation in the regions of negative materials parameters via surface plasmon states on the interface (Ramakrishna and Pendry 2003), which has been shown experimentally for surface plasmons propagating on a metal surface in contact with a laser dye (Seidel et al. 2005, Noginov et al. 2007) and for metallic nanoparticles (Noginov et al. 2006). Consequently, it appears that nonlinearity and amplification may become an integral feature of future metamaterials.

We shall subsequently present here two metamaterials: a self-switched SRR medium exhibiting bistability and wave mixing, and an actively switched metamaterial via photoconductivity. Control of the capacitance through the embedded dielectric is the key to control the resonance of the metamaterial units. In thin wire arrays, one can make similar use of atomic or molecular magnetic media at microwave frequencies to control the magnetic permeability of the medium and hence the inductance of the wires.

3.6.1 Nonlinear split-ring resonators

The split-ring resonator shows large local field enhancements at frequencies near the resonance and there is a tremendous concentration of the electric fields in the small capacitive gaps. Consider the single ring SRR with two splits of Fig. 3.12. The electric field across the capacitive gaps is $E_c = V_c/d_c$, where V_c is given by Eq. (3.60), yielding

$$E_c(\omega_0) = \frac{\mu_0 \omega_p^2 r \tau_c}{2\gamma L_c}. \tag{3.78}$$

For the incident radiation, the energy is equally distributed between the electric and magnetic fields. We can estimate an enhancement factor of

$$Q = \frac{1}{2} \frac{1/2\varepsilon_0 |E_c(\omega_0)|^2}{1/2\mu_0 |H_{ext}(\omega_0)|^2} = \frac{1}{8c^2} \left[\frac{r\tau_c\omega_p^2}{2\gamma L_c} \right]^2 \tag{3.79}$$

for the energy stored in the capacitive gaps at resonance, where the factor of 2 accounts for the fact that there are two gaps in the system. Typical numbers for this factor range from 10^4 to 10^6 (Pendry et al. 1999, O'Brien and Pendry 2002).

Hence the material in the capacitive gaps can be crucially utilized to change the properties of the SRR. Indeed it has been pointed out that changing the substrate properties of planar metallic SRRs can be used effectively to control the metamaterial characteristics (Sheng and Varadan 2007). Any small nonlinearity in the material placed in the gaps can drastically affect the performance of the system as the nonlinear effects are amplified many times due to large field enhancements. Consider then filling the capacitive gaps with a material exhibiting a Kerr nonlinearity. Regardless of the frequency of interest, the nonlinear Kerr effect in a dielectric is always achievable (Boyd 2003). This is a typical nonlinearity where the refractive index of the dielectric depends on the electromagnetic fields as

$$n = \sqrt{\varepsilon} = n_0 + n_2 I, \tag{3.80}$$

where $I = 1/2(\varepsilon/\mu)^1/2\varepsilon_0 c |E|^2$ is the intensity of light. The sign of n_2 can be positive or negative in which case it is called a focusing or defocusing nonlinearity, respectively. Suppose that the capacitive gap in the SRR is filled with a Kerr nonlinear dielectric. The capacitance of the SRR depends on the value of the embedded dielectric in-between, and hence the resonant frequency becomes a function of the incident field strengths.

Consider a metamaterial of SRR with a nonlinear dielectric in the capacitive gaps. A relation between the incident field strength and nonlinear resonance frequency can be obtained in a quasi-static calculation as (Zharov et al. 2003, O'Brien et al. 2004)

$$|H_{ext}|^2 = \frac{Z_d n^2 d_c^2}{4n_2 L_c^2 \omega_0^2} \frac{(1-x^2)[(x^2 - \Omega^2)^2 + \Omega^2 \Gamma'^2]}{\Omega^2 x^6}, \tag{3.81}$$

where $\Omega = \omega/\omega_0$, $x = \omega_{NL}/\omega_0$, $\Gamma' = \Gamma/\omega_0$, ω_0 is the resonance frequency for the SRR embedded in the linear material, and ω_{NL} is the resonance frequency for the SRR embedded in the nonlinear material. The plot of the nonlinear

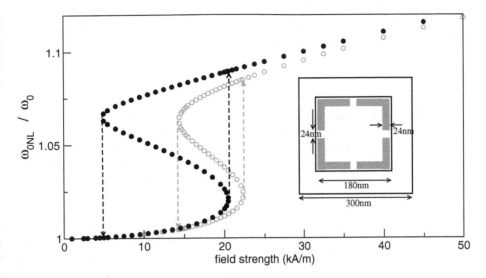

Figure 3.31 Nonlinear resonant frequency vs. the field strength for two values of the dissipation rate: filled circles are for the γ of silver and the open circles for 3γ. The inset shows the geometry of SRR structure with the relevant dimensions. The inner box shows the nonlinear medium just enclosing the SRR structure. (Reproduced with permission from (Ramakrishna 2005). © 2005, Institute of Physics Publishing, U.K.)

resonance frequency with the field strength is shown in Fig. 3.31. It is obvious that the system is bistable. The bistable behavior is reasonably stable against dissipation, which was checked by increasing the dissipation parameter by as much as three times (shown in Fig. 3.31). One can see that the resonant frequency switches from a lower frequency (ω_L) to a higher frequency (ω_H). Thus, for frequencies $\omega_L < \omega < \omega_H$, the medium appears as a negative permeability medium ($\mu < 0$, $\varepsilon > 0$) at low intensity, while at higher intensities, the medium appears as a positive permeability medium ($\mu > 0$, $\varepsilon > 0$). The material can switch between a negative magnetic medium with high reflectivity to a positive medium that can almost be transparent depending on the incident intensity. Use of this mechanism to switch the behavior of the metamaterial of thin wires and SRR composite from negative refractive index to positive index or to plasma-like medium has also been suggested (Zharov et al. 2003). The SRR with Kerr media inclusion shows a nonlinear magnetization with a $\chi^{(3)}$ nonlinearity. This is somewhat unique at optical or near-infrared frequencies where even magnetic activity is special, let alone nonlinear magnetism.

Metamaterials with embedded nonlinear media or devices would have non-linear magnetization, motivating the description of an SRR with an embedded nonlinear diode in a similar manner to Lapine et al. (2003), where a general lumped resonant circuit approach is described. Consider a nonlinear diode embedded in series in the legs of the single SRR with two splits. The voltage-current characteristic of the nonlinear diode can be taken to be

$$I = \frac{1}{R_d}(V + \alpha V^2), \tag{3.82}$$

where R_d is the Ohmic resistance and α is the nonlinearity coefficient (per unit length of the cylinder if the system has invariance along the axis). Such a system is entirely implementable for the SRR designed for microwave operations. Due to the presence of the quadratic term, driving currents at different frequencies get coupled. In the Fourier domain, the current voltage response can be seen to be

$$V(\omega) = Z_d(\omega)I(\omega) + \frac{1}{2}\int_{-\infty}^{\infty} \alpha(\omega;\omega',\omega-\omega')Z_d(\omega')Z_d(\omega-\omega')I(\omega')I(\omega-\omega')\frac{d\omega'}{2\pi}, \tag{3.83}$$

where the kernel $\alpha(\omega;\omega',\omega-\omega')$ is complex and Z_d is the linear impedance of the diode. Hence, in the SRR there is an additional potential drop across the serial diode to account for in Eq. (3.37). Clearly, we have a $\chi^{(2)}$ nonlinearity in this system and can have three wave mixing processes at the frequencies ω, ω', and $\omega \pm \omega'$. Note that the insertion of the diode breaks the centrosymmetry of the unit cell. The magnetization that develops at any one frequency depends on the wave amplitudes at other frequencies. This is a well-studied process (Boyd 2003) and can be utilized for the resonant enhancement of nonlinear processes such as sum and difference wave generation, second harmonic generation, and parametric wave amplification at microwave frequencies.

Let us explicitly estimate the nonlinear susceptibility for a three-wave mixing process between three waves at frequencies ω_1 (pump), ω_2 (idler), and ω_3 (probe). We assume a small signal gain regime, no pump depletion due to the nonlinearity, and we consider only the linear impedance for the waves with frequency ω_1 and ω_2. Consider Eq. (3.37) and write the net linear impedance (per unit length) of the circuit as

$$Z_L(\omega) = -i\omega(1-f)\mu_0\pi r^2 + \frac{1}{-i\omega C(\omega)} + 2\pi r\rho + Z_d(\omega), \tag{3.84}$$

such that for waves with frequency ω_1 and ω_2, we can write the corresponding currents in the loop as

$$j(\omega_1) = \frac{i\omega_1\mu_0\pi r^2}{Z_L(\omega_1)}H(\omega_1), \tag{3.85a}$$

$$j(\omega_2) = \frac{i\omega_2\mu_0\pi r^2}{Z_L(\omega_2)}H(\omega_2). \tag{3.85b}$$

Including the nonlinear diode term for the currents at frequency ω_3, we obtain for the currents at ω_3:

$$j(\omega_3) = \frac{i\omega_3\mu_0\pi r^2}{Z_L(\omega_3)}H(\omega_3) - \frac{\alpha(\omega_3;\omega_1,\omega_2)}{Z_L(\omega_3)}\frac{i\omega_1\mu_0\pi r^2}{Z_L(\omega_1)}\frac{i\omega_2\mu_0\pi r^2}{Z_L(\omega_2)}H(\omega_1)H(\omega_2).$$

(3.86)

Noting that the magnetization $M(\omega) = fj(\omega)$, we can see that the magnetization at frequency ω_3 is proportional to the product of the magnetic fields at the other two frequencies so that we identify the nonlinear magnetic susceptibility as

$$\chi_m^{(2)}(\omega_3;\omega_1,\omega_2) = -\frac{\alpha(\omega_3;\omega_1,\omega_2)Z_d(\omega_1)Z_d(\omega_2)}{Z_L(\omega_3)}\frac{i\omega_1\mu_0\pi r^2}{Z_L(\omega_1)}\frac{i\omega_2\mu_0\pi r^2}{Z_L(\omega_2)},$$

$$= \frac{\alpha(\omega_3;\omega_1,\omega_2)Z_d(\omega_1)Z_d(\omega_2)}{i\omega_3\mu_0\pi r^2 f^2}\chi_m^{(1)}(\omega_1)\chi_m^{(1)}(\omega_2)\chi_m^{(1)}(\omega_3),$$

(3.87)

where $\chi_m^{(1)}$ is the linear susceptibility. There is a resonant enhancement of the nonlinear susceptibility whenever there is a resonance for any of the three waves involved. This is to be expected in view of the large local field enhancement caused by the metamaterial. It can be straightforwardly seen that the nonlinear material will support second harmonic generation ($\omega_1 = \omega_2$ and $\omega_3 = \omega_1 + \omega_2$).

3.6.2 Actively controllable metamaterials

One would like to actively control the behavior of the metamaterials by externally applied inputs. For example, the dielectric permittivity of the medium in which the metallic structure is embedded could be easily modified by applying external electric fields using the electro-optic effect. In this way, one could easily switch the system in and out of resonance. In general, for such active control, the key consideration is the switching speed required. Any electronic means would limit the bandwidth and clearly high speed operations in excess of terabits per second would necessarily be optical in nature.

A more drastic approach was taken in Chen et al. (2006b) where an actively switchable metamaterial was demonstrated at about 1 THz where the embedding medium was a highly doped n-type semiconductor (GaAs). The metamaterial consisted of single split resonators that have a resonant electric polarizability for radiation with the electric field oriented along the capacitive gap. The entire planar metamaterials were deposited on an n-type semiconductor which in turn was deposited on a semi-insulating substrate (Si-GaAs). Using a bias voltage between the metamaterial and the substrate, one can control the charge carrier density in the semiconductor in the regions of the capacitive gaps. The bias voltage essentially controls the ability of the n-type semiconductor to conduct (low field) or not (high field). Thus, with no bias field, the capacitive gaps are shorted out and the metamaterial appears like a conducting surface. With a bias field, one can deplete the charge carriers in the gaps and thus the metamaterial becomes resonant with a dielectric permittivity following a Lorentz-like dispersion. Such active switches in the THz band represent an important development for the control of terahertz radiation.

4

Negative refraction and photonic bandgap materials

Photonic crystals are inhomogeneous materials that consist of a periodic arrangement of (usually) dielectric scatterers which create a periodic modulation of the dielectric function in space. Photonic crystals are one of the very few inhomogeneous materials that can be treated exactly within the framework of the electromagnetic theory due to their inherent periodicity. The latter is at the heart of their working principle and is responsible for some unusual properties, such as the appearance of a photonic bandgap (John 1987, Yablonovitch 1987), i.e., a frequency band within which propagating waves are excluded from the crystal.*

Most photonic crystals, studied theoretically and realized experimentally, are two-dimensional, where the inclusions are infinitely long (very long compared to the transverse dimensions in experiments) cylinders of circular cross-section. One-dimensional photonic crystals or periodically layered materials have been extensively studied as well, especially since the work of Lord Rayleigh (see Chapter 1), but have been less popular because of the fewer degrees of freedom they offer. Nonetheless, due to such well-established grounds, one-dimensional photonic crystals have benefited from the resurgence of interest in the search for negative indices of refraction. Three-dimensional photonic crystals are less common because they are more complex to analyze theoretically and more challenging to realize experimentally. Here also, however, the regained interest in negative refraction phenomena and in particular in flat lens imaging has given the study of three-dimensional photonic crystals a fillip and various studies have been reported in the literature.

Within the framework of this book, we shall primarily discuss one- and two-dimensional periodic materials. The one-dimensional level is used to illustrate how a bandgap appears when a wave propagates through the medium whose index of refraction is modulated periodically. It is intended to show that just an elementary knowledge of transmission line theory is sufficient to understand the appearance of a bandgap, which is a key property in photonic crystals. The specific theory of photonic crystals based on the Maxwell equa-

---

*This property is the reason for the terminology *photonic crystals*: a periodic modulation of the index of refraction similar to the periodic arrangement of atoms in natural crystals, which defines a frequency band of forbidden propagation of photons.

tions is then applied to two-dimensional structures with cylindrical inclusions of circular cross-section. The theory is used within the framework of left-handed media, with specific attention to negative refraction, collimation, and lensing effects. Theoretical details are provided in order to help the reader follow the concepts and reproduce the results. For a more detailed discussion of photonic crystals specifically, with in-depth analysis and explanations, we refer the reader to Joannopoulos et al. (1995), whose notation we follow in Section 4.1.4. For more information on natural crystals and solid-state physics, the reader is referred to Ashcroft and Mermin (1976) and Kittel (1996). For a detailed description on how photonic crystals can be utilized to tailor the photonic density of modes and radiation properties of embedded emitters, we refer the reader to Sakoda (2005).

4.1 Photonic crystals and bandgap materials

4.1.1 One-dimensional photonic crystals: transmission lines approach

Note: As is customary in transmission line theory, we use the symbol j to indicate the imaginary number $\sqrt{-1}$ in this section alone, instead of i which is used in the rest of this book.

Although photonic crystals are most commonly realized in two or three dimensions (as a periodic arrangement of dielectric inclusions in a host medium), it is still beneficial to first simplify the configuration to a one-dimensional one. The one-dimensional photonic crystal is then equivalent to a layered dielectric medium and can be analyzed with one of the many theories offered in this regard. We choose here to follow the transmission line theory since it is a common and well-known method. We therefore consider a layered medium made of two families of media, characterized by their permittivity, thickness, wave-vector, and impedance: $(\varepsilon_1, d_1, k_1, \eta_1)$ and $(\varepsilon_2, d_2, k_2, \eta_2)$. The one-dimensional succession of these layers can be viewed as a series of transmission lines, as shown in Fig. 4.1.

The voltages and currents at the input and output of each transmission line are related via the matrices $\bar{\bar{M}}_i$ $(i = 1, 2)$ (Pozar 2005):

$$\begin{bmatrix} V_{n+\frac{1}{2}} \\ I_{n+\frac{1}{2}} \end{bmatrix} = \bar{\bar{M}}_1 \cdot \begin{bmatrix} V_n \\ I_n \end{bmatrix} \quad \text{where} \quad \bar{\bar{M}}_1 \begin{bmatrix} \cos k_1 d_1 & -j\eta_1 \sin k_1 d_1 \\ -j \sin k_1 d_1 / \eta_1 & \cos k_1 d_1 \end{bmatrix},$$
(4.1a)

$$\begin{bmatrix} V_{n+1} \\ I_{n+1} \end{bmatrix} = \bar{\bar{M}}_2 \cdot \begin{bmatrix} V_{n+\frac{1}{2}} \\ I_{n+\frac{1}{2}} \end{bmatrix} \quad \text{where} \quad \bar{\bar{M}}_2 \begin{bmatrix} \cos k_2 d_2 & -j\eta_2 \sin k_2 d_2 \\ -j \sin k_2 d_2 / \eta_2 & \cos k_2 d_2 \end{bmatrix}.$$
(4.1b)

For the purpose of illustration, we take $k_1 d_1 = k_2 d_2 = kd$, while the interested reader might carry on the treatment with the general variables. Combining Eqs. (4.1):

$$\begin{bmatrix} V_{n+1} \\ I_{n+1} \end{bmatrix} = \bar{\bar{M}}_2 \cdot \bar{\bar{M}}_1 \cdot \begin{bmatrix} V_n \\ I_n \end{bmatrix},$$

$$= \begin{bmatrix} \cos^2 kd - (\eta_2/\eta_1)\sin^2 kd & -j(\eta_1 + \eta_2)\cos kd \sin kd \\ -j(1/\eta_1 + 1/\eta_2)\sin kd \cos kd & \cos^2 kd - (\eta_1/\eta_2)\sin^2 kd \end{bmatrix} \cdot \begin{bmatrix} V_n \\ I_n \end{bmatrix}.$$

$$(4.2)$$

For forward propagating waves, we look for periodic solutions of the type $V_{n+1} = V_n \exp(-j\theta)$, $I_{n+1} = I_n \exp(-j\theta)$ in the infinite lattice. Introducing these expressions into Eqs. (4.2), multiplying the two equations, and simplifying by $V_n I_n$, we finally obtain the dispersion relation

$$\cos\theta = \cos^2 kd - \frac{\eta_1/\eta_2 + \eta_2/\eta_1}{2}\sin^2 kd. \qquad (4.3)$$

Interesting points are at $kd = m\pi$ and $kd = m\pi/2$ where the value of the right-hand side term (shown as the black line in Fig. 4.2) is 1 and $-(\eta_1/\eta_2 + \eta_2/\eta_1)/2 < -1$, respectively. Hence, at $kd = \pi/2$ and within a small band around, the right-hand side term of Eq. (4.3) is smaller than (-1) and θ does not admit real values as a solution. Since θ represents the propagating vector of the voltage wave in the transmission line (which is immediately related to the propagation of the electromagnetic wave through the layered medium), this is tantamount to saying that the wave-vector does not take real values in a certain frequency band, and the wave is evanescent. The frequency band where this phenomenon occurs is called a frequency bandgap, for obvious reasons, and is represented by the gray areas in Fig. 4.2. Note that from Eq. (4.3), we see that a frequency bandgap appears in the dispersion as soon as there is a mismatch between the layers, however small. In the case when $\varepsilon_1 = \varepsilon_2 = \varepsilon_0$, real solutions for θ always exist, the dispersion relation is linear as expected, and no bandgap is formed.

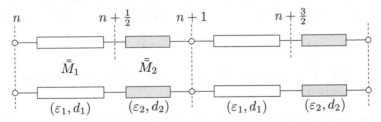

Figure 4.1 Transmission line model of a one-dimensional photonic crystal composed of two media with permittivities and thicknesses (ε_1, d_1) and (ε_2, d_2). Nodes are numbered with integers n to denote the periodicity with respect to medium 1; half-integers denote the periodicity with respect to medium 2. $\bar{\bar{M}}_i$ are the transmission line matrices of each medium $(i = 1, 2)$.

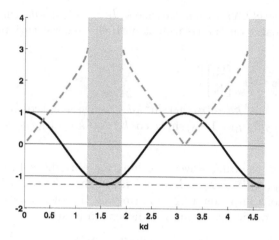

Figure 4.2 Illustration of the bandgap opening in a one-dimensional photonic crystal (equivalent to a layered medium). The governing equation is Eq. (4.3): the function on the right-hand side is shown in black line (with a minimum of $-(\eta_1/\eta_2 + \eta_2/\eta_1)/2$), while real solutions for θ are shown in gray line. The bandgap corresponds to values where $\cos\theta < -1$ and is shown as the gray areas.

Despite its simplicity, one-dimensional photonic crystals have had important applications in the design of highly reflective coatings, for example in laser applications. More recently, within the framework of left-handed media, it was shown that a structure with material parameters alternating between a regular medium and a left-handed medium can exhibit a transmission coefficient with an unusual angular dependency when the average refractive index of the structure is close to zero (Li et al. 2003a, Shadrivov et al. 2003), or can even create a bandgap for both TE and TM polarizations, therefore being effectively opaque to any in-plane propagating waves (Shadrivov et al. 2005). Additional interesting properties have also been presented in Section 2.5.2 on 52.

4.1.2 Two-dimensional photonic crystals: definitions and solution

The concepts presented in the previous section in one-dimension are generalizable to two and three dimensions, although the theory itself becomes more involved than the simple transmission line approach. Yet, dispersion relations and frequency bandgaps are concepts generalizable to any number of dimensions, and are fundamental to the study of general photonic crystals. We shall present the theory for two-dimensional photonic crystals next.

There is a close analogy between the electromagnetic problem under consideration (to find the electromagnetic fields inside a periodic two-dimensional photonic crystal) and the problem of finding the potential function in natural crystals. Consequently, the electromagnetic community has borrowed many concepts and terminology from solid-state physics, useful for the solution of the Maxwell equations in periodic media. Some of this terminology is briefly reviewed hereafter, to the extend that it is necessary for a self-contained reading of the sections to follow. For more details, the reader is referred to Ashcroft and Mermin (1976) and Kittel (1996).

4.1.2.1 Direct lattice

The photonic crystal we consider is two-dimensional: invariant along the $\hat{z}$ direction and periodic in the (xy) plane. In this plane, we can therefore define a translational vector $\mathbf{R}$ and express any vector $\mathbf{r}'$ in the (xy) plane as

$$\mathbf{r}' = \mathbf{r} + \mathbf{R}, \tag{4.4}$$

where $\mathbf{r}$ is a vector within the unit cell of reference arbitrarily chosen. Mathematically, if $(\mathbf{u}_x, \mathbf{u}_y)$ are the lattice vectors, then $\mathbf{R} = \mathbf{u}_x a_x + \mathbf{u}_y a_y$, where a_x and a_y are integers, positive or negative. Two of the most common lattices are the square and the triangular ones (also referred to as hexagonal), shown in the left panels of Fig. 4.3.

4.1.2.2 Reciprocal lattice

The reciprocal lattice can be thought of as defined in the Fourier transform space of the (xy) plane. In electromagnetic parlance, this corresponds to the spectral domain, defined by the wave-vector $\mathbf{k}$. Since the reciprocal space is also periodic, we can define a reciprocal translational vector $\mathbf{G}$ which, similarly to $\mathbf{R}$, can be written as $\mathbf{G} = \mathbf{v}_x b_x + \mathbf{v}_y b_y$, where (b_x, b_y) are integers, positive or negative. The reciprocal lattice vectors $(\mathbf{v}_x, \mathbf{v}_y)$ are obtained from the direct ones $(\mathbf{u}_x, \mathbf{u}_y)$ by enforcing the periodicity in the permittivity function. Indeed, being a periodic function, $\varepsilon(\mathbf{r})$ can be Fourier decomposed as:

$$\varepsilon(\mathbf{r}) = \sum_{\mathbf{G}} \tilde{\varepsilon}(\mathbf{r}) \, e^{i\mathbf{G}\cdot\mathbf{r}} \quad \text{where } \tilde{\varepsilon}(\mathbf{G}) = \frac{1}{V_{\text{cell}}} \iiint d\mathbf{r} \, \varepsilon(\mathbf{r}) \, e^{-i\mathbf{G}\cdot\mathbf{r}}, \tag{4.5}$$

where the $\sim$ sign denotes the Fourier component and where V_{cell} is the area (volume) of the two-dimensional periodic unit cell. Using this expansion and imposing $\varepsilon(\mathbf{r} + \mathbf{R}) = \varepsilon(\mathbf{r})$, we write

$$\varepsilon(\mathbf{r} + \mathbf{R}) = \sum_{\mathbf{G}} \tilde{\varepsilon}(\mathbf{G}) \, e^{i\mathbf{G}\cdot\mathbf{r}} \, e^{i\mathbf{G}\cdot\mathbf{R}}, \tag{4.6}$$

which imposes the condition $\exp(i\mathbf{G} \cdot \mathbf{R}) = 1$ or $\mathbf{G} \cdot \mathbf{R} = 2m\pi$, where $m \in \{\ldots, -1, 0, 1, 2, \ldots\}$. This condition is immediately satisfied by defining the

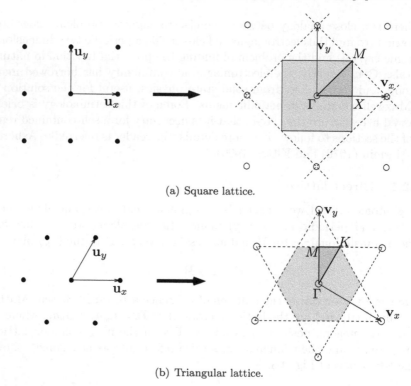

(a) Square lattice.

(b) Triangular lattice.

Figure 4.3 Square and triangular lattices represented in both the direct space (left) and reciprocal space (right). The Brillouin zones are highlighted in gray and the irreducible Brillouin zones are highlighted in darker gray. The symmetry points in each case are indicated.

following relation between the vectors of the direct lattice and those of the reciprocal lattice

$$\mathbf{u}_i \cdot \mathbf{v}_j = 2\pi\delta_{ij}, \quad i, j = x, y. \tag{4.7}$$

More explicitly,

$$\mathbf{v}_1 = 2\pi \frac{\mathbf{u}_2 \times \mathbf{u}_3}{\mathbf{u}_1 \cdot \mathbf{u}_2 \times \mathbf{u}_3}, \tag{4.8}$$

while the other vectors can be obtained by index permutation. The right panels in Fig. 4.3 show the reciprocal lattices for the square and triangular direct lattices, respectively. It can be seen that the square lattice transforms into another square lattice, while the triangular lattice transforms into a rotated triangular lattice. The definitions of the vectors $(\mathbf{v}_x, \mathbf{v}_y)$ are provided in Tab. 4.1 for both cases.

4.1.2.3 Brillouin zone and irreducible Brillouin zone

The spatial periodicity in the spectral plane is generated by the repetition of a unit cell, or tile, which can be generated by choosing a lattice point as reference, drawing all the connections to its closest neighbors, and drawing the medians to all the segments thus defined. The smallest area delimited by the medians is called the first Brillouin zone, as shown in Fig. 4.3.

This zone is, however, not the smallest self-repeating tile: because of its symmetry, the first Brillouin zone can be folded into an even smaller region, called the irreducible Brillouin zone. The latter is represented in Fig. 4.3 for the two examples of square and triangular lattice and is important for a few reasons. First, it defines the smallest area that needs to be examined for the reciprocal vectors: the behavior of the whole crystal can be inferred by the study of the irreducible Brillouin zone only. Second, its edges correspond to points of maximum diffraction. Third, its symmetry points (identified by $\Gamma - X - M$ and $\Gamma - K - M$ for the respective cases of a square and triangular lattices) correspond to zero group velocities of the waves: the tangents to the dispersion contours at these points are horizontal, as it is clear by a direct inspection of the band diagrams. Hence, when studying the wave propagation inside the photonic crystal, the wave-vectors need to be varied so as to span the regions defined by the symmetry points of the irreducible zone:

- For the square lattice:

$$\Gamma \to (k_x = 0, k_y = 0), \tag{4.9a}$$

$$X \to (k_x = \frac{\pi}{a}, k_y = 0), \tag{4.9b}$$

$$M \to (k_x = \frac{\pi}{a}, k_y = \frac{\pi}{a}). \tag{4.9c}$$

Table 4.1 Definitions of $\mathbf{u}_i$ and $\mathbf{v}_j$ vectors for square and triangular (or hexagonal) lattices ($i, j = x, y$).

Lattice	Fractional vol.	Direct vectors	Reciprocal vectors
Square	$f_r = \dfrac{\pi R_c^2}{a^2}$	$\mathbf{u}_x = \hat{x}a$ $\mathbf{u}_y = \hat{y}a$	$\mathbf{v}_x = \hat{x}\, 2\pi/a$ $\mathbf{v}_y = \hat{y}\, 2\pi/a$
Triangular	$f_r = \dfrac{2\pi R_c^2}{\sqrt{3}a^2}$	$\mathbf{u}_x = \hat{x}a$ $\mathbf{u}_y = a(\hat{x} + \sqrt{3}\hat{y})/2$	$\mathbf{v}_x = 2\pi/a\,(\hat{x} - \sqrt{3}/3\,\hat{y})$ $\mathbf{v}_y = \hat{y}\,(2\pi/a)\,(2/\sqrt{3})$

- For the triangular (hexagonal) lattice:

$$\Gamma \rightarrow (k_x = 0, k_y = 0), \tag{4.10a}$$

$$K \rightarrow (k_x = \frac{2\pi}{3a}, k_y = \frac{2\pi}{\sqrt{3}a}), \tag{4.10b}$$

$$M \rightarrow (k_x = 0, k_y = \frac{2\pi}{\sqrt{3}a}). \tag{4.10c}$$

4.1.3   Bloch theorem and Bloch modes

Intuitively, the periodicity of the medium induces a periodicity in the electromagnetic fields, which can therefore be written as the product of two functions: one governing the propagation of the wave itself, the other one reflecting the periodicity. For example, the magnetic field is written as

$$\mathbf{H_k}(\mathbf{r}) = e^{i\mathbf{k}\cdot\mathbf{r}}\mathbf{\Phi_k}(\mathbf{r}), \qquad \text{where } \mathbf{\Phi_k}(\mathbf{r}+\mathbf{R}) = \mathbf{\Phi_k}(\mathbf{r}). \tag{4.11}$$

These are often referred to as Bloch waves. Note that we have anticipated the notation and have added an index $\mathbf{k}$ to denote the dependency of the field with the wave-vector. A justification of this is provided subsequently. Since $\mathbf{\Phi_k}$ is periodic, it can be Fourier transformed:

$$\mathbf{\Phi_k}(\mathbf{r}) = \sum_{\mathbf{G}} \tilde{\mathbf{\Phi}}_{\mathbf{G}}\, e^{i\mathbf{G}\cdot\mathbf{r}}, \tag{4.12}$$

where the sum spans all the reciprocal vectors $\mathbf{G}$ (i.e., the integers b_x and b_y are summed from $-\infty$ to $+\infty$). This expansion is important subsequently for the expression of the field in the photonic crystal.

4.1.4   Electromagnetic waves in periodic media

Based on the previous brief review of fundamental concepts, we proceed with the study of electromagnetic fields per se in the photonic crystal. The governing equations are of course the Maxwell equations and the Helmholtz wave equation. Since the treatment based on the magnetic field yields a Hermitian operator when only the dielectric permittivity is inhomogeneous and thus an eigen-system simpler to solve than the one based on the electric field (Joannopoulos et al. 1995), we choose to proceed with the former and write

$$\nabla \times \left[\frac{1}{\varepsilon(\mathbf{r})}\nabla \times \mathbf{H}(\mathbf{r})\right] = \left(\frac{\omega}{c}\right)^2 \mu_r \mathbf{H}(\mathbf{r}), \tag{4.13a}$$

$$\nabla \cdot \mathbf{H}(\mathbf{r}) = 0. \tag{4.13b}$$

The whole problem is therefore to find the modes of $\mathbf{H}(\mathbf{r})$ that satisfy the eigenvalue problem

$$\Theta\mathbf{H}(\mathbf{r}) = \left(\frac{\omega}{c}\right)^2 \mu_r\mathbf{H}(\mathbf{r}), \qquad (4.14a)$$

where the linear operator

$$\Theta = \nabla \times \left[\frac{1}{\varepsilon(\mathbf{r})}\nabla\times\right]. \qquad (4.14b)$$

The solution of this system yields a series of eigenvectors that correspond to functions that describe the field distribution, and a series of eigenvalues that correspond to the squared frequencies of each mode. Since the eigenvalue problem is restricted to a limited domain in space (defined by the irreducible Brillouin zone), the eigenvalues are discrete. In other words, for each value of $\mathbf{k}$, a discrete number of modes of the magnetic field are allowed. The band diagram of the crystal is therefore constructed by letting $\mathbf{k}$ span the edge of the irreducible Brillouin zone and looking at the evolution of each eigenvalue (only the necessary first few are important in practice).

We proceed by using the Bloch theorem of Eq. (4.11), based on which the operator Θ is rewritten as

$$\Theta = (i\mathbf{k} + \nabla) \times \left[\frac{1}{\varepsilon(\mathbf{r})}(i\mathbf{k} + \nabla)\times\right] (i\mathbf{k} + \nabla) \times [\kappa(\mathbf{r})(i\mathbf{k} + \nabla)\times], \qquad (4.15)$$

where we have defined $\kappa(\mathbf{r}) = 1/\varepsilon(\mathbf{r})$. The inverse permittivity $\kappa(\mathbf{r})$ is also a periodic function which, similarly to Eq. (4.5), can be Fourier expanded as

$$\kappa(\mathbf{r}) = \sum_{\mathbf{G}} \tilde{\kappa}(\mathbf{G})\, e^{i\mathbf{G}\cdot\mathbf{r}}, \qquad (4.16)$$

and introduced into the operator of Eq. (4.15). By the virtue of Bloch theorem, the solutions of the magnetic field that are sought are of the form

$$\mathbf{H}_{\mathbf{k}}(\mathbf{r}) = \sum_{\mathbf{G}} \mathbf{h}_{\mathbf{G}} e^{i(\mathbf{k}+\mathbf{G})\cdot\mathbf{r}} \qquad (4.17)$$

and the eigensystem is written as

$$-\sum_{\mathbf{G}''}\sum_{\mathbf{G}'} \tilde{\kappa}_r(\mathbf{G}'')(\mathbf{k}+\mathbf{G}')\times[(\mathbf{k}+\mathbf{G}')\times\mathbf{h}_{\mathbf{G}'}]\, e^{i(\mathbf{k}+\mathbf{G}'+\mathbf{G}'')\cdot\mathbf{r}}$$

$$= \left(\frac{\omega}{c}\right)^2 \mu_r \sum_{\mathbf{G}'} \mathbf{h}_{\mathbf{G}'} e^{i(\mathbf{k}+\mathbf{G}')\cdot\mathbf{r}}. \qquad (4.18)$$

Upon substituting $\mathbf{G} = \mathbf{G}' + \mathbf{G}''$, simplifying by $\exp(i\mathbf{k}\cdot\mathbf{r})$, multiplying by $\exp(-i\mathbf{G}''\cdot\mathbf{r})$, integrating over the whole space, and finally substituting $\mathbf{G}''$ by $\mathbf{G}$ (since the indices are dummy), we obtain

$$-\sum_{\mathbf{G}'} \tilde{\kappa}_r(\mathbf{G}-\mathbf{G}')(\mathbf{k}+\mathbf{G}')\times[(\mathbf{k}+\mathbf{G}')\times\mathbf{h}_{\mathbf{G}'}] = \left(\frac{\omega}{c}\right)^2 \mu_r\mathbf{h}_{\mathbf{G}} \qquad (4.19)$$

for all $\mathbf{G}$. The vector $\mathbf{h_G}$ has in principle three components. Making use of Eq. (4.13b), however, the number of components can be reduced to two in the appropriate coordinate system. From Eq. (4.17),

$$(\mathbf{k} + \mathbf{G'}) \cdot \mathbf{h_{G'}} = 0, \qquad (4.20)$$

so that we can define three vectors $(\hat{e}_1, \hat{e}_2, \hat{e}_3)$ such that

$$\mathbf{k} + \mathbf{G'} = |\mathbf{k} + \mathbf{G'}| \, \hat{e}_3 \,, \qquad (4.21a)$$

$$\hat{e}_1 \cdot \hat{e}_3 = \hat{e}_2 \cdot \hat{e}_3 = 0, \qquad (4.21b)$$

and $(\hat{e}_1, \hat{e}_2, \hat{e}_3)$ form a right-handed orthonormal basis. Note that this decomposition is standard in other areas of electromagnetics: it is for example used in the definition of the horizontal and vertical polarizations of waves (Tsang et al. 2000a), and is also at the basis of the (kDB) system for the study of wave propagating in anisotropic and bianisotropic homogeneous media (Kong 2000). Within the basis defined by the vectors $(\hat{e}_1, \hat{e}_2, \hat{e}_3)$, the vector $\mathbf{h_{G'}}$ has only two components and is written as

$$\mathbf{h_{G'}} = h_{1_{\mathbf{G'}}} \hat{e}_1 + h_{2_{\mathbf{G'}}} \hat{e}_2 = \sum_{\lambda=1,2} h_{\lambda_{\mathbf{G'}}} \hat{e}_\lambda \,, \qquad (4.22)$$

where again we follow the notation of Joannopoulos et al. (1995). The wave equation (4.18) becomes

$$-\sum_{\mathbf{G}} \sum_{\mathbf{G'}} \sum_\lambda h_{\lambda_{\mathbf{G'}}} \tilde{\kappa}_r(\mathbf{G} - \mathbf{G'}) \left[(\mathbf{k} + \mathbf{G}) \times [(\mathbf{k} + \mathbf{G'}) \times \hat{e}_\lambda] \right] e^{i(\mathbf{k}+\mathbf{G})\cdot\mathbf{r}}$$

$$= \left(\frac{\omega}{c} \right)^2 \mu_r \sum_{\mathbf{G'}} \sum_\lambda h_{\lambda_{\mathbf{G'}}} \, e^{i(\mathbf{k}+\mathbf{G'})\cdot\mathbf{r}} \, \hat{e}_\lambda. \quad (4.23)$$

The exponential term can be disposed of in the usual manner, i.e., by simplifying by $\exp(i\mathbf{k} \cdot \mathbf{r})$, multiplying by $\exp(-i\mathbf{G''} \cdot \mathbf{r})$, and integrating over the entire space, leaving a delta function. Finally, dot multiplying both terms of the wave equation by $\hat{e}'_\lambda$ and using the identity $\mathbf{C} \cdot (\mathbf{A} \times \mathbf{B}) = \mathbf{B} \cdot (\mathbf{C} \times \mathbf{A})$, we obtain the governing equation as

$$\sum_{\mathbf{G}} \sum_\lambda \left\{ \left[(\mathbf{k} + \mathbf{G}) \times \hat{e}_\lambda \right] \cdot \left[(\mathbf{k} + \mathbf{G'}) \times \hat{e}'_\lambda \right] \right\} \tilde{\kappa}_r(\mathbf{G'} - \mathbf{G}) h_{\lambda_{\mathbf{G}}} = \left(\frac{\omega}{c} \right)^2 \mu_r h_{\lambda'_{\mathbf{G'}}} .$$

$$(4.24)$$

This equation can be simply cast into a matrix form yielding the operator of Eq. (4.14a):

$$\Theta = \tilde{\kappa}_r(\mathbf{G'} - \mathbf{G}) \, |(\mathbf{k} + \mathbf{G})| \, |(\mathbf{k} + \mathbf{G'})| \begin{pmatrix} \hat{e}_2 \cdot \hat{e}_2 & -\hat{e}_2 \cdot \hat{e}_1 \\ -\hat{e}_1 \cdot \hat{e}_2 & \hat{e}_1 \cdot \hat{e}_1 \end{pmatrix} . \qquad (4.25)$$

The diagonal elements of this system yield the TE and TM propagating modes in the photonic crystal, which correspond to the magnetic field and the electric field being parallel to the two-dimensional inclusions, respectively.

At this stage, only the Fourier components $\tilde{\kappa}_r$ need to be detailed. In the case of infinitely long dielectric rods of circular cross-section and permittivity ε_a embedded in a background of permittivity ε_b, an analytical expression for $\tilde{\kappa}_r$ can be derived (Cassagne 1998). Since it is more intuitive to discuss the permittivity contrast rather than the inverse permittivity contrast, we only present the derivation for the permittivity. The inverse $\kappa(\mathbf{r})$ is obtained in an exactly similar way.

As a reminder, we write the permittivity as

$$\varepsilon(\mathbf{r}) = \sum_{\mathbf{G}} \tilde{\varepsilon}(\mathbf{G})\, e^{i\mathbf{G}\cdot\mathbf{r}}, \qquad \tilde{\varepsilon}(\mathbf{G}) = \frac{1}{V_{cell}} \int_{V_{cell}} \mathrm{d}\mathbf{r}\, \varepsilon(\mathbf{r})\, e^{-i\mathbf{G}\cdot\mathbf{r}}, \qquad (4.26)$$

where V_{cell} denotes the surface of the elementary cell and where it should be kept in mind that all vectors lie in the (xy) plane. The idea is to write the permittivity as

$$\varepsilon(\mathbf{r}) = \varepsilon_b + (\varepsilon_a - \varepsilon_b) \sum_{\mathbf{R}} S(R_c - |\mathbf{r} - \mathbf{R}|), \qquad (4.27)$$

where R_c is the radius of the dielectric rods and S denotes the step function. Merging these two equations, we obtain:

$$\tilde{\varepsilon}(\mathbf{G}) = \frac{\varepsilon_b}{V_{cell}} \int_{V_{cell}} \mathrm{d}\mathbf{r}\, e^{-i\mathbf{G}\cdot\mathbf{r}} + \frac{\varepsilon_a - \varepsilon_b}{V_{cell}} \int_{V_{cell}} \mathrm{d}\mathbf{r} \sum_{\mathbf{R}} S(R_c - |\mathbf{r} - \mathbf{R}|)\, e^{-i\mathbf{G}\cdot\mathbf{r}}.$$
$$(4.28)$$

The first integral is immediately evaluated as ε_b if $\mathbf{G} = 0$ and zero otherwise. The second integral is evaluated by introducing the change of variable $\mathbf{r}' = \mathbf{r} - \mathbf{R}$. Since $\mathbf{r}$ spans the periodic domain V_{cell} and $\mathbf{R}$ is the translational vector, $\mathbf{r}'$ spans the whole space. We can therefore replace the sum of integrals over V_{cell} by a single integral over the whole two-dimensional space. Separating the cases when $\mathbf{G} = 0$ and $\mathbf{G} \neq 0$, we obtain

$$\frac{\varepsilon_a - \varepsilon_b}{V_{cell}} \iint \mathrm{d}\mathbf{r}'\, S(R_c - |\rho - \mathbf{R}_\rho|) = f_r(\varepsilon_a - \varepsilon_b) \qquad (4.29a)$$

for $\mathbf{G} = 0$, where f_r is the fractional volume (see Tab. 4.1), and for $\mathbf{G} \neq 0$,

$$\frac{1}{V_{cell}}(\varepsilon_a - \varepsilon_b) \iint_{-\infty}^{\infty} \mathrm{d}\mathbf{r}'\, S(R_c - |\mathbf{r} - \mathbf{R}|)\, e^{-i\mathbf{G}\cdot\mathbf{r}'}$$

$$= \frac{\varepsilon_a - \varepsilon_b}{V_{cell}} \int_0^{R_c} \mathrm{d}r'\, r' \int_0^{2\pi} \mathrm{d}\phi\, e^{-i\mathbf{G}r'\cos(\phi-\theta)}$$

$$= \frac{\varepsilon_a - \varepsilon_b}{V_{cell}} 2\pi \int_0^{R_c} \mathrm{d}r'\, r'\, J_0(r'\mathbf{G}), \qquad (4.29b)$$

where we have used the standard change of variable $x' = r' \sin\phi$, $y' = r' \cos\phi$, $G_x = G\sin\theta$, $G_y = G\cos\theta$, and the well-known identity for the Bessel function. The integral is evaluated using $\int x J_0(\alpha x)\mathrm{d}x = x/\alpha\, J_1(\alpha x)$, yielding the final result

$$\tilde{\varepsilon}(\mathbf{G}) \begin{cases} \varepsilon_a f_r + \varepsilon_b (1 - f_r) & \text{if } \mathbf{G} = 0 \\ (\varepsilon_a - \varepsilon_b) f_r \frac{2J_1(GR_c)}{GR_c} & \text{elsewhere.} \end{cases} \qquad (4.30)$$

The reconstructed permittivity in the two cases of a square and a triangular lattice is shown in Fig. 4.4. The expected Gibbs phenomenon due to the Fourier transformation cannot be entirely removed, but has a minimal influence on the solution of the eigensystem and can therefore be ignored.

4.2 Band diagrams and iso-frequency contours

4.2.1 Free-space and standard photonic crystal

The previous section presented the mathematical equations necessary to solve the eigensystem (4.14a) and highlighted the following methodology:

- Geometry:

    1. Set the basis and the background (e.g. circular rods of radius R_c and permittivity ε_a in a background ε_b).

    2. Set the direct lattice (e.g. a square lattice with vectors $|\mathbf{u}_x| = |\mathbf{u}_y| = a$).

    3. Compute the vectors of the reciprocal lattice $(\mathbf{v}_x, \mathbf{v}_y)$.

    4. Determine the edge of the irreducible Brillouin zone (e.g. the coordinates of Γ, X, and M).

    5. Compute a set of vectors $\mathbf{G}$ by spanning (b_x, b_y). Sort the vectors according to their magnitude (the most important vectors being close to the origin).

- Electromagnetics:

    1. Compute $\tilde{\kappa}(\mathbf{G})$ using the equivalent of Eq. (4.30).

    2. Compute a set of $\mathbf{k}$ vectors spanning the edge of the irreducible Brillouin zone. The coordinates of the symmetry points are given in Eqs. (4.9) and Eqs. (4.10).

    3. For each incidence $\mathbf{k}$ from the set, build and solve the eigensystem: for each $\mathbf{G}$ from the set, span $\mathbf{G}'$ from the same set and compute $(\mathbf{k} + \mathbf{G}; \hat{e}_1; \hat{e}_2)$, $(\mathbf{k} + \mathbf{G}'; \hat{e}_1'; \hat{e}_2')$, $\tilde{\kappa}(\mathbf{G} - \mathbf{G}')$ (where $\mathbf{G} - \mathbf{G}'$ is also a vector of the reciprocal lattice), build the system of Eq. (4.25), and solve for the eigenvalues.

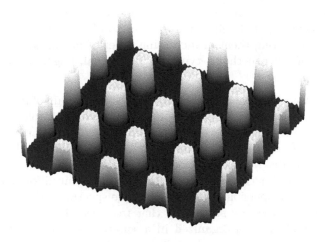

(a) Square lattice.

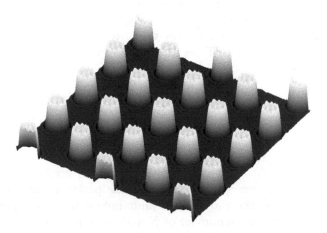

(b) Triangular lattice.

Figure 4.4 Reconstruction of the permittivity profile from Eq. (4.30) for cylindrical rods in a square and triangular lattice ($\varepsilon_a = 10$, $\varepsilon_b = 1$, $R_c/a = 0.2$).

Note that this method has the computational drawback of requiring a convergence check because of the direct solution of the eigensystem. As is customary in direct solution schemes, the eigenvalues depend on the size of the system chosen and they eventually converge to their final values once enough modes have been included. When plotting the band diagram of photonic crystals, one is usually interested in the first few eigenvalues (ten or so), which might require the inclusion of the first hundreds or so of reciprocal vectors (hence the necessity of sorting them in increasing order of magnitude). In that regard, the use of other methods, typically variational, might be of interest to the reader.

As a first illustration, we shall compute the band diagram of free-space with a double purpose in mind. First, the solution is of course simple and analytic, but should also be obtained by the method presented in the previous section. Free-space therefore constitutes a first good check of validity of the algorithm. Second, the band-diagram already looks very much like the band-diagrams of real photonic crystals so that free-space provides a first intuition of the folding mechanism of the dispersion relation due to the spectral periodicity.

We therefore proceed by supposing that free-space is a periodic medium with inclusions of $\varepsilon_a = \varepsilon_0$ organized in a square lattice in a background of $\varepsilon_b = \varepsilon_0$. We suppose without loss of generality that the vectors $\mathbf{u}_{x,y}$ are orthonormal so that the vectors $|\mathbf{v}_{x,y}| = 2\pi$. The irreducible Brillouin zone is therefore delimited by the points (Γ, X, M). The electromagnetic aspect of the problem can be treated in closed form. The dispersion relation of free-space is of course given by $k_x^2 + k_y^2 = (\omega/c)^2$ where the left-hand side depends on the values of k_x and k_y, i.e., on how we move in the reciprocal plane. The first way of walking the path $\Gamma \to X \to M$ is to start from the origin and follow (k_x, k_y) as

$$\Gamma \to X : k_x = 0, k_y = 0 \to \frac{\pi}{a}, \tag{4.31a}$$

$$X \to M : k_x = 0 \to \frac{\pi}{a}, k_y = \frac{\pi}{a}, \tag{4.31b}$$

$$M \to \Gamma : k_x = \frac{\pi}{a} \to 0, k_y = \frac{\pi}{a} \to 0. \tag{4.31c}$$

This path is, however, one of the many possible paths: because of the periodicity of the medium, there are in fact an infinite number of Γ points in the crystal, as well as an infinite number of X and M points. We could therefore, for example, decide to walk the same path, but translated by $2\pi/a$ in the k_x direction, or similarly in the k_y direction, or even jump between cells. All those paths are possible and yield increasingly larger values of (ω/c) as k_x and k_y move away from the origin (arbitrarily chosen). The lower values of (ω/c) are of course the most influential, and it is clear that the higher the frequency, the more possibilities of walking along $(\Gamma - X - M - \Gamma)$. In other words, the higher the frequency, the more modes are obtained. Eventually, keeping the first few modes, the dispersion relation of free-space expressed in a *photonic crystal language* is shown in Fig. 4.5. The figure does look more complicated than the free-space dispersion, but it is only due to the folding mechanism along the directions suggested by the assumption of a periodic medium. It can be seen that as expected, the density of curves increases with increas-

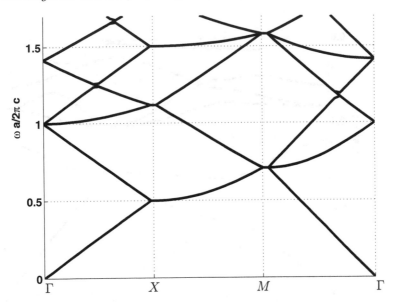

Figure 4.5 Band diagram of free-space when modelled as a periodic medium with $\varepsilon_a = \varepsilon_b = \varepsilon_0$.

ing frequency (or normalized frequency). In addition, at low frequencies, the dispersion along the paths $\Gamma \to X$ and $X \to M$ are straight lines, character-istic of free-space. Finally, the band structure does not exhibit a bandgap, like expected for free-space (this is the same situation as the one-dimensional photonic crystal studied in Section 4.1.1: no modulation in the permittivity does not produce a bandgap).

The band diagram of a realistic photonic crystal is shown in Fig. 4.6 (the parameters are given in the caption of the figure). At low frequencies, the similarity with Fig. 4.5 is clearly seen, with the major difference appearing at the symmetry points of the crystal where the tangent to the curves becomes horizontal. At higher frequencies, the coupling between the modes renders the band diagram very different from the one of free-space, as expected. Fig. 4.6 also illustrates the property of a partial bandgap: it is seen that a TM mode exists at all frequencies, while the TE modes have no solutions for normalized frequencies in $[0.23, 0.28]$ and $[0.35, 0.39]$, approximately. These frequencies therefore correspond to bandgaps for TE modes but not for TM modes. A total bandgap would be obtained if both modes had coinciding bandgaps, which can be obtained by properly adjusting the parameters of the crystal (i.e., the size of the inclusions and the permittivity contrast).

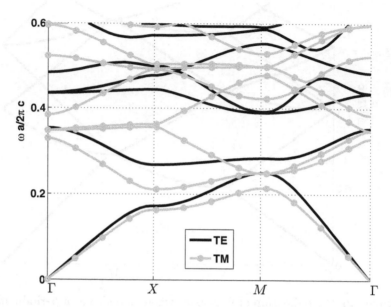

Figure 4.6 Band diagram of a photonic crystal with the parameters from Luo et al. (2002): $\varepsilon_a = \varepsilon_0$, $\varepsilon_b = 12\varepsilon_0$, $R_c/a = 0.35$, organized in a square lattice.

4.2.2 Iso-frequency contours

Like in the previous case of band diagrams, let us first start by a brief discussion on the iso-frequency contours in free-space, as a way of relating new concepts to well-established grounds.

The iso-frequency contours of free-space (or any homogeneous isotropic dielectric if the permittivity is included in the equation) are straightforwardly obtained from the dispersion relation

$$k_x^2 + k_z^2 = k^2 = \left(\frac{\omega}{c}\right)^2 , \tag{4.32}$$

where we have supposed that $k_y = 0$. The iso-frequency contours are therefore circles of increasing radii with frequency, and along with the phase matching condition (i.e., the continuity of the tangential component of the wave-vector across an interface) can be used to derive Snell's law at the boundary between two media. Although elementary, this simple picture contains a subtlety that has been usually overlooked because refraction was always assumed to be positive: upon phase-matching the wave-vector, two solutions appear in the transmitted medium, corresponding to two possible transmitted waves (see Fig. 5.1 and the corresponding discussion for more details). Yet, only one is chosen based on power flow argument: the power flow normal to the interface has to always be of the same sign, i.e., the power has to always flow away from the source. From the iso-frequency diagram, the power flow direction, or group velocity direction, is obtained from the relation $\mathbf{v}_g = \nabla_{\mathbf{k}}\omega$ (see Section 1.4). Being vectorial, this relation defines a direction and orientation that can be both obtained from the iso-frequency diagram. Let us imagine the circles obtained from Eq. (4.32) at a frequency ω and at a slightly higher frequency $\omega + \delta\omega$: the second circle is slightly larger than the first one and centered at the same point $(k_x, k_z) = (0,0)$. Hence the direction of the gradient is in the radial direction (a straightforward conclusion in the case of a circle) while the orientation is given by the change due to frequency, outward in this case. The power at a given phase-matched point is therefore perpendicular to the iso-frequency contour and pointing outward. The same methodology can be applied to more irregular shapes of iso-frequency contours, an example of which is given in Chapter 5.

Let us also remark that the iso-frequency contour defined by Eq. (4.32) can be plotted in the (k,ω) space rather than the (k_x, k_y) space, and yields a straight line, referred to as the light-line. It is now interesting to look back at Fig. 4.6, specifically around the Γ point at low frequencies: the dispersion relation is seen to be very well approximated by the light line if the permittivity of the background is accounted for (and it is of course the free-space light line in the case of Fig. 4.5). This indicates that at low frequency, the photonic crystal appears as a periodic medium where the various dielectric inclusions are not coupled: the iso-frequency contours look like the ones shown in the top left panel of Fig. 4.7: a series of independent circles obeying (in the first approximation) Eq. (4.32) centered at the symmetry points of the reciprocal lattice. The obvious question is: what happens as the frequency increases? Within the simple picture of Eq. (4.32), the circles would continue to expand, intersect each other, and remain circles of increasing radii. This picture is obviously not correct since it neglects the interactions between the dielectric inclusions in the photonic crystal, but it is still inspiring to some extent as we shall illustrate subsequently.

The formalism to obtain the correct iso-frequency contours has been presented in Section 4.1.2 and yields the patterns shown in Fig. 4.7(b): the circular shapes centered at Γ points disappear at higher frequencies and square-like shapes (in the case shown here of a square lattice) appear centered at M points. Different from the initial circles, these squares are larger at low frequencies than at higher frequencies. In addition, gaps appear since the contours do not touch each other (showing that the simple vision of increasing and intersecting circles is indeed incorrect). These gaps in the iso-frequency contours are the fundamental explanation of the bandgaps appearing in the band diagrams, for example the one shown in Fig. 4.6: the appearance of a gap prevents a solution to the phase-matching condition to exist at that specific incident angle and frequency. If the gap is large enough (obtained by a larger contrast in permittivities inside the photonic crystal), no solution exists for any incident angle, and a bandgap appears (Notomi 2000). Fig. 4.7(c) shows a

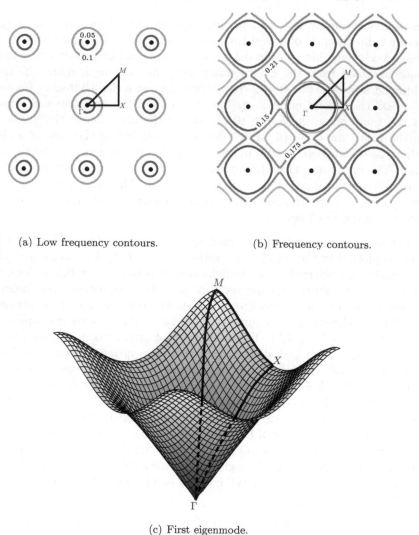

(a) Low frequency contours.   (b) Frequency contours.

(c) First eigenmode.

Figure 4.7 Frequency contours at a normalized ω for a photonic crystal defined by the parameters of Fig. 4.6. The superposed circle is a fictitious iso-frequency contour corresponding to an isotropic medium.

three-dimensional view of the first eigenvalue in the space (k_x, k_y, ω). The iso-frequency contours shown in Figs. 4.7(a)−4.7(b) are therefore different cuts of this curve at various heights. It can be seen that the three-dimensional surface starts from the Γ point and expands as a cone corresponding to a homogeneous dielectric medium. At higher frequencies, the cone is deformed due to the interaction between the dielectric inclusions inside the photonic

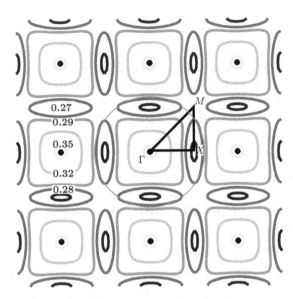

Figure 4.8 Iso-frequency contours at higher ω for the same configuration as in Fig. 4.7. The superposed circle is a fictitious iso-frequency contour corresponding to an isotropic medium.

crystal. It can also be seen that as frequency increases, the iso-frequency contours shrink toward the M point.

Coming back to our simple vision of expanding circles, it should be noted that the gaps appear near those points where we were expecting the circles to intersect. This is where the coupling between all the Bloch diffracted waves is the strongest, related to the maximum diffraction at the edge of the Brillouin zone. Although not completely accurate, the simple picture of expanding circles is nonetheless instructive. Indeed, looking away from the gap, we see that the apexes of the squares are not linked by straight lines but by inward curved ones. These lines in fact correspond to the circles centered at Γ which keep expanding, as illustrated by the fictitious circular iso-frequency contour in Fig. 4.7. Hence, as frequency increases, the circles of Eq. (4.32) keep increasing, making the square shrink until they ultimately disappear. As frequency continues to increase, the circles continue to expand and other gaps appear at their new crossing points. The influence of the increasing isotropic circle can still be seen as shown in Fig. 4.8, although the visualization of the circles is much less obvious at higher frequencies. This defines higher modes and more complex iso-frequency contours in the photonic crystal, yet always based on the same principle.

Let us finally remember the previous discussion on the group velocity in free-space: we have argued that the group velocity is pointing outward because the circles are expanding with frequency. Following the same reasoning, it is seen that the group velocity obtained on the shrinking squares may point inward (we remain cautious since this conclusion needs to be checked by a thorough examination of the curvature of the iso-frequency contours, and because the notion of "inward pointing" is not precisely defined). Like in the case of homogeneous dielectrics with $\varepsilon < 0$ and $\mu < 0$ treated in Chapter 5, this property can potentially yield negative refraction.

4.3 Negative refraction and flat lenses with photonic crystals

4.3.1 Achieving negative refraction

Unlike metamaterial structures discussed in other chapters of this book, photonic crystals do no have unit cells that are small compared to the wavelength and therefore cannot be viewed as homogenizable media. It is remarkable, however, that to some extent, they can be described based on their iso-frequency contours and in that regard, one is left wondering if they can exhibit a negative refraction when they operate at a frequency where the contour shrinks around a point[†]

We have seen in Fig. 4.7 that the photonic crystal whose parameters are given in Fig. 4.6 does exhibit a shrinking iso-frequency contour around the M point. This contour, however, is not circular at first (see the shape of the contour at the normalized frequency $a/\lambda = 0.173$), and becomes more and more circular as frequency increases, until the mode disappears and the second mode takes over. During this process of "becoming more and more circular," the curvature of the contour changes from concave to convex, with the important difference that the group velocity changes from pointing away from M to pointing toward M. The notion of "pointing toward" is here defined as the tendency of the normal directions to the iso-frequency contour to converge, as opposed to being divergent with concave contours, as illustrated in Fig. 4.9. In the ideal situation of a circular contour, the convergence happens toward the same point, instead of being smeared out like in the present case which is non-ideal. Fig. 4.9, however, proves that negative refraction is possible provided that the wave-vectors on the proper contour can be excited, i.e., can be phase-matched with an incident wave-vector.

This issue of excitation by phase matching is directly related to the orientation of the crystal at the interface with an incident medium or, in other

---

[†]Homogenizable media such as metamaterials exhibit iso-frequency contours that shrink or expand about the origin in the spectral plane, whereas photonic crystals exhibit expanding and shrinking contours about any symmetry point of the reciprocal lattice.

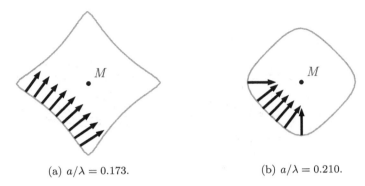

(a) $a/\lambda = 0.173$.   (b) $a/\lambda = 0.210$.

Figure 4.9 Illustration (zoom) on the iso-frequency contour at the normalized frequencies $a/\lambda = 0.173$ and $a/\lambda = 0.210$ for the photonic crystal of Fig. 4.6. The normal directions at various points along the contour show a divergent direction at $a/\lambda = 0.173$ and a convergent direction at $a/\lambda = 0.210$.

words, to how the crystal is cut. As a matter of fact, a rotation of the direct lattice induces a rotation in the reciprocal lattice, while the interface with the incident medium defines the direction of phase matching. Two situations are represented in Fig. 4.10, where the spectral and spatial representations are shown on the left and right, respectively. The iso-frequency contours are shown at the normalized frequency $a/\lambda = 0.210$, where a negative refraction is expected (Luo et al. 2002). In the first case, the crystal is cut along a line parallel to a direct lattice vector so that the spectral plane looks like the one in Fig. 4.7. Two wave-vectors are considered in the homogeneous incident medium, $\mathbf{k}_{i1}$ and $\mathbf{k}_{i2}$. Note that we have not assumed any specific values for the constitutive parameters of the incident medium so that the radii of the circular iso-frequency contours are chosen arbitrarily. Of course, as soon as the permittivity and the permeability are chosen, the radii must be obtained from the specific frequency at which the iso-frequency contour of the photonic crystal has been obtained (corresponding here to $a/\lambda = 0.210$). The medium for the first incident vector $\mathbf{k}_{i1}$ is selected such that the circular dispersion relation has a small radius and does not yield any real solution to the phase matching condition. The wave is therefore totally reflected from the crystal. The medium for the second incidence is chosen with a higher index of refraction, increasing the radius of the circle and offering real solutions to the phase matching condition beyond a certain incident angle. Choosing the phase-matched solution that corresponds to an up-propagating power, it is seen that the wave experiences a positive refraction, also represented in the right panel. In Fig. 4.10(b), the crystal is cut at a 45° angle, as shown in the right panel. The corresponding spectral domain representation is shown in the left panel, where it is seen that not only a phase-matched solution exists, but also that it corresponds to a negative refraction of the power. Note,

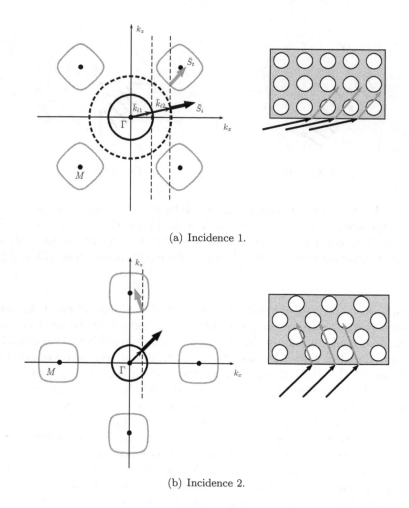

(a) Incidence 1.

(b) Incidence 2.

Figure 4.10 Correspondence between the spectral domain (left) and spatial domain (right) for two crystal orientations, and illustration of the associated phase matching. Note that the radii of the circular iso-frequency contours of the homogeneous media are chosen arbitrarily for the purpose of illustration.

however, that as already mentioned, this negative refraction is not isotropic because of the non-circular shape of the iso-frequency contour.

The above discussion therefore illustrates that negative refraction is indeed possible with a photonic crystal, provided that the following conditions are met:

1. At the operating frequency, portions or the whole iso-frequency contour is convex in order to induce converging energy refraction.

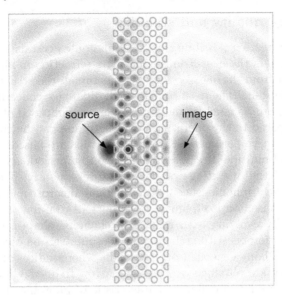

Figure 4.11 Focusing a point source with a slab of photonic-crystal. (Courtesy of Prof. J. Joannopoulos, MIT, Cambridge, USA.)

2. The crystal is properly cut in order to allow phase matching from the incident medium.

3. The frequency is chosen such that the iso-frequency contours of the photonic crystal and of the incident homogeneous medium yield real solutions to the phase matching condition.

It should be mentioned that Luo et al. (2002) added the condition that the frequency should be below $\pi c/a_s$, where a_s is the surface parallel period. In addition, the authors also generalized the concept of negative refraction to an *all angle* negative refraction, by which all incident rays are negatively refracted by the photonic crystal. For this to happen, the third condition needs to be updated in the sense that all the incident wave-vectors need to be included in the iso-frequency surface of the photonic crystal.

Developing upon the concept of negative refraction, Luo et al. (2002) showed an important analogy with the negative refraction obtained in metamaterials: after identifying two frequency bands in which all the above conditions are met, the authors showed that a slab of such photonic crystal is able to image a point source placed in front of it, very much like the imaging capabilities of slabs of left-handed media (see Chapter 8). An illustration of such imaging capability is offered in Fig. 4.11.

4.3.2   Image quality and stability

Despite the interesting possibility of using photonic crystals as a flat lens for imaging a source in the near-field, the negative refraction in photonic crystals is very different from the one obtained with metamaterial structures described in the other chapters. The most fundamental difference concerns the homogenization condition: as much as metamaterials are composed of unit cells that are small compared to the operating wavelength and can be characterized by effective constitutive parameters, photonic crystals are based on the complex superposition of multiple Bragg scattering due to a periodicity and a unit cell on the order of the operating wavelength. Consequently, photonic crystals are still inherently inhomogeneous, even if their macroscopic properties can be understood based on iso-frequency contours. As we have seen, these contours are rapidly evolving with frequency, changing shapes, curvature, and even symmetry points. Although circular contours can be obtained, yielding similarities with homogeneous isotropic media where indices of refraction can be defined, they are obtained over a very narrow frequency band and therefore cannot be used as a defining characteristic of photonic crystals. In fact, photonic crystals cannot be viewed as homogeneous media, even though they exhibit some of the properties of homogeneous media.

An immediate consequence of these considerations is that the wave propagation inside the crystal strongly depends on the orientation of its constituents for example, which would be unexplained if the crystal was truly homogenizable. Upon designing a configuration that supports negative refraction (based on the same principle enunciated above), Martinez and Marti (2005) showed that an incident beam is indeed negatively refracted for some particular lattice orientations only. For other ones, the beam is either strongly attenuated[‡] or scattered to multiple Bloch modes. Similarly, Moussa et al. (2005) and later Decoopman et al. (2006) showed that surface termination has a direct and important impact on the transmission level through a slab of photonic crystal as well as on the values of the associated effective constitutive parameters, despite the fact that the crystal was optimized to exhibit a circular iso-frequency contour. It has been postulated by Decoopman et al. (2006) that these effects are due to the interplay between the influence of the impedance and the influence of the index of refraction on the wave propagating inside the crystal. As a matter of fact, the iso-frequency contour provides information on n only (which is here understood to be an effective parameter), but not on the permittivity and permeability. Hence, any values of effective ε and μ that satisfy $\sqrt{\varepsilon\mu} = n$ comply with the iso-frequency contour and in particular, once n is fixed, any value of ε that satisfies $\varepsilon = n^2/(\mu c^2)$ is acceptable. However, ε and μ are also strongly dependent on the location of the boundary (i.e., the cut)

[‡]It should be emphasized that the iso-frequency contours drawn in the spectral plane only give an indication on the propagation direction of waves, while no information is provided on their amplitude, which can potentially be very low.

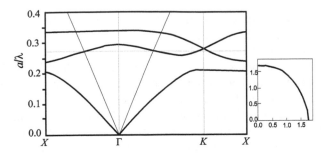

Figure 4.12 Dispersion diagram for a triangular array of circular rods $r = 0.45a$. Horizontal axis: Bloch vector describing the first Brillouin zone $\Gamma - K - X$. Vertical axis: Normalized frequencies a/λ, with λ the wavelength. Right: Iso frequency contour at the intersection of the light line in vacuum with optical band showing negative group velocity at $\lambda = 3.66a$. (Redrawn from the data in Ramakrishna et al. (2007b) © 2007, American Physical Society.)

within a unit cell: while maintaining a constant product, the ratio between them is not preserved so that the surface impedance of the crystal varies, and so does the transmission level. These results therefore confirm the fact that the iso-frequency contours are not sufficient to conclude on the isotropy of a photonic crystal.

Despite these words of caution, iso-frequency contours can still be used as an intuitive tool to understand some aspects of propagation in photonic crystals, especially for frequencies corresponding to the higher bands and for crystals cut along their symmetry directions (Foteinopoulou and Soukoulis 2005). Using these principles, both negative refraction and imaging by a flat slab have been consistently reported, in a one-dimensional structure (Monzon et al. 2006), in a two-dimensional hexagonal lattice (Wang et al. 2004), in a three-dimensional structure (Lu et al. 2005b), and even in a metallic photonic crystal (Parimi et al. 2004) and a photonic crystal with silver nanowires operating at optical frequencies (Ao and He 2005).

For the sake of further illustration, we subsequently present the calculated results for the case of a triangular lattice of cylindrical dielectric rods with a refractive index of $n = 4$ in air. The band structure for this configuration is shown in Fig. 4.12 where it is seen that the second band has a negative slope or a negative group velocity. Thus, the iso-frequency surface shrink around the Γ point as the frequency increases. The point of intersection of this band with the dispersion curve for free light in vacuum with $(n = +1)$ corresponds to $\lambda = 3.66a$. The iso-frequency surface is shown on the right side of Fig. 4.12 at this wavelength and it is seen to be almost circular, which is ideal for focusing applications. Note that the ratio of the crystal lattice period and the free-space wavelength is still smaller than unity and that an incident plane wave couples to a single Bragg mode inside the photonic crystal. In Fig. 4.13, we show the results of a calculation where a Gaussian beam of finite width is incident upon a slab made of this photonic crystal (16 × 70 cylinders) and undergoes a negative refraction. A multipole method (Tayeb and Maystre 1997) has been used to calculate the fields with the magnetic field along the cylindrical axis, and the magnitude of the electric field has been plotted as a gray scale in the figure. It is clear that the beam experiences a negative transverse shift upon transmission across the slab, which is one of the hallmarks of negative refraction. One can also see the negative refraction onto the same side of the normal inside the photonic crystal although that is quite obscured by the interference pattern inside the slab. The interference pattern on the incident side arises due to interference between the incident and reflected beams. Although we have negative refraction in this case which almost appears like $n = -1$, we do not have impedance matching which sensitively depends on the surface termination of the photonic crystal (the plane at which the slab is cut out of the crystal). The finite reflectivity arises from this impedance mismatch. Note that one has to keep the beam width reasonably large to avoid a large beam divergence which would otherwise not allow the clear demonstration of the negative refraction effect. One can clearly see the negative refraction inside the photonic crystal by considering the refraction of a finite beam through a prism made of the photonic crystal. An equilateral triangular prism formed out of this photonic crystal by placing cuts along the symmetry directions is shown in Fig. 4.14. A Gaussian beam of light of wavelength $\lambda = 3.6798a$ is incident on the prism from the bottom left, making an angle of 30° with the vertical axis. The beam is centered on the middle of the prism side consisting of 60 rods, and its width is $20a$. It is clear that the beam refracts negatively across the prism and bends toward the apex. In this figure, due to the spatial separation between the two beams inside the photonic crystal, there are no interference effects that obscure the negative refraction, which is thus clearly visible. Note that the beam on the top right side of the prism arises from the reflection at the second interface.

It should also be emphasized that whenever it is obtained, the imaging phenomenon with a photonic crystal-based flat lens usually achieves some amount of sub-wavelength resolution due to the larger iso-frequency contours than those of the incident medium at the same frequency. As a matter of fact, the reciprocal effect of total reflection occurs, by which an evanescent wave in the incident medium is coupled to a propagating wave inside the crystal. For sources in the near field of the first slab interface, the evanescent waves have not decayed enough to be negligible and the corresponding propagating waves inside the crystal carry meaningful information that can be transmitted over a large distance, typically to the other side of the slab. The resulting image, if also in the near field of the flat lens, therefore contains information from the evanescent waves from the source and can thus have sub-wavelength resolution to some extent.

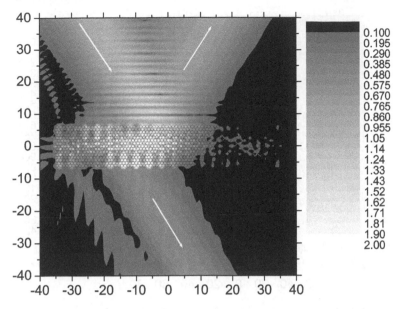

Figure 4.13 Negative refraction of a Gaussian beam across a slab of a two-dimensional photonic crystal with a triangular lattice. The radiation has a wavelength of $\lambda = 3.6798a$ and is incident from the top right as shown by the arrow. The beam negatively refracts inside the photonic crystal and has a negative transverse shift upon transmission across the slab. (This figure has been kindly provided by Gérard Tayeb, Institut Fresnel, Marseilles, France.)

4.4 Negative refraction vs. collimation or streaming

The ambiguity in defining homogeneous parameters for photonic crystals is at the origin of some seemingly contradictory results reported in the literature. As we have argued in the previous section, defining the index of refraction of a photonic crystal solely based on a circular iso-frequency curve at a given frequency can be justified within some very specific conditions, but is not exact in general and is quickly bound to be misinterpreted. Hence, simultaneously to the reports of flat lens imaging based on the negative refraction property, other authors interpreted apparently similar imaging results based on a different wave propagation mechanism, thus questioning the relative importance of negative refraction.

The argument to question the homogeneity and the importance of the negative refraction in a photonic crystal is based on a simple property of the imaging configuration using a flat lens: a truly isotropic and homogeneous medium with a negative index of $n = -1$ creates an image of a point source

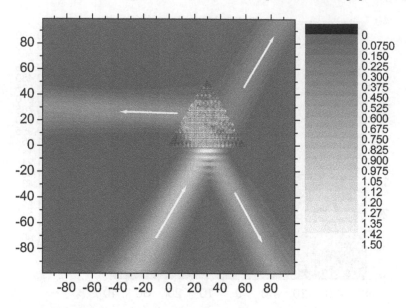

Figure 4.14 Negative refraction of a Gaussian beam in an equilateral triangular prism of two-dimensional photonic crystal with a triangular lattice. The radiation has a wavelength of $\lambda = 3.6798a$ and is incident from the bottom left making an angle of 30° with the vertical axis. The beam is centered on the middle of the triangle side consisting of 60 rods and has a width of $20a$. (Taken from Ramakrishna et al. (2007b) © 2007, American Physical Society.)

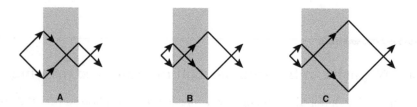

Figure 4.15 Ray tracing illustration of the flat lens made of an isotropic homogeneous medium of index of refraction $n = -1$. The distance between the image point and the second interface of the lens varies as function of the distance between the source and the first interface (situation A to B), as well as function of the lens thickness (situation B to C).

whose position can be varied by two parameters, the distance between the source and the first interface of the lens as well as the lens thickness. These observations come directly from the simple ray diagram illustrated in Fig. 4.15.

Upon choosing a configuration that exhibits a negative refraction, however, Li and Lin (2003) showed using a multiple scattering approach that the image does not follow the expected behavior reported in Fig. 4.15, but instead that it is consistently located in the near field of the second interface. In addition, no focusing inside the slab was visible either, but a seemingly collimated beam that was propagating from the first to the second interface of the lens. In fact, returning to Fig. 4.9, one can see that even though the iso-frequency contour is convex, the negative refraction is somewhat weak, and the power is more directed toward a single direction rather than being focused to a point. The reason of this behavior is of course the shape of the contour, which is not circular but more square-like, with approximately flat portions whose normal vectors point in very similar directions. This effect has been termed *collimation* due to the confined lateral extent of the beam as it propagates through the crystal. Consequently, a Gaussian beam which, contrary to a Bessel beam for example, naturally spreads, can be collimated in a photonic crystal by using its inherent anisotropy at a frequency where the iso-frequency contour has a low curvature. Interestingly, this collimation is a purely linear effect, unlike for example solitons which are created by a nonlinear compensation for the beam spread due to diffraction.

A theoretical study of the collimation effect in photonic crystals can be performed by studying the curvature of the iso-frequency contour, as suggested in Shin and Fan (2005). Supposing that the relationship between k_z and (k_x, k_y) is known, $k_z = f(k_x, k_y)$, one can compute the radius of curvature as

$$R = \frac{1}{\partial^2 f / \partial u^2}, \tag{4.33}$$

where $\hat{u} = \hat{x} \cos\theta + \hat{y} \sin\theta$ is an arbitrary tangential direction. Hence, the photonic crystal exhibits collimation in a given direction if R is infinite in that direction. Of course, in order to realize a good collimation, it is necessary to obtain an infinite radius over as wide portion of the iso-frequency contour as possible. It has been suggested that such a wide, flat portion of the contour can be obtained at frequencies for which the contour mimics the form of the first Brillouin zone (Chigrin et al. 2003). Applications for example include the reshaping of the radiation pattern of a dipole source (or an antenna in general), with enhanced directivity in specific directions (Chigrin 2004, Guven and Ozbay 2007).

5

Media with $\varepsilon < 0$ and $\mu < 0$: theory and properties

This chapter presents various theoretical aspects related to left-handed media, i.e., media whose permittivity ε and permeability μ can achieve negative values simultaneously. As we have already mentioned in previous chapters, the sole possibility of achieving $\varepsilon < 0$ and $\mu < 0$ prompts us to revisit even the most basic phenomena of electromagnetics and optics, one such example discussed at length being the sign of the index of refraction.

We thus start by identifying the possible origins of negative refraction, showing that the negative refraction of the power is not a new phenomenon. The specificity of left-handed media is shown to come from the negative refraction of both the power and of the wave-vector $\mathbf{k}$, which prompts us to examine the choices of $\mathbf{k}$ and its consequences. A few key properties are then discussed, such as the reversal of Snell's law, of the Doppler shift, of Čerenkov radiation, and of the Goos-Hänchen shift. The modified Mie scattering is also discussed, providing a vivid illustration of the impact of negative refraction on the field distribution: it is shown that the usual forward focusing obtained with standard dielectric Mie spheres is modified, yielding a clear focus of the field inside the sphere.

The chapter continues by expanding the discussion to anisotropic media and indefinite media, where the signs of the diagonal terms in the permittivity and permeability are allowed to be independently chosen, either positive or negative. The various iso-frequency contours thus generated are presented, and their impact on the amphoteric refraction* of waves, on the inversion of the critical angle and the Brewster angle, and on the realization of flat lenses is discussed. The chapter terminates by a generalization to bianisotropic media, showing how a strong bianisotropy can flip the refraction from positive to negative.

---

*By amphoteric refraction, it is implied that the refraction can be either positive or negative. For example, one could have a situation where the refraction of the phase is positive while the one of the power is negative.

5.1   Origins of negative refraction

Wave refraction at an interface between two media is one of the most fundamental phenomena of optics and electromagnetics, and is quantified by Snell's refraction law which stipulates that the transmitted angle θ_t is related to the incident angle θ_i by the relation $n_i \sin \theta_i = n_t \sin \theta_t$, where n_i and n_t are the refractive indices of the incident and transmitted media, respectively. Within a ray optics approximation, Snell's law can be obtained as a direct application of Fermat principle and yields a positive refraction for media with positive index of refractions, which represent all the natural homogeneous media known to date. As the wave theory of light matured with Fresnel and later with Maxwell's electromagnetic theory, Snell's law was shown to be the consequence of the more fundamental concept of phase matching. Furthermore, Maxwell's theory also pointed out the necessary distinction between refraction of the phase, related to the propagation vector **k** of the wave, and the refraction of the power flow associated with the wave. Within this more general framework, it appeared that both positive and negative transmission angles were solutions of the refraction equations, but that a negative refraction of the power would also require a negative refraction of the phase. Since no media were known to exhibit this property, negative refraction was sought within more complex media such as anisotropic or inhomogeneous.

In the case of uniaxial anisotropic crystals for example, the dispersion relation, which represents the variation of the components of the wave-vector **k** with frequency (ω), yields elliptic iso-frequency contours that can be rotated by cutting the crystal appropriately with respect to its axes in order to achieve a negative refraction of the power flow (but a positive refraction of the phase) within a specific range of incident angles. In the case of inhomogeneous media, intense efforts by the photonic crystal community have led to the successful demonstration of photonic crystal structures exhibiting negative refraction, as discussed in the previous chapter. Unlike the anisotropic situation, the negative refraction in this case is most often due to shrinking iso-frequency contours around a given point in the spectral space as a function of increasing frequency.

An important conceptual generalization was to reconcile isotropy and shrinking dispersion relation, which was done by postulating the existence of substances exhibiting negative values of the permittivity ε and the permeability μ within a certain frequency band (Veselago 1968). A few interesting properties of such media were immediately identified, among others the negative refraction, but also the left-handed triad formed by the electric field vector, the magnetic field vector, and the wave-vector, which led to the terminology "left-handed media" (LHM) for these substances, the reversal of Doppler shift and the reversal of Čerenkov radiation. All these phenomena were later experimentally confirmed: negative refraction has been measured in a prism experiment (Shelby et al. 2001b) using an artificial metamaterial based on small unit cells containing split rings and rods discussed in Chapter 3 (Pendry et al. 1998; 1999), backward phase waves were shown by simulating real metamaterials Lindell et al. (2001), while reversed Čerenkov radiation and Doppler shift have been reported in Lu et al. (2003), Luo et al. (2003), and Stancil et al. (2004), respectively.

A few facts can be highlighted from these considerations. First, negative refraction is not unusual and has been known and studied for decades already. However, left-handed media have an interesting property of exhibiting a negative refraction of both the power and the phase, which had not been achieved before. Second, negative refraction does not imply $\varepsilon < 0$ and $\mu < 0$ while the implication holds in the other direction. Third, negative refraction (of the power and/or the phase) can be attributed to one of the following four origins:

1. An elliptic dispersion relation properly oriented in the spectral domain, typically achieved by uniaxial media.

2. A shrinking iso-frequency contour with frequency like in photonic crystals.

3. A negative index of refraction due to negative values of the permittivity ϵ and the permeability μ, which effectively also yield a shrinking dispersion relation but within an isotropic and homogeneous medium.

4. The motion of an otherwise isotropic and homogeneous medium.

It should be mentioned that negative refraction can be obtained by yet other methods, for example the proper design of a series of parallel plate waveguides. However, unlike the four properties listed above, most of these other methods cannot be associated with effective material properties and we shall therefore not consider them here in detail. In the following, we describe how the negative refraction phenomenon is generated by the four aforementioned properties, and we start by discussing the fundamental concept of dispersion relation.

5.1.1   Dispersion relation

The dispersion relation is fundamental in order to understand the refraction phenomenon at the boundary between two homogeneous media. The dispersion relation relates the components of the wave-vector of the propagating electromagnetic wave to the properties of the medium and to the frequency. In the case of a homogeneous isotropic dielectric of relative constitutive parameters ε and μ for example, the dispersion relation is well known and is written as

$$k = \sqrt{k_x^2 + k_z^2} = \frac{\omega}{c}\sqrt{\varepsilon\mu}, \tag{5.1}$$

where c is the velocity of light and where the wavenumber k is the magnitude of the wave-vector $\mathbf{k} = \hat{x}k_x + \hat{z}k_z$, written here in two dimensions for the sake of simplicity. The parameters ϵ and μ are understood to be relative values.

A systematic way of obtaining the dispersion relation of a medium is to combine the Maxwell equations and the constitutive relations to yield an equation for the electric field only (a generalization of the well-known Helmholtz equation for homogeneous media). The dispersion relation is obtained upon setting the determinant of the matrix operator to zero, a condition that can be written as

$$\left| \omega^2 \varepsilon_0 \bar{\bar{\varepsilon}} + \mu_0^{-1} \bar{\bar{k}} \cdot \bar{\bar{\mu}}^{-1} \cdot \bar{\bar{k}} \right| = 0. \tag{5.2}$$

Note that this method can be generalized to bianisotropic media as well, as shown in Chapter 2, Eq. (2.81) on page 59. In Eq. (5.2), $\bar{\bar{\varepsilon}}$ and $\bar{\bar{\mu}}$ are second-rank tensors and can be potentially fully populated. Of course, for highly complex constitutive tensors, it may be difficult to obtain an analytical solution to Eq. (5.2).

The graphical representation of the dispersion relation in the spectral plane yields information on both the refraction of the wave-vector and of the power when an interface is present. The refraction of the wave-vector is a direct consequence of phase matching (i.e., the continuity of the tangential component of the wave-vector across the interface). This situation is represented in Fig. 5.1: Fig. 5.1(a) shows the physical situation while Fig. 5.1(b) shows the spectral representation. A wave impinging from a homogeneous medium of parameters (ε_0, μ_0) onto another homogeneous medium of parameters $(\varepsilon_1, \mu_1) = (\varepsilon\varepsilon_0, \mu\mu_0)$ experiences both a reflection and a refraction phenomenon. In the first medium, the wave-vectors are $\mathbf{k}_i$ and $\mathbf{k}_r$ for the incident and reflected waves, respectively, while in the second medium the transmitted wave propagates with a wave-vector $\mathbf{k}_t$. The components of these wave-vectors satisfy the dispersion relations in the respective media (supposing again for simplicity that $k_y = 0$):

$$k_{ix}^2 + k_{iz}^2 = \omega^2 \varepsilon_0 \mu_0 \,, \tag{5.3a}$$

$$k_{rx}^2 + k_{rz}^2 = \omega^2 \varepsilon_0 \mu_0 \,, \tag{5.3b}$$

$$k_{tx}^2 + k_{tz}^2 = \omega^2 \varepsilon_1 \mu_1 \,, \tag{5.3c}$$

as well as the phase matching condition

$$k_{ix} = k_{rx} = k_{tx} \,, \tag{5.4}$$

so as to satisfy the continuity of the tangential components of the fields across the interface between the media. These relations are represented in Fig. 5.1(b): Eqs. (5.3) yield circles while Eq. (5.4) yields the dashed line, and the intersection between the two indicates the wave-vectors that are solutions of both conditions. From this picture, it is obvious that both the incident and transmitted media support two solutions. In the incident medium, both solutions

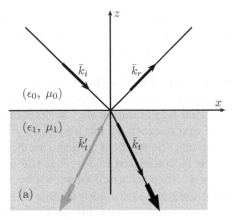

(a) Spatial representation.

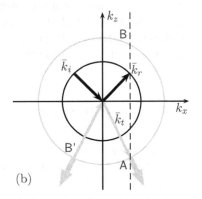

(b) Spectral representation.

Figure 5.1 Spatial and spectral representation of a plane wave impinging onto a dielectric half-space, generating a reflected and a transmitted wave. In the spectral domain, the wave-vectors satisfy Eqs. (5.3) and Eq. (5.4). Solution A corresponds to a regular medium where the wave-vector (thin arrow) and power flow (thick arrow) are parallel whereas solution B corresponds to a left-handed medium where the wave-vector and the power flow are anti-parallel. The direction of the power flow is obtained from the gradient of the iso-frequency curve with respect to the frequency.

are excited: one is the incident wave and the other one is the reflected wave. The situation in the transmitted medium deserves further discussion. It has been customary to choose solution A, which yields the regular positive refraction phenomenon, but one is left wondering if the second solution B could be chosen as well.

The resolution of this question lies in the physical requirement that the power always flows away from the source. In the situation of Fig. 5.1, this requires the $\hat{z}$ component of the Poynting vector to be always negative (the source being located at $z \to \infty$ and the wave propagating downward in the figure).

The direction of the power flow can be obtained from the gradient of the dispersion relation with respect to frequency $\nabla_{\mathbf{k}}\omega(\mathbf{k})$ (if the dispersion $\omega(\mathbf{k})$ does not vary fast – see Section 6.5). Graphically, we can draw an iso-frequency contour at frequency f and another one at frequency $f + \delta f$ (where δf is a small frequency increment), the gradient giving the direction of the power. For a standard dielectric as considered in Fig. 5.1, the iso-frequency contour expands with frequency, yielding the gradient shown by the thicker arrows at point A, while a similar outgoing arrow corresponds to point B (not shown). Clearly, only one solution is permitted (only one solution has a negative z component), which corresponds to the solution intuitively chosen. One could imagine, however, that if the iso-frequency contour were shrinking with frequency, the gradient would point inward and the other phase-matched component would have to be chosen, corresponding to the situation at point B'. This is precisely what happens in left-handed media, as we further discuss in Section 5.1.4.

5.1.2   Anisotropic media with positive constitutive parameters

When the permittivity and the permeability tensors are not scalar quantities, the medium is said to be anisotropic, i.e., the material properties are different in different directions of propagation. In this case, the iso-frequency contours are more complex than the simple sphere of a homogeneous medium, and typically take elliptic shapes. We consider here a biaxial medium: the relative permittivity and permeability tensors are simultaneously diagonalizable and the three values might be different:

$$\bar{\bar{\varepsilon}} = \begin{pmatrix} \varepsilon_x & 0 & 0 \\ 0 & \varepsilon_y & 0 \\ 0 & 0 & \varepsilon_z \end{pmatrix} \quad \text{and} \quad \bar{\bar{\mu}} = \begin{pmatrix} \mu_x & 0 & 0 \\ 0 & \mu_y & 0 \\ 0 & 0 & \mu_z \end{pmatrix}. \tag{5.5}$$

Note that an isotropic medium is just the limiting case of an anisotropic biaxial medium where $\varepsilon_x = \varepsilon_y = \varepsilon_z$ and $\mu_x = \mu_y = \mu_z$.

For a plane wave in the xy plane with an electric field polarized along the $\hat{y}$ direction and a magnetic field in the xz plane (corresponding to an S polarization), the dispersion relation equivalent to Eq. (5.1) is written as:

$$\frac{k_x^2}{\varepsilon_y \mu_z} + \frac{k_z^2}{\varepsilon_y \mu_x} = k_0^2. \tag{5.6}$$

The components $(\varepsilon_x, \varepsilon_z, \mu_y)$ influence the other polarization. When all the parameters of Eq. (5.6) are positive (we shall consider the case when they

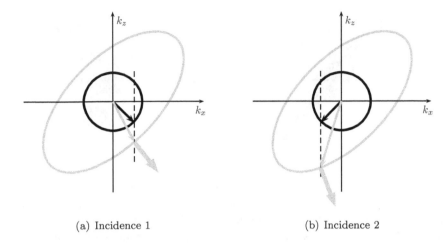

(a) Incidence 1   (b) Incidence 2

Figure 5.2 Illustration of the amphoteric refraction between free-space and a rotated anisotropic medium. The incidences on the left and right yield a positive and a negative refraction of the power, respectively. In both cases, the refraction of the phase is positive.

can be negative in Section 5.3), the iso-frequency curves are elliptic with the principal axes aligned with the $\hat{x}$ and $\hat{z}$ directions. These directions, however, may not coincide with the principal axes of the crystal. This is to say, the medium might be biaxial in some coordinate system different from the $(\hat{x}, \hat{y}, \hat{z})$ system defined *a priori*. When this happens the iso-frequency contours in the xz plane become rotated ellipses, and this rotation can induce a negative refraction of the power.

The above situation is illustrated in Fig. 5.2, while the mathematical treatment of this case is postponed to Section 5.3. Upon performing the phase matching and gradient considerations as described above (all the parameters here being positive, the iso-frequency curves are expanding with frequency), we see that for a proper choice of parameters and rotation angle, some incidences (Fig. 5.2(a)) have a positive refraction, whereas some other incidences (Fig. 5.2(b)) have a negative refraction of the power but positive refraction of the phase. When using a beam of light as illustrated in Fig. 5.3, one would therefore witness a negative refraction of the beam and could conclude on a negative refraction phenomenon with a natural material, for example a properly cut crystal of $CaCO_3$ or of YVO_4 (Chen et al. 2005c, Du et al. 2006). This negative refraction, however, has some fundamental differences when compared to the negative refraction due to left-handed media:

1. It holds only for the power, whereas the phase (i.e., the wave-vector), is positively refracted. These media therefore do not support backward waves.

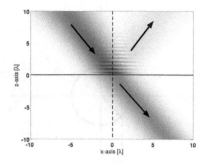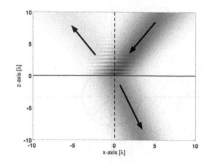

Figure 5.3 Amphoteric refraction of a beam of light between free-space and a rotated anisotropic medium. Parameters are: $\varepsilon_y = 3\varepsilon_0$, $\mu_x = \mu_0$, $\mu_z = 3\mu_0$, and a rotation of 45 degrees. The incidence is from top to bottom in both cases and the weaker beam represents the reflection from the interface.

2. It holds only for a few incident angles, whereas the refraction is positive for most other angles. This would, for example, preclude the use of these media for lensing applications as described in Chapter 8.

5.1.3   Photonic crystals

Photonic crystals have been treated in detail in Chapter 4 and, more specifically, the phenomenon of negative refraction has been treated in Section 4.3. It is therefore known that in this case as well, despite the very inhomogeneous nature of the medium, negative refraction can still be understood with iso-frequency contours and their evolution as function of frequency. Thus a wave incident from free-space would couple to a single Bragg mode with a negative phase vector in the photonic crystal. It then becomes a designing procedure: the proper permittivity of the background and that of the inclusions as well as the distance between them and the operating frequency need to be determined in order to obtain a shrinking iso-frequency contour around a symmetry point of the crystal. Additionally, circular iso-frequency contours are often sought in order to mimic an isotropic medium and to borrow the terminology of index of refraction. If these conditions can be met at a low enough frequency, it is possible to cut the crystal in such a way to obtain an all-angle negative refraction (Luo et al. 2002).

Photonic crystals, however, are inherently inhomogeneous media, and circular iso-frequency contours can be obtained only within a very narrow frequency band. In addition, they depend very sensitively on the relative placement of the scatterers and the lattice. We therefore prefer to refrain ourselves from defining an index of refraction in this case, and consider the negative refraction produced by photonic crystals as a special case of a much more complex wave propagation phenomenon governed by the theory of Bragg diffraction.

5.1.4   Left-handed media

We consider here left-handed media as they were originally defined (Vese-lago 1968): homogeneous and isotropic substances with a negative dielectric permittivity and a negative magnetic permeability. The case of anisotropic left-handed media, fundamentally related to their physical implementation as a succession of metal split rings and thin wires, is postponed until Section 5.3.

We have discussed earlier the properties of the dispersion relations and their influence on the iso-frequency contours of standard dielectrics, which has en-abled us to explain the well-known phenomenon of refraction at an interface in a systematic manner. This approach was necessary in order to understand the refraction between free-space and a left-handed medium: since our intu-ition is of no help in this new situation, we need to resort to a mathematical approach and examine its consequences.

We have seen in the previous chapter that left-handed media exhibit a negative permittivity and a negative permeability within a certain frequency range, as a result of a frequency dispersive phenomenon (typically a Drude-like model for the permittivity and a Lorentz-like model for the permeability). A typical evolution of these parameters in a homogeneous medium is shown in Fig. 3.21, where the index of refraction $n = \sqrt{\epsilon\mu}$ is directly proportional to the magnitude of the wave-vector k in Eq. (5.1). From this picture, it is therefore clear that when the index of refraction is negative, its magnitude decreases with increasing frequency, thus yielding a smaller circle in the (k_x, k_z) plane (since $k^2 = k_x^2 + k_z^2$ and k decreases with frequency). The iso-frequency contour therefore shrinks with the frequency, changing the direction of the gradient. Consequently, for the same incidence as in Fig. 5.1, the solution B needs to be chosen (the one with a negative $\hat{z}$ component of the power), which is equivalently represented by the point B' and yields a negative refraction. In order to preserve the phase matching condition (k_x continuous across the interface), k_z needs to be in the opposite direction with respect to the case in Fig. 5.1. This inversion of the wave-vector in the propagation direction is at the origin of the backward waves supported by left-handed media. An example of negative refraction of a finite beam of light at the interface between free-space and a left-handed medium is shown in Fig. 5.4, while Fig. 5.5 illustrates the motion of a pulse through a negative index medium where the phase of the wave propagates backward while the envelope of the pulse propagates forward.

5.1.5   Moving media

The electrodynamics of moving media represent an entire topic per se whose extensive discussion would merit a chapter by itself. This being out of the

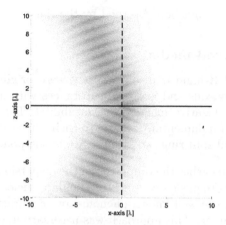

Figure 5.4 Illustration of the refraction of a Gaussian beam between free-space and a left-handed medium characterized by the constitutive parameters $(-\varepsilon_0, -\mu_0)$. Note the simultaneous negative refraction of the power and of the phase.

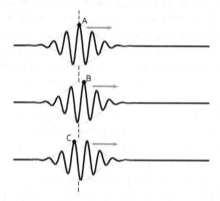

Figure 5.5 Illustration of the propagation of a pulse through a standard medium and through a left-handed medium. Top: reference position. Middle: position at a later time when the pulse propagates through a standard medium. Bottom: position at a later time when the pulse propagates through a left-handed medium.

scope of this book, we refer the reader to Kong (2000) as well as to the open literature for more details, while we shall only present this topic within the scope of negative refraction.

The motion of a medium is potentially a fourth cause of negative refraction (Grzegorczyk and Kong 2006) which, like in the previous situations, can be directly understood from the dispersion relation given by

$$k_x^2 + \frac{1}{\alpha}\left(k_z - k_z^+\right)\left(k_z - k_z^-\right) = 0, \tag{5.7}$$

where it is assumed that the motion is in the $\hat{z}$ direction,

$$k_z^{\pm} = k_0 \frac{n \pm \beta}{n\beta \pm 1}, \quad \alpha = \frac{1 - \beta^2}{1 - n^2\beta^2}, \tag{5.8}$$

$k_0 = \omega/c$, $n^2 = c^2\varepsilon'\mu' > 1$, c is the velocity of light in free-space, $\beta = \hat{z}\beta = \mathbf{v}/c$, $\beta < 1$, and ε' and μ' are the isotropic permittivity and permeability of the medium in its rest frame, respectively. The iso-frequency contours are then either ellipses (if $\beta < 1/n$) or hyperbolae (if $\beta > 1/n$), both centered at $k_{z0} = k_0\,\beta(n^2 - 1)/(n^2\beta^2 - 1)$. We note that $k_{z0} < 0$ if $\beta < 1/n$, which indicates that the ellipses are displaced toward negative k_z values, while $k_{z0} > 0$ if $\beta > 1/n$ (the Čerenkov zone), which indicates that the hyperbolae are displaced toward positive k_z values. Consequently, the phase of the wave can only be positively refracted, while the refraction of the power can be either positive or negative. Two regimes can be distinguished: one below and one above the Čerenkov limit of $\beta = 1/n$. For velocities below the limit, the displaced ellipse can produce a negative refraction for the proper quadrant of incident angles, as illustrated in Fig. 5.6(a). All the angles in the quadrant are included if $\beta = 1/n^2$. For velocities above the Čerenkov limit, the displaced hyperbola produces a negative refraction for an entire quadrant of incident angles, as shown in Fig. 5.6(b), and explains the standard high-velocity Fizeau-Fresnel drag (Censor 1969, Parks and Dowell 1974).

Note, however, that this negative refraction holds only for the power and not for the wave-vector. Consequently, a backward phase vector is not generated here. Additionally, the motion of the medium can be combined with the natural negative refraction of an isotropic left-handed medium for example. The combined effect is either an enhanced refraction or an attenuated refraction with a possible sign change, as function of frequency (the constitutive parameters of left-handed media being frequency dispersive), incident angle, and velocity (Grzegorczyk and Kong 2006).

5.2 Choice of the wave-vector and its consequences

A fundamental consequence of the inversion of the wave-vector in the phase matching diagram is the inversion of the sign of the index of refraction. This concept is so deeply related to the topic of left-handed medium and metamaterials that the terminology "negative index media" (or an equivalent) has often

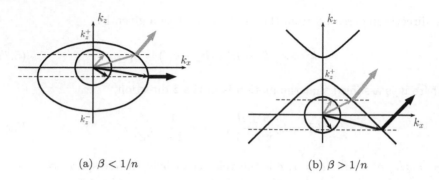

(a) $\beta < 1/n$ (b) $\beta > 1/n$

Figure 5.6 Iso-frequency contours for free-space (circle) and for a moving medium as function of its velocity. The wave-vectors (thin arrows), power (thick arrows) and phase matching condition show the possibility of obtaining a negative refraction at some incident angles. In all cases, the refraction of the k-vector is positive. Note that the curves are given only for the purpose of illustration and are not drawn to scale.

been used to characterize them. In this section, we examine more rigorously why a negative dielectric permittivity and a negative magnetic permeability induce a negative index of refraction.[†]

Mathematically, the index of refraction n is expressed as function of the product of the permittivity and the permeability as (the Maxwell relation)

$$n = \sqrt{\varepsilon\mu}. \tag{5.9}$$

Hence, it is not immediately obvious that $\varepsilon < 0$ and $\mu < 0$ imply $n < 0$. Like often in electromagnetics, one has to look at the lossy situation in order to extrapolate the conclusions to lossless media.

Upon including losses, the permittivity and the permeability are written in the polar coordinate system (in the complex plane) as

$$\varepsilon = |\varepsilon|e^{i\theta_\varepsilon}, \qquad \mu = |\mu|e^{i\theta_\mu}. \tag{5.10}$$

The index of refraction thus becomes

$$n = \sqrt{|\varepsilon|\,|\mu|}\; e^{i\frac{1}{2}(\theta_\varepsilon + \theta_\mu)}. \tag{5.11}$$

Our convention (using $i = \sqrt{-1}$ to denote the imaginary number) imposes that $\varepsilon'' > 0$ (the imaginary part of the permittivity to be positive[‡]) and $\mu'' > 0$, so that $\theta_\varepsilon \in [0, \pi]$ (illustrated in Fig. 5.7 in the case of a negative permittivity) and $\theta_\mu \in [0, \pi]$. Consequently, the angle of the index of refraction is $(\theta_\varepsilon +$

[†]We shall limit our discussion to media where it makes sense to talk about an index of refraction, i.e., isotropic and homogeneous media in our case.
[‡]We denote by single primes the real part operator and double primes the imaginary operator, so that $\mathrm{Re}(\varepsilon) = \varepsilon'$ and $\mathrm{Im}(\varepsilon) = \varepsilon''$.

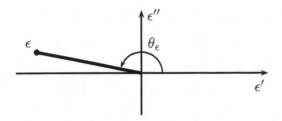

Figure 5.7 Representation of the permittivity in the complex plane.

$\theta_\mu)/2 \in [0, \pi]$. This range is reduced to $[\pi/2, \pi]$ in the situation when $\varepsilon' < 0$ and $\mu' < 0$, so that the real part of the index of refraction is negative ($n' < 0$). In particular, the lossless limit is obtained as $\theta_\varepsilon \to \pi$ and $\theta_\mu \to \pi$ which produces:

$$n = \sqrt{|\varepsilon| \, |\mu|} \, e^{i\pi} = -\sqrt{|\varepsilon| \, |\mu|}. \tag{5.12}$$

Under the assumption of an isotropic medium, this index of refraction can be directly introduced into Snell's law and is seen to reverse the refraction direction, as already explained based on phase matching of the wave-vector.

Another situation inducing $n' < 0$ should be pointed out, theoretically justified but of practical limited interest. As a matter of fact, a negative index of refraction is obtained as soon as $(\theta_\varepsilon + \theta_\mu) > \pi$. This can be obtained when $\theta_\varepsilon > \pi/2$ and $\theta_\mu > \pi/2$, which is the situation of a standard left-handed medium described above, or for example if we let $\theta_\varepsilon = \pi/2 + \delta_\varepsilon$, where $\delta_\varepsilon \in [0, \pi/2]$. In this latter situation, a negative n' is obtained when $\theta_\mu > \pi/2 - \delta_\varepsilon$:

1. When $\delta_\varepsilon \to \pi/2$, a situation can be selected where the permittivity has a negative real part and small losses ($\theta_\varepsilon \to \pi/2$) and the permeability has a positive real part and small losses ($\theta_\mu \to 0$). With the proper combination of angles, a negative refraction can be obtained.

2. When $\delta_\varepsilon \to 0$, the permittivity has a small negative real part and high losses which, combined with a permeability with a small positive real part and high losses, can also yield a negative refraction.

In both cases, however, the resulting complex n has a negative real part as desired, but also a very large imaginary part. This configuration therefore corresponds to a medium with high losses where electromagnetic waves are strongly evanescent, and has a limited practicality except for some very specific near-field applications.

5.2.1   Modified Snell's law of refraction

Let us examine the modified Snell's law of refraction more closely. Consider an interface in the xy plane between (for example) vacuum and a medium with some arbitrary homogeneous, isotropic medium with complex material parameters ($\varepsilon = \varepsilon' + i\varepsilon''$ and $\mu = \mu' + i\mu''$). While the medium could also be amplifying in general, we confine ourselves to the discussion of passive, dissipative media only: $\varepsilon'' \geq 0$ and $\mu'' \geq 0$. If we define the incident plane as being xz, the incident wave has a parallel component of the wave-vector k_x, which is conserved upon refraction due to phase matching. The discussion on obtaining the refractive index presented above is a special case (for $k_x = 0$) of the more general problem of choosing the sign of the square root for the normal component of the wave-vector in the medium:

$$ k_z = \pm \left(\varepsilon\mu k_0^2 - k_x^2\right)^{1/2} = \pm \left([\varepsilon'\mu' k_0^2 - \varepsilon''\mu'' k_0^2 - k_x^2] + i[\varepsilon'\mu'' + \varepsilon''\mu']k_0^2\right)^{1/2}, \tag{5.13} $$

where $k_0 = \omega/c$. In order have a propagation in the medium, we assume that the levels of dissipation are low and subsequently can extrapolate to large values if required. In the case of small values of ε'' and μ'':

$$ k_z = \pm\{ [\varepsilon'\mu' k_0^2 - \varepsilon''\mu'' k_0^2 - k_x^2]^{1/2} + i\frac{[\varepsilon'\mu'' + \varepsilon''\mu']k_0^2}{2[\varepsilon'\mu' k_0^2 - \varepsilon''\mu'' k_0^2 - k_x^2]^{1/2}}. \tag{5.14} $$

Note that we have non-propagating evanescent waves with a predominantly imaginary wave-vector if $(\varepsilon'\mu' - \varepsilon''\mu'')k_0^2 - k_x^2 < 0$.

Physical boundary conditions have to be imposed to obtain the correct sign of the wave-vector. The relevant condition here is that the field amplitudes and the energy flow should go to zero as $z \to \infty$ since the medium is dissipative. Hence for propagating waves and $\varepsilon' > 0$ and $\mu' > 0$, it is obvious that the positive sign for the wave-number has to be chosen since it results in the decay of the wave amplitude with z. Similarly, for propagating waves and $\varepsilon' < 0$ and $\mu' < 0$, the negative sign has to be chosen in order to yield waves of decaying amplitude with z if $[\varepsilon'\mu'' + \varepsilon''\mu'] < 0$. It is necessary to look at the limit of zero dissipation to obtain the physically sensible wave-vector in a non-dissipative medium. Obviously in a medium where ε' and μ' have opposite signs, the waves are all evanescent and we again have decaying waves into the medium. Since it is meaningless to associate a direction of propagation to evanescent waves, we do not discuss the choice of the sign for the square root for this case, except to remark that the sign should always be chosen such that the wave decays to zero at the infinities in a dissipative medium.

We note in passing that physically relevant quantities such as the transmission coefficient and the reflection coefficient of a slab of a material are all invariant under the transformation $k_z \to -k_z$. A second point is that a slab can support both evanescent decaying and evanescent growing waves. In the case of media with negative material parameters, it is possible to have some cases where the amplifying waves dominate the solution, which is a scenario

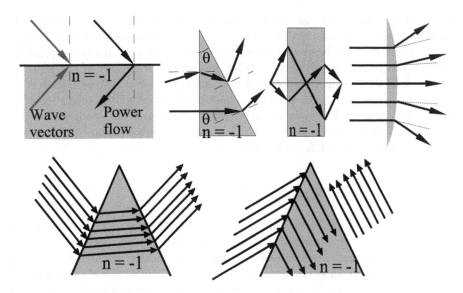

Figure 5.8 Pictorial examples of the modified refraction process for a negative index medium where it is assumed $n = -1$ (Top, from left to right): A planar interface (the phase vectors are shown in grey), a prism, a flat slab and a plano-convex lens. Bottom: Refraction across a wedge. In the second case (right) which can be obtained by decreasing the angle in the first case (left), the rays become tangential to the second interface and the output rays appear as if associated with a source at infinity.

that usually arises due to the resonant excitation of surface waves on the far surface.[§] However, note that the solution in a semi-infinite medium cannot be obtained from the solutions of a finite slab by taking the limit of infinite slab thickness $d \rightarrow \infty$ unless the light has a finite coherence length (or a finite frequency bandwidth).

Finally, Fig. 5.8 presents graphically a few interesting cases of negative refraction. In the first case we have the usual negative refraction at an interface: note the opposite directions of the phase vector and the energy flow (ray). The second panel on the top shows a prism made of a negative material: the negative refraction causes the ray to deflect toward the apex, in contrast to the usual case of refraction toward the base in normal positive materials. The third panel (top) shows a flat lens or imaging device that can focus a source located on one side of the lens to an image on the other side. A convex lens, shown in the fourth panel, made of the negative material behaves as a diverging lens and, by analogy, a concave lens causes rays from infinity

---

[§]See Fig. 8.5 on page 295, for example.

to converge. Monzon et al. (2005) also presented an intriguing case of light incident on a triangular wedge with $n = -1$. One can see that as the angle of incidence changes beyond a point, the incident bundle of rays approaches the tangent to the lower interface. The rays on the other side of the interface appear as if they could be associated with another source located infinitely far away. This scenario arises because negative refraction at the other interface happens at a point infinitely far away from the apex of the wedge. Actually the power flow in this example associated with the rays is similar to the power flow associated with surface plasmons (see Chapter 7).

5.2.2 Reversed Doppler shift

The Doppler effect refers to the frequency shift between the frequency of a source emitting a time harmonic radiation and the one measured by a receiver, when the source and the receiver are in relative motion with respect to one another. The Doppler effect can be easily measured with both electromagnetic waves as well as acoustic waves. In the first case, it is for example responsible for the red shift measured when observing a receding star, whereas in the second case, it is responsible for the change of pitch heard before and after the passing of an ambulance.

Fig. 5.9 gives a simple illustration of the origin of the frequency shift. A source S is moving at a velocity $\mathbf{v}$ while radiating. At subsequent times t_i, the source is at r_i, $i \in \{0, 1, 2, 3\}$. The figure shows the time history of the positions as well as the wave-fronts at the specific time t_4. Hence, for example, the circle marked C_0 is centered at the position r_0 along the path of the source (from which the radiation was originated) with a radius corresponding to the time step t_4. From this simple figure, it is clear that the receiver R_1, from which the source is moving away, measures a decrease in frequency, whereas the receiver R_2, toward which the source is moving, measures an increase in frequency.

This illustration supposes that all the circles are in phase, i.e., that the phase is propagating forward. One can therefore intuitively understand that a reversal of the phase velocity induces a reversal of the Doppler shift: the receivers R_1 and R_2 would measure an increase and a decrease in frequency, respectively. It is a simple exercise to repeat the intuitive illustration of Fig. 5.9 in the case where the phase is propagating backward in the background medium, and to track the circles of identical phase in order to see the reversal of the Doppler shift.

The illustration of Fig. 5.9 is, however, a non-relativistic intuitive explanation, whereas the Doppler shift is a fundamental effect for electromagnetic radiation that is based on the relativistic invariance of the phase. Within this formulation, and for an emission frequency of ω in a background medium characterized by the index of refraction n, the frequency measured by a detector

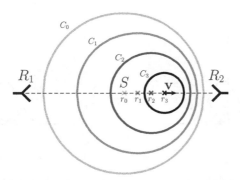

Figure 5.9 Illustration of the Doppler shift: a source S moves at a velocity $\mathbf{v}$ while radiating. The receiver R_1 measures an apparent smaller frequency, whereas the receiver R_2 measures an apparent higher frequency.

considered to be at rest in the frame of the negative medium is

$$\omega' = \gamma\left(\omega + \mathbf{k}\cdot\mathbf{v}\right) = \gamma\left(\omega + \frac{nv\omega}{c}\cos\theta\right), \tag{5.15}$$

where $\gamma = (1 - v^2/c^2)^{-1/2}$ is the relativistic factor and θ is the angle between the wave-vector and the velocity vector of the source. For an emission along the direction of the motion ($\theta = 0$) in a medium with $n = -1$, we therefore obtain

$$\frac{\omega'}{\omega} = \sqrt{\frac{c - v}{c + v}} < 1, \tag{5.16}$$

so that the measured frequency by a detector is confirmed to be smaller when the source is moving toward it. This should be compared to the frequency increase that is measured in a normal medium for which $n > 0$. Again it is the reversed phase vector in the negative refractive index medium that is responsible for this reversed Doppler shift.

Although this effect is often quoted as a fundamental property of left-handed media, it has not been measured within this context because it requires the unrealistic configuration of the entire background supporting a backward wave (the inhabitants of an anti-world, where the normal vacuum would be anti-free-space, would measure a reversed Doppler shift compared to ours). However, as we have seen, an inverse Doppler shift only requires the background medium to support backward waves, which does not limit this effect to left-handed media only. For example, a reversed shift has been measured in the context of negative phase velocity of dipolar spin waves in a Yttrium Iron-Garnet (YIG) magnetic material (Stancil et al. 2004).

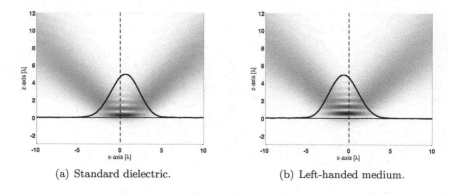

(a) Standard dielectric.   (b) Left-handed medium.

Figure 5.10 Illustration of the Goos-Hänchen shift when a beam impinges onto a standard dielectric and a left-handed medium. The incidence is initially centered at $(x, z) = (0, 0)$. The beam represents the total fields, whereas the one-dimensional curve represents the reflected field only.

5.2.3   Reversed Goos-Hänchen shift

The phenomenon of total reflection occurs for incident angles greater than the critical angle when a wave impinges from a medium onto a less dense medium. In this case, the reflection coefficient is a complex number with unit amplitude (corresponding to the total reflection phenomenon) and a certain phase:

$$R = e^{2i\phi}, \qquad \text{where} \quad \phi = -\tan^{-1}\left(\frac{\mu_0\sqrt{k_x^2 - k_t^2}}{\mu_t k_z}\right), \qquad (5.17)$$

where the subscript "t" indicates the transmitted medium, (k_x, k_z) are the components of the incident wave-vector, and k_t is the transmitted wave-number. The phase ϕ introduces a spatial shift in the reflection of a finite beam, which is known as the Goos-Hänchen shift. For angles of incidence sufficiently away from the critical angle and the grazing angle, the displacement of the beam was shown to be (Artmann 1948)

$$\delta_{\mathbf{gh}} = -\frac{\partial\phi}{\partial k_x} \qquad (5.18)$$

and is illustrated in Fig. 5.10(a) for the case of an incident Gaussian beam onto a standard dielectric half-space. The incident beam is taken centered at $(x, z) = (0, 0)$ whereas it is clear that the reflected field (one-dimensional curve) and the total field (two-dimensional plot) are both shifted to the right. Fig. 5.10(b) illustrates the Goos-Hänchen shift obtained when the beam is impinging onto a left-handed medium half-space. Both in the figure and analytically, it is straightforward to see that the shift is negative in this case.

It is, however, not correct to conclude that left-handed media are necessarily associated with a negative Goos-Hänchen shift. This conclusion holds only for half-spaces, whereas the situation in the case of a slab is more complex because of the presence of the second interface (multi-layer configurations with alternating positive index and negative index media have also been considered in Kim (2005)). The complexity comes from the fact that left-handed media slabs can support growing evanescent waves, so that a surface plasmon resonance can be excited at the second interface, with high field magnitudes that strongly influence the field distribution back in the incident medium. The two limiting cases are as follows:

- When the left-handed medium slab is strongly mismatched to free-space, the growing evanescent waves are only weakly excited, thus not influencing the field distribution in the incident medium. The case is therefore similar to a half-space situation, and the Goos-Hänchen shift is negative.

- When the left-handed medium slab is exactly matched to free-space (i.e., its constitutive parameters are $-\varepsilon_0$ and $-\mu_0$), the surface plasmon is excited at its resonance and the very strong field amplitude at the second interface dominates the field distribution. The Goos-Hänchen shift is in this case positive, as shown in Fig. 5.11 (Grzegorczyk et al. 2005a).

For parameters in-between these two limiting cases, the Goos-Hänchen shift can take either positive or negative values as a function of the incident angle of the impinging beam (Chen et al. 2005a). Note also that the magnitude of the shift can be much enhanced if the operating frequency is close to a resonance of the medium, as pointed out in Wang and Zhu (2005).

5.2.4 Reversed Čerenkov radiation

The reversal of Čerenkov radiation is one of the key properties of left-handed media, already mentioned in 1968 (Veselago 1968). The Čerenkov effect is a relativistic effect whereby a charged particle emits electromagnetic radiation when it travels at a velocity $\mathbf{v}$ larger than the velocity of light in the surrounding medium:

$$|\mathbf{v}| > \frac{c}{|n|}, \tag{5.19}$$

where c is the velocity of light in free-space and n is the refractive index of the medium. This radiation was first observed by Čerenkov in 1934 (although the most commonly cited English papers appeared only in 1937 (Čerenkov 1934)) and later explained theoretically in Frank and Tamm (1937): the wave-vector was shown to be expressed as (Kong 2000) $\mathbf{k} = \hat{\rho}k_\rho + \hat{z}\omega/v$ where $k_\rho = \sqrt{k^2 - \omega^2/v^2}$, and the electric field was shown to be perpendicular to the plane defined by the radiation wave-vector and the velocity vector of

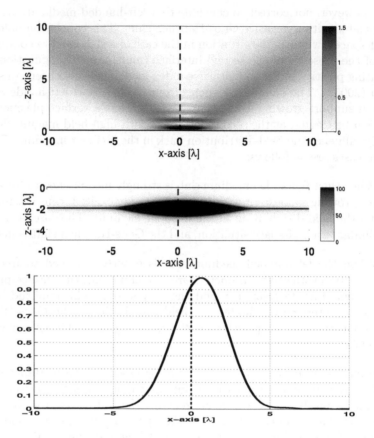

Figure 5.11 Top: Total reflection of a Gaussian beam impinging on a slab of metamaterial above critical angle ($\varepsilon_{2ry} = -0.5$, $\mu_{2rx} = \mu_{2rz} = -0.5$). Center: Surface plasmon excited at the second interface of the slab when the third medium is exactly matched ($\varepsilon_{3ry} = 0.5$, $\mu_{3rx} = \mu_{3rz} = 0.5$). Bottom: Amplitude of the reflected electric field at the first interface showing a displacement to the right, illustrating a positive Goos-Hänchen shift.

the particle. In addition, this radiation exhibits a cylindrical symmetry and creates the well-known Čerenkov cone whose angle θ is given by

$$\cos \theta = \frac{c}{nv}. \tag{5.20}$$

We re-examine these conclusions here within the specific framework of left-handed media, where the index of refraction n is negative. The analysis starts by expressing the current defined by the motion of the charged particle as (Kong 2000)

$$\mathbf{J}(\mathbf{r}, t) = \hat{z}qv\delta(z - vt)\,\delta(x)\,\delta(y), \tag{5.21}$$

where q is the charge of the particle that is assumed to move with a speed v along the $\hat{z}$ direction (hence $\mathbf{v} = \hat{z}v$), and where δ is the Dirac function. The current density thus defined can be transformed into the frequency domain and used in the vectorial wave equation for the electric field. The latter is readily solved using a standard Green's function technique and the separation between the variables ρ and z. In the cylindrical coordinate system, the equation in ρ reduces (still in the frequency domain) to the Poisson equation:

$$\left[\frac{1}{\rho} \frac{\partial}{\partial \rho} \left(\rho \frac{\partial}{\partial \rho} \right) + k_\rho^2 \right] g(\rho) = -\frac{\delta(\rho)}{2\pi\rho}, \tag{5.22}$$

where $g(\rho)$ is yet to be determined and where

$$k_\rho = \frac{\omega}{c} \sqrt{n^2 - \frac{c^2}{v^2}}. \tag{5.23}$$

Eq. (5.22) admits two solutions that are directly related to the two-dimensional Green's function. Since no justification on the choice of the sign of the solution has been given yet, the two solutions are kept and are written as:

$$1) \quad g(\rho) = \frac{i}{4} H_0^{(1)}(k_\rho \rho), \qquad \mathbf{k} = \hat{\rho}k_\rho + \hat{z}k_z, \tag{5.24a}$$

$$2) \quad g(\rho) = -\frac{i}{4} H_0^{(2)}(k_\rho \rho), \qquad \mathbf{k} = -\hat{\rho}k_\rho + \hat{z}k_z. \tag{5.24b}$$

As can be seen, the first set of solutions correspond to outgoing waves whereas the second set correspond to ingoing waves. In order to be able to choose a particular solution, it is necessary to compute the energy W radiated in both the $\hat{\rho}$ and the $\hat{z}$ directions, keeping in mind that the energy radiated in the $\hat{\rho}$ direction is subject to the Sommerfeld radiation condition.

After some manipulations, one finds that the two directed energies can be expressed as follows for the first set of solutions (Lu et al. 2003):

$$W_\rho^{(1)} = \int_{-\infty}^{+\infty} dt\, S_\rho(\mathbf{r},t) = \frac{q^2}{8\pi^2\rho} \int_0^\infty d\omega \frac{k_\rho^2}{\omega\varepsilon}, \tag{5.25a}$$

$$W_z^{(1)} = \int_{-\infty}^{+\infty} dt\, S_z(\mathbf{r},t) = \frac{q^2}{8\pi^2\rho v} \int_0^\infty d\omega \frac{k_\rho}{\varepsilon}, \tag{5.25b}$$

whereas for the second set of solutions:

$$W_\rho^{(2)} = \int_{-\infty}^{+\infty} dt\, S_\rho(\mathbf{r},t) = -\frac{q^2}{8\pi^2\rho} \int_0^\infty d\omega \frac{k_\rho^2}{\omega\varepsilon}, \tag{5.26a}$$

$$W_z^{(2)} = \int_{-\infty}^{+\infty} dt\, S_z(\mathbf{r},t) = \frac{q^2}{8\pi^2\rho v} \int_0^\infty d\omega \frac{k_\rho}{\varepsilon}, \tag{5.26b}$$

where S_ρ and S_z are the $\hat{\rho}$ and $\hat{z}$ components of the Poynting vector, respectively. In a regular medium with $\varepsilon > 0$ and $\mu > 0$, we see that $W_\rho^{(1)} > 0$ and

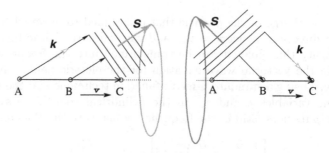

Figure 5.12 The Čerenkov radiation cone angle is modified for a reversed phase vector. There is constructive interference for emission only in one direction as the phase of the wavefront emitted by the particle at A, B, and C has to be identical. The case on the left represents the Čerenkov radiation in a positive medium with positive **k**, while the case on the right shows the emission in a medium with negative **k**. (Reproduced with permission from (Ramakrishna 2005) © 2005, Institute of Physics Publishing, U.K.)

$W_\rho^{(2)} < 0$. Since an energy $W_\rho < 0$ would violate the Sommerfeld radiation condition mentioned above, we conclude that in regular media, the first set of solutions has to be chosen, yielding the Green's function $(i/4)H_0^{(1)}(k\rho)$ as expected. In addition, this situation corresponds to $W_z^{(1)} < 0$, so that the Poynting power and the wave-vector are directed toward the same direction. In a left-handed medium, the situation is reversed: $\varepsilon < 0$ and $\mu < 0$ correspond to $W_\rho^{(1)} < 0$ and $W_\rho^{(2)} > 0$ so that the second set of solutions needs to be chosen, corresponding to the Green's function $(-i/4)H_0^{(2)}(k\rho)$.¶ The energy in the $\hat{z}$ direction is in this case negative ($W_z^{(2)} < 0$), and yields a Poynting vector in the opposite direction to the wave-vector. The Čerenkov cone is in this case still defined by the angle of Eq. (5.20) in which the index of refraction n is negative, as illustrated in Fig. 5.12.

We can attempt to follow the same intuitive approach as in the previous section for the Doppler shift to explain the Čerenkov radiation. As a matter of fact, the starting point is very similar in both cases: a source is moving with a certain velocity and radiates an electromagnetic wave as it progresses. The major difference of course is that in order to have a Čerenkov radiation, the source needs to move faster than the speed of light in the medium. Let us return to Fig. 5.9: it is clear that the source is moving at a velocity slower than the velocity at which the radii of the wave-fronts increase. It is therefore not a relativistic situation with consequently no Čerenkov radiation. Let us then repeat this figure for a source that is moving faster than the rate of

---

¶Note that we still require $\mu < 0$ in order to have a real wavenumber and, thus, a propagating electromagnetic radiation.

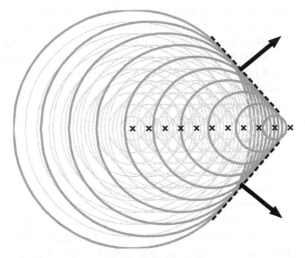

(a) Family of wave-fronts at a given time when the phase is increasing with time (as in a regular medium).

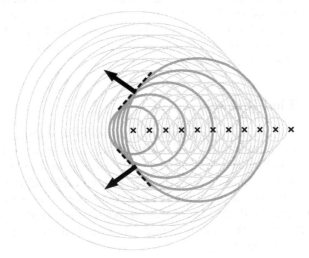

(b) Family of wave-fronts at a given time when the phase is decreasing with time (as in a left-handed medium).

Figure 5.13 Time history of the wave-front as emitted by a source moving at a velocity larger than the velocity of light in the medium.

increase of the radii of the wave-front, which yields Fig. 5.13(a) (note that we also show the wave-fronts at each time step for each discretized position of the source). A family of wave-fronts at a given time is highlighted with darker lines, under the assumption of a positive phase increase as the wave propagates. It is seen that the wave-fronts define a coherent radiation and provide a vivid illustration of the Čerenkov cone. In Fig. 5.13(b), we have repeated the exact same figure as in Fig. 5.13(a) but the family of wave-fronts is selected this time under the assumption of a backward phase propagation. It is clearly seen that the cone is reversed but that it otherwise maintains the same absolute value of its angle. Fig. 5.13 therefore provides a simple and intuitive illustration of the Čerenkov radiation and its reversal depending on the background medium.

Next, we need to examine the effect of frequency dispersion, inherent in left-handed media, on the previous conclusions. Upon examining the spectrum of frequencies and the regions where $n(\omega)^2 > 1$ (a necessary condition to have the possibility of Čerenkov radiation), it is an easy exercise to realize that the previous conclusions indeed hold so that the Čerenkov radiation is indeed reversed in a left-handed medium, as originally predicted in Veselago (1968).

Finally, it is known that frequency dispersion necessarily implies dissipation so that these conclusions should be generalized to lossy media as well. A detailed treatment of the Čerenkov radiation in dispersive and dissipative regular media can be found in Saffouri (1984), and a similar methodology can be reproduced in the case of left-handed media. The derivations are left as an exercise to the interested reader.

5.2.5 Modified Mie scattering

We conclude this section by considering a standard scattering problem generalized to left-handed media: the scattering of a plane wave by a sphere. This problem is well known to be analytically solvable by the Mie theory and has been presented in numerous textbooks (Bohren and Huffman 1983, Kong 2000). The motivation of including it here stems from the necessity of carefully re-examining the derivation of the scattering coefficients, which usually cannot be taken from the standard texts and applied to negative values of permittivity and permeability.

We therefore consider the standard problem of a sphere of radius a and of constitutive parameters (ε_s, μ_s) illuminated by a plane wave which we write in the spherical coordinate system as

$$\mathbf{E} = \hat{x}E_0 e^{ikz} = \hat{x}E_0 e^{ikr\cos\theta} = \hat{x}E_0 \sum_{n=0}^{\infty} (-i)^{-n}(2n+1)j_n(kr)P_n(\cos\theta), \quad (5.27)$$

where j_n is the spherical Bessel function and P_n is the Legendre function. Following the Mie scattering theory, the incident plane wave is further decomposed into Debye potentials π_e and π_m depending on the polarization

(TE to **r** or TM to **r**) so that

$$\mathbf{A} = \hat{r}\pi_e, \tag{5.28a}$$

$$\mathbf{H} = \nabla \times \mathbf{A} = \hat{\theta}\frac{1}{\sin\theta}\frac{\partial}{\partial\phi}\pi_e - \hat{\phi}\frac{\partial}{\partial\theta}\pi_e, \tag{5.28b}$$

for TM waves and

$$\mathbf{Z} = \hat{r}\pi_m, \tag{5.28c}$$

$$\mathbf{E} = \nabla \times \mathbf{Z} = \hat{\theta}\frac{1}{\sin\theta}\frac{\partial}{\partial\phi}\pi_m - \hat{\phi}\frac{\partial}{\partial\theta}\pi_m, \tag{5.28d}$$

for TE waves. The Debye potentials are given by

$$\pi_e = \frac{E_0\cos\phi}{\omega\mu r}\sum_{n=1}^{\infty}\frac{(-i)^{-n}(2n+1)}{n(n+1)}\psi_n(kr)P_n^1(\cos\theta), \tag{5.29a}$$

$$\pi_m = -\frac{E_0\sin\phi}{kr}\sum_{n=1}^{\infty}\frac{(-i)^{-n}(2n+1)}{n(n+1)}\psi_n(kr)P_n^1(\cos\theta), \tag{5.29b}$$

where ψ_n is the Riccati-Bessel function defined as $\psi_n(kr) \equiv krj_n(kr)$.

In order to match the boundary conditions for the tangential electric and magnetic fields, the internal and scattered fields (denoted by the subscripts i and s, respectively) are expanded in a similar fashion:

$$\pi_e^s = \frac{E_0\cos\phi}{\omega\mu r}\sum_{n=1}^{\infty}a_n\xi_n(kr)P_n^1(\cos\theta), \tag{5.30a}$$

$$\pi_m^s = -\frac{E_0\sin\phi}{kr}\sum_{n=1}^{\infty}b_n\xi_n(kr)P_n^1(\cos\theta), \tag{5.30b}$$

$$\pi_e^i = \frac{E_0\cos\phi}{\omega\mu_s r}\sum_{n=1}^{\infty}c_n\psi_n(k_s r)P_n^1(\cos\theta), \tag{5.31a}$$

$$\pi_m^i = -\frac{E_0\sin\phi}{k_s r}\sum_{n=1}^{\infty}d_n\psi_n(k_s r)P_n^1(\cos\theta), \tag{5.31b}$$

where the Riccati-Bessel function $\xi_n(kr) \equiv krh_n(kr)$. In the above definition of the potentials, the Hankel function has been used for the scattered field since the potential has to decay at infinity in order to obey the radiation condition. In the interior domain of the sphere, however, the potential needs to be regular at the origin $r = 0$ so that the Bessel functions are chosen. Note that the potentials in Eqs. (5.29) are entirely determined since they represent the incident field which is a known quantity. The potentials of Eqs. (5.30) and Eqs. (5.31), however, are written as function of unknown coefficients

(a_n, b_n, c_n, d_n) that need to be solved for by applying the boundary conditions for the tangential fields at the boundary of the sphere. Upon doing so, the coefficients are obtained as (Pacheco, Jr. 2004)

$$a_n = \frac{(-i)^{-n}(2n+1)}{n(n+1)} \frac{k\varepsilon_s \psi_n(k_s a)\psi_n'(ka) - k_s \varepsilon \psi_n(ka)\psi_n'(k_s a)}{k_s \varepsilon \xi_n(ka)\psi_n'(k_s a) - k\varepsilon_s \psi_n(k_s a)\xi_n^{(1)'}(ka)}, \quad (5.32a)$$

$$c_n = \frac{(-i)^{-n}(2n+1)}{n(n+1)} \frac{ik\varepsilon_s \mu_s}{k\varepsilon_s \mu \psi_n(k_s a)\xi_n^{(1)'}(ka) - k_s \varepsilon \mu \xi_n(ka)\psi_n'(k_s a)} \quad (5.32b)$$

for the TM waves and

$$b_n = \frac{(-i)^{-n}(2n+1)}{n(n+1)} \frac{k\mu_s \psi_n(k_s a)\psi_n'(ka) - k_s \mu \psi_n(ka)\psi_n'(k_s a)}{k_s \mu \xi_n(ka)\psi_n'(k_s a) - k\mu_s \psi_n(k_s a)\xi_n^{(1)'}(ka)}, \quad (5.33a)$$

$$d_n = \frac{(-i)^{-n}(2n+1)}{n(n+1)} \frac{ik_s \mu_s}{k\mu_s \psi_n(k_s a)\xi_n^{(1)'}(ka) - k_s \mu \xi_n(ka)\psi_n'(k_s a)} \quad (5.33b)$$

for the TE waves.

The importance of these expressions is revealed by examining their symmetry properties with respect to the sign of k_s. Using the properties of the Riccati-Bessel functions, one can show that

$$a_n(-k_s) = a_n(k_s), \quad (5.34a)$$
$$b_n(-k_s) = b_n(k_s), \quad (5.34b)$$
$$c_n(-k_s) = (-1)^{n+1} c_n(k_s), \quad (5.34c)$$
$$d_n(-k_s) = (-1)^n d_n(k_s). \quad (5.34d)$$

From these relations, it can be seen that the potentials of Eqs. (5.30) and Eqs. (5.31) are independent of the sign of k_s (use the fact that $J_n(-x) = (-1)^n J_n(x)$). Consequently, Eqs. (5.32)–(5.33) avoid the necessity of choosing a specific sign for k_s depending on whether the medium is a regular medium or a left-handed medium.

Fig. 5.14 illustrates the difference between the Mie scattering from a regular dielectric sphere and a left-handed medium sphere. Both spheres have the same radius of $1.5\lambda_0$ where λ_0 is the free-space wavelength, with the respective relative constitutive parameters equal to $(2, 1)$ and $(-1, -1)$. In the case of the regular dielectric, the pattern shows the well-known strong forward scattering, illustrating the lensing effect of the sphere. In the case of the left-handed medium sphere, however, the focus point is seen to lie inside the sphere, due to the negative refraction of the rays at the interface of the sphere.

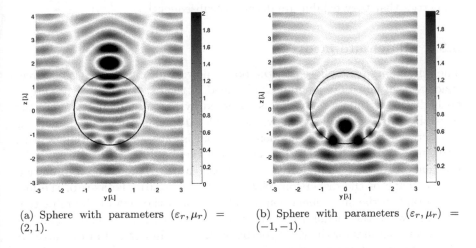

(a) Sphere with parameters $(\varepsilon_r, \mu_r) = (2, 1)$.

(b) Sphere with parameters $(\varepsilon_r, \mu_r) = (-1, -1)$.

Figure 5.14 $|E_x|$ of the total field due to the Mie scattering of a plane wave $\mathbf{E} = \hat{x}e^{ikz}$ by two different spheres. In both cases the sphere has a radius of $1.5\lambda_0$, where λ_0 is the wavelength in free-space. Note the forward focusing with the regular medium and the focusing inside the sphere for the left-handed medium.

5.3 Anisotropic and chiral media

The topic of anisotropic media has already been approached in Section 5.1.2 while discussing the origins of negative refraction. In the present section, we want to generalize this discussion, allowing for some or all of the components of the permittivity and permeability tensors to take negative values. The media considered are characterized by the following constitutive tensors:

$$\bar{\bar{\varepsilon}} = \begin{pmatrix} \varepsilon_x & 0 & 0 \\ 0 & \varepsilon_y & 0 \\ 0 & 0 & \varepsilon_z \end{pmatrix}, \qquad \bar{\bar{\mu}} = \begin{pmatrix} \mu_x & 0 & 0 \\ 0 & \mu_y & 0 \\ 0 & 0 & \mu_z \end{pmatrix}, \qquad (5.35)$$

where the sign of each component is arbitrary (positive or negative). We consider in addition that the electric field is polarized along the $\hat{y}$ direction and the magnetic field is in the xz plane, so that the only relevant parameters are ε_y, μ_x, and μ_z, governed by the dispersion relation of Eq. (5.6). The possibility of having negative constitutive components offers the possibility of achieving hyperbolic dispersion relations, a unique feature for stationary media. Due to the fact that these media are described by tensors that are neither definite positive nor definite negative, they have been termed indefinite media (Smith and Schurig 2003, Smith et al. 2004a). Some of the unique properties that they exhibit are discussed subsequently.

5.3.1 Indefinite media

The refraction of the phase and the power flow at an interface between free-space and an indefinite medium is obtained in the exact same way as previously discussed: phase refraction is governed by phase matching whereas the refraction of the power flow is governed by $\nabla_{\mathbf{k}}\omega(\mathbf{k})$. As usual, it is important to ascertain whether the iso-frequency contour is expanding or shrinking with frequency, and which are the proper axes of the medium with respect to the boundary. Tab. 5.1 summarizes the various cases by providing information on the shape of the dispersion relation from Eq. (5.6), the axes of the foci (important in the subsequent considerations), and the sign of the refraction obtained when the medium boundary coincides with the xy plane. The iso-frequency contours for the two hyperbolic cases are illustrated in Fig. 5.15. It is seen that the standard left-handed medium, corresponding to case (*viii*), is not the only configuration that yields a negative refraction. Yet, as is seen in Fig. 5.15, the other situations do not induce a negative refraction of both the phase and the power, thus emphasizing the uniqueness of left-handed media in this regard.

Table 5.1 Shapes of iso-frequency curves and refraction properties of the power for anisotropic media governed by Eq. (5.6) (Thomas et al. 2005).

Case	ε_y	μ_x	μ_z	Shape	Refraction
i	+	+	+	elliptical	positive
ii	+	+	−	z^*-hyperbolic	negative
iii	+	−	+	x-hyperbolic	positive
iv	+	−	−	imaginary	
v	−	+	+	imaginary	
vi	−	+	−	x-hyperbolic	negative
vii	−	−	+	z-hyperbolic	positive
viii	−	−	−	elliptic	negative

* *Indicates the axis on which the foci lie.*

Indefinite media are directly realizable using the proper orientation of split ring resonators and rods, as discussed in Chapter 3, to achieve all the situations listed in Tab. 5.1. Supposing a direction of propagation along the $\hat{z}$ axis, in the xz plane:

- A negative ε_y can be achieved by introducing a medium of wires along the $\hat{y}$ direction and impinging a polarized electric field along $\hat{y}$ as well. As described in Chapter 3, the medium of wires can be modeled as an

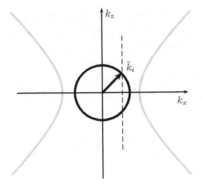

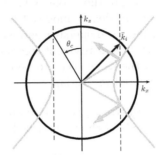

(a) x-hyperbolic case with no phase matched solution.

(b) x-hyperbolic case with phase matched solutions at large incident angles. This case illustrates the inversion of critical angle discussed in Section 5.3.3.

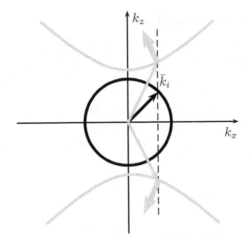

(c) z-hyperbolic case.

Figure 5.15 Iso-frequency contours for x-hyperbolic and z-hyperbolic left-handed media, corresponding to cases *(ii, iii, vi, vii)* in Tab. 5.1.

effective medium with a frequency dispersive permittivity that obeys a Drude model. Thus, below the plasma frequency, the dielectric permittivity takes negative values. Note also that it was described in Section 2.5.2 how a layered stack of silver and silica (layered along the $\hat{z}$ direction) can produce $\varepsilon_x < 0$ and $\varepsilon_z > 0$ or $\varepsilon_x > 0$ and $\varepsilon_z < 0$ in different frequency ranges. The magnitudes can be tuned by the layer thicknesses.

- A negative μ_x is achieved by introducing split rings whose axes are parallel to the $\hat{x}$ axis. Chapter 3 described how such a metamaterial can be equivalently modeled by an effective medium exhibiting a Lorentz frequency dispersive permeability, yielding a negative μ_x between the plasma and resonant frequencies. A proper design for the rod spacing and the rings needs to be found if the regions of $\varepsilon_y < 0$ and $\mu_x < 0$ are expected to overlap.

- A negative μ_z is achieved in a similar fashion as a negative μ_x but with the axis of the rings parallel to the $\hat{z}$ axis.

The combination of the three metamaterial elements above yields all the cases listed in Tab. 5.1. For example, case (*iii*) is obtained by having one set of rings only with axis aligned with $\hat{x}$, while case (*vi*) is obtained by combining wires and rings with axis along $\hat{z}$.

5.3.2 Amphoteric refraction

The previous sections have presented qualitative results on the wave propagation in indefinite media, which are media whose components in the permittivity and permeability tensors can take negative values. In the present section, we are interested in the quantitative study of the same phenomenon, with the purpose of being able to exactly compute the numerical values of the angles of refraction of both the phase and the power.

The systematic study of the iso-frequency contours is obtained by studying the general form of Eq. (5.6) written as

$$\text{Ellipse:} \qquad \frac{k_z^2}{\alpha^2} + \frac{k_x^2}{\beta^2} = 1, \qquad (5.36\text{a})$$

$$\text{Hyperbola:} \qquad \frac{k_z^2}{\alpha^2} - \frac{k_x^2}{\beta^2} = 1, \qquad (5.36\text{b})$$

where

$$\alpha^2 = k_0^2|\varepsilon_y\mu_x|, \quad \beta^2 = k_0^2|\varepsilon_y\mu_z|, \qquad (5.37)$$

the permittivity and the permeabilities being given as relative values. At first, we assume that k_x and k_z are aligned with $\hat{x}$ and $\hat{z}$, which correspond to the situation where the interface of the metamaterial coincides with the direction of its lattice. The rotated case is treated subsequently.

In order to obtain the refraction angles of the phase and the power, we need to generalize Snell's law to the cases of hyperbolic contours (elliptic contours share the same refraction laws since the two cases are related by the transformation $\beta^2 \rightarrow -\beta^2$). Using the convention that angles are defined with respect to the $\hat{z}$ axis and are counted positive if the $\hat{x}$ component of the wave-vector is positive, one obtains after some simple algebra (Grzegorczyk et al. 2005c):

$$\sin \theta_k = \frac{\sin \theta_i}{\sqrt{\varepsilon_y \mu_x + (1 - \mu_x/\mu_z) \sin^2 \theta_i}}, \tag{5.38a}$$

$$\sin \theta_s = \frac{\sin \theta_i}{\sqrt{\varepsilon_y \mu_z - (1 - \mu_x/\mu_z) \sin^2 \theta_i}} \sqrt{\frac{\mu_x}{\mu_z}}. \tag{5.38b}$$

where θ_i is the incident angle, θ_k is the refraction angle of the phase obtained from the phase matching condition, and θ_s is the refraction angle of the power, obtained from the gradient of the iso-frequency contours. Eq. (5.38a) is the refraction law of the wave-vector for positive (i.e., expanding with frequency) elliptic and hyperbolic dispersion relations. The angles for the dual cases are obtained from the transformation $\theta_k(\theta_i) \to \pi - \theta_k(\theta_i)$. Similarly, Eq. (5.38b) is the power refraction laws for the positive elliptic and negative hyperbolic cases while the dual cases are obtained from the transformation $\theta_s(\theta_i) \to -\theta_s(\theta_i)$.

The case of rotated dispersion relations is accounted for by redefining the permittivity and permeability as

$$\bar{\bar{\varepsilon}}' = \bar{\bar{T}}(\phi) \cdot \bar{\bar{\varepsilon}} \cdot \bar{\bar{T}}^{-1}(\phi), \tag{5.39a}$$

$$\bar{\bar{\mu}}' = \bar{\bar{T}}(\phi) \cdot \bar{\bar{\mu}} \cdot \bar{\bar{T}}^{-1}(\phi), \tag{5.39b}$$

where ϕ is the rotation angle illustrated in Fig. 5.16 and $\bar{\bar{T}}$ is the rotation matrix about the $\hat{y}$ axis. Under the limitation of studying ε_y and (μ_x, μ_z), the rotation leaves the permittivity unchanged ($\varepsilon'_y = \varepsilon_y$) but $\bar{\bar{\mu}}$ loses its diagonal property if $\mu_x \neq \mu_z$. If the two components of the permeability are equal ($\mu_x = \mu_z$), however, the dispersion relation reduces to a circle and the refraction is obviously not affected by any rotation.

The determination of the refraction laws is a mathematical exercise that can be approached from various directions, either geometrical or analytical. The choice of the demonstration method is left to the discretion of the reader, and the final result is written as (Grzegorczyk et al. 2005c)

$$\tan \theta_k = \frac{\sin \theta_i (\mu_x \sin^2 \phi + \mu_z \cos^2 \phi)}{(\mu_z - \mu_x) \sin \theta_i \sin \phi \cos \phi + \Delta \sqrt{\mu_x \mu_z}}, \tag{5.40a}$$

$$\tan \theta_s = \frac{(\mu_x - \mu_z) \Delta \sin \phi \cos \phi + \sqrt{\mu_x \mu_z} \sin \theta_i}{\Delta (\mu_x \sin^2 \phi + \mu_z \cos^2 \phi)}, \tag{5.40b}$$

$$\Delta = \sqrt{\varepsilon_y (\mu_x \sin^2 \phi + \mu_z \cos^2 \phi) - \sin^2 \theta_i}. \tag{5.40c}$$

The numerical computation of the angles depends on their domains of definition, either $[0, 2\pi]$ or $[-\pi, \pi]$, the latter being chosen here. In addition, the square roots need to be carefully chosen: like in the case of homogeneous left-handed media, the square root should be taken in the complex plane and in

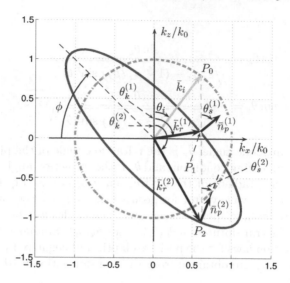

Figure 5.16 Illustration of the iso-frequency contour when it corresponds to a rotated ellipse by an angle ϕ. The two rotation angles of the wave-vector and of the Poynting vector are identified by $\theta_k^{(i)}$ and $\theta_s^{(i)}$, respectively, $i = 1, 2$. (From Grzegorczyk et al. (2005c).© 2005 IEEE).

particular, $\sqrt{\mu_x \mu_z}$ yields a negative number if $\mu_x < 0$ and $\mu_z < 0$. Moreover, one should note that depending on the numerical evaluation of the arctangent function, the numerators and denominators in Eqs. (5.40) might have to be simultaneously multiplied by i or (-1). Despite these numerical adjustments, Eqs. (5.40) can be viewed as the generalized Snell's laws of the phase and power for the indefinite media and provide the value of the refracted angles of the phase and of the power, as function of the incident angle and the material parameters. Consequently, the negative refraction qualitatively explained previously on the basis of iso-frequency curves is now exactly quantified. Some examples are provided in Fig. 5.17 for elliptic contours and in Fig. 5.18 for hyperbolic contours. In both cases, it is seen that the phase is always refracted positively as expected from the iso-frequency arguments (i.e., θ_k is positive/negative when θ_i is positive/negative, respectively), whereas the power may be refracted negatively (i.e., θ_s is positive/negative when θ_i is negative/positive, respectively). In particular, all the cases of Tab. 5.1 can be verified, which is left as an exercise.

We conclude this section by considering a prism configuration, which has been historically used for the first demonstration of the negative refraction obtained with metamaterials (Shelby et al. 2001b). The prism configuration is essentially a two-interface problem: the first interface coincides with the principal axes of the metamaterial, whereas the second interface is cut at an

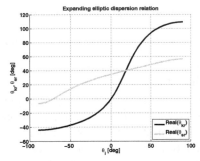

(a) Expanding contour with frequency.   (b) Shrinking contour with frequency.

Figure 5.17 Curves of $\theta_k(\theta_i)$ and $\theta_s(\theta_i)$ for an elliptic contour, either expanding or shrinking. Parameters are: $\alpha = 0.5k_0$, $\beta = 1.5k_0$, $\phi = -40°$.

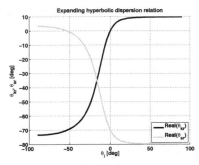
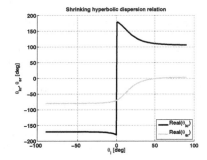

(a) Expanding contour with frequency.   (b) Shrinking contour with frequency.

Figure 5.18 Curves of $\theta_k(\theta_i)$ and $\theta_s(\theta_i)$ for a hyperbolic contour either expanding or shrinking. Parameters are: $\alpha = 0.5k_0$, $\beta = 0.5k_0$, $\phi = 35°$.

angle. In terms of iso-frequency contours, this implies no rotation at the first interface and a rotation at the second (Smith et al. 2004a, Thomas et al. 2005). Nonetheless, this rotation is not arbitrary: by construction, the angle of incidence of the wave at the second interface is equal to the physical angle of the prism (see Fig. 5.8), thus exactly opposite to the rotation of the iso-frequency contour: $\theta_k = \theta_s = -\phi$.[||] This very special incidence on the iso-

---

[||] Note that in Eqs. (5.40), the incidence is from free-space to the left-handed medium, so that θ_i is in free-space, whereas θ_k and θ_s are inside the left-handed medium. At the second interface of the prism, however, the incidence is from the left-handed medium into free-space. The known angles are therefore θ_k and θ_s, whereas θ_i (the transmitted angle to free-space) needs to be solved for.

frequency contour drastically simplifies the refraction law which becomes

$$\sin\theta_i = \pm\sqrt{\varepsilon_y\mu_x}\,\sin\phi, \tag{5.41}$$

where the $\pm$ sign refers to expanding and shrinking contours. Eq. (5.41) is simple enough to be assimilated to a modified Snell's law with a redefined index of refraction $n = \sqrt{\varepsilon_y\mu_x}$. Remarkably, the refraction law does not depend on μ_z, which explains the similarity between the results obtained with one-dimensional and two-dimensional prism configurations. This independence is directly visible from the iso-frequency contour: the very specific incidence maps onto the lower point of the hyperbola (a similar argument holds for the ellipse) which depends on α only, thus on ε_y and μ_x only.

5.3.3   Reversal of critical angle and Brewster angle

This section focuses on cases (*iii*) and (*vi*) in Tab. 5.1, where the iso-frequency curves are hyperbolic with foci on the $\hat{x}$ axis. Interestingly, the phenomenon of refraction is inverted compared to standard dielectrics:

1. At small incidences (close to normal), there is no real solution for the transmitted wave-vector and the field is totally reflected.

2. At high incidences (close to grazing), a phase matched solution exists and the field is transmitted with a certain refraction angle.

This phenomenon is in drastic contrast with the refraction at interfaces with standard dielectrics, where the incident field is transmitted into the second medium for low incidences and may reach a critical angle at higher incidences beyond which the field is totally reflected. If we think of this latter situation as a high-angle filter, the former is a low-angle filter.

In addition to the inversion of critical angle, the Brewster angle (corresponding to no reflection) is also inverted: when it exists, it occurs beyond the critical angle, whereas it occurs of course below the critical angle in standard dielectrics. This inversion of the critical angle and the Brewster angle strongly depends on the proper combination of positive and negative signs in the constitutive tensors.

A quantitative analysis of this phenomenon requires the analytical expressions of the Fresnel reflection (and transmission) coefficients, for example using the method presented in Section 2.6. In the case of a one-interface problem (i.e., a half-space problem denoted by the subscript "hs"), these coefficients are given by Eqs. (2.98) on page 63, rewritten here as:

$$R_{\text{hs}} = \frac{k_{z1}\mu_{x2} - k_{z2}\mu_1}{k_{z1}\mu_{x2} + k_{z2}\mu_1}, \quad T_{\text{hs}} = \frac{2k_{z1}\mu_{x2}}{k_{z1}\mu_{x2} + k_{z2}\mu_1}, \tag{5.42a}$$

$$k_{z1} = \sqrt{\omega^2/c^2 - k_x^2}, \quad k_{z2} = \sqrt{\left(\frac{\omega^2}{c^2}\varepsilon_{y2}\mu_{z2} - k_x^2\right)\frac{\mu_{x2}}{\mu_{z2}}}, \tag{5.42b}$$

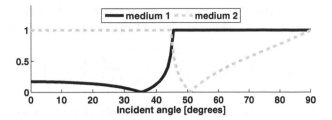

Figure 5.19 Absolute value of the Fresnel reflection coefficient at the interface between free-space and two different media defined by *Medium 1*: $(\varepsilon_{2y} = 1, \mu_{2x} = \mu_{2z} = 0.5)$ and *Medium 2*: $(\varepsilon_{2y} = -1, -\mu_{2x} = \mu_{2z} = -0.5)$.

where the subscripts "1" and "2" refer to the free-space and the biaxial medium, respectively. Setting the reflection to 1 or 0 yield the critical angle θ_c and the Brewster angle θ_b, respectively:

$$\theta_c = \sin^{-1}\left(\sqrt{\frac{\epsilon_{y2}\mu_{z2}}{\epsilon_1\mu_1}}\right), \tag{5.43a}$$

$$\theta_b = \sin^{-1}\left(\sqrt{\frac{\mu_{z2}(\epsilon_{y2}\mu_1 - \epsilon_1\mu_{x2})}{\epsilon_1(\mu_1^2 - \mu_{x2}\mu_{z2})}}\right). \tag{5.43b}$$

When these two angles exist, the analytical formulae confirm their relationship (greater or smaller): $\theta_c > \theta_b$ when all the parameters are positive, $\theta_c < \theta_b$ when the appropriate parameters are negative. This inversion is illustrated in Fig. 5.19. The first medium represented is described by only positive constitutive components, yielding a medium less dense than the incident free-space. In addition, the parameters have been chosen so that a Brewster angle exists as well. As can be seen, the Brewster angle occurs for an incidence of about 35.3 degrees, whereas the critical angle occurs beyond, at 45 degrees. The second medium represented in Fig. 5.19 is an x-hyperbolic medium that is also less dense than the incident free-space. The Brewster angle is seen to occur at about 50.8 degrees, whereas the critical angle occurs before, again at 45 degrees.

These considerations can be extended to a slab configuration, which is more suitable for experimental measurements (the wave does not need to be measured inside the metamaterial but can be measured outside). The reflection and transmission coefficients, identified by the subscript "slab," are given in this case by

$$R_{\text{slab}} = \frac{R_{\text{hs}}(e^{2\Phi} - 1)}{R_{\text{hs}}^2 e^{2\Phi} - 1}, \qquad T_{\text{slab}} = \frac{R_{\text{hs}}^2 - 1}{R_{\text{hs}}^2 e^{2\Phi} - 1}, \tag{5.44}$$

where $\Phi = ik_{z1}d$. The condition $R_{\text{slab}} = 0$ is still fulfilled with $R_{hs} = 0$, indicating that the Brewster angle for the half-space is also a Brewster angle for the slab. Additional Brewster angles are obtained by letting $(e^{2\Phi} - 1) = 0$, which produces the family of angles defined by Grzegorczyk et al. (2005d):

$$\theta_{b_m} = \sin^{-1}\left(\sqrt{\frac{\epsilon_{y2}\mu_{z2}}{\epsilon_1\mu_1} - \frac{\mu_{z2}}{\mu_{x2}}\frac{(m\pi/d)^2}{(\omega^2/c^2)\epsilon_1\mu_1}}\right), \tag{5.45}$$

where m = 1,2,3,... (note that m = 0 is excluded since it yields the critical angle). A study of Eq. (5.45) easily reveals that when it exists, θ_{b_m} occurs after θ_c in the case shown in Fig. 5.15(b).

Note that the experimental verification of the inversion of critical angle can be performed on a simple metamaterial based on wires to produce $\epsilon_y < 0$ and on rings perpendicular to the direction of propagation to produce $\mu_z < 0$. As no rings are present along the other direction, the material has a $\mu_x > 0$. Experimental results showing the inversion of the critical angle have been demonstrated, for example, in Grzegorczyk et al. (2005d).

5.3.4   Negative refraction due to bianisotropic effects

As has been seen and demonstrated in the previous sections, the recipe to obtain a negative refraction of the power is to design a metamaterial whose dispersion relation presents a proper gradient as function of frequency, within a certain frequency band. The knowledge of the dispersion relation of a metamaterial is therefore fundamentally important and holds the key to the understanding of the refraction phenomena.

With this understanding and mathematical tool, let us look back at the material constituted by the well-known square split ring resonators of Fig. 2.8 on page 66. It has been argued in Chapter 2 that this material can be described by bianisotropic constitutive parameters of the form of Eqs. (2.101) on page 64, also repeated here for convenience (Marques et al. 2002):

$$\bar{\bar{\epsilon}} = \begin{bmatrix} \epsilon_x & 0 & 0 \\ 0 & 1 & 0 \\ 0 & 0 & \epsilon_z \end{bmatrix}, \qquad \bar{\bar{\mu}} = \begin{bmatrix} 1 & 0 & 0 \\ 0 & \mu_y & 0 \\ 0 & 0 & 1 \end{bmatrix}, \tag{5.46a}$$

$$\bar{\bar{\xi}} = \begin{bmatrix} 0 & 0 & 0 \\ 0 & 0 & 0 \\ 0 & -i\xi & 0 \end{bmatrix}, \qquad \bar{\bar{\zeta}} = \begin{bmatrix} 0 & 0 & 0 \\ 0 & 0 & i\xi \\ 0 & 0 & 0 \end{bmatrix}. \tag{5.46b}$$

The dispersion relation inside such medium has been derived as Eq. (2.103) on page 65 and is written as

$$\frac{\epsilon_z}{\epsilon_x}k_{z2}^2 = \frac{\omega^2}{c^2}(\epsilon_z\mu_y - \xi^2) - k_x^2. \tag{5.47}$$

This equation is similar to Eq. (5.6) for biaxial media with an additional ξ factor. The discussion on biaxial media can therefore be extended to the present case, and the iso-frequency contours are expected to be ellipses or hyperolae as function of the signs of the various terms. An important difference, however, is that we need to look at the sign of $(\varepsilon_z\mu_y - \xi^2)$ instead of the sign of $\varepsilon_z\mu_y$ only as in the biaxial case. It is immediately seen that if the sign of $\varepsilon_z\mu_y$ is positive, the sign of $(\varepsilon_z\mu_y - \xi^2)$ is governed by the bianisotropic parameter: a low bianisotropy does not change the shape of the iso-frequency curve while a strong bianisotropy does.

This situation is depicted in Fig. 5.20 for a hypothetical case where $\varepsilon_x = 1.5$, $\varepsilon_z = -1.5$, and $\mu_y = -1.5$. The cutoff value for ξ is therefore 2.25. Fig. 5.20(a) shows the iso-frequency contour when $\xi = 0$, which reduces to the known case of a biaxial medium. The iso-frequency contour is similar to the cases of Fig. 5.15(a) and Fig. 5.15(b). Taking an interface along the $\hat{x}$ axis, low incidences yield no phase-matched solutions, while high incidences might yield a solution if the free-space circle is sufficiently large. If, however, the metamaterial is realized with split rings that exhibit the bianisotropy of Eqs. (5.46) and if this bianisotropy is strong enough, the refraction phenomenon is totally different. Fig. 5.20(b) illustrates the iso-frequency contour for a medium with the same parameters as before but with $\xi = 2$. The contour is seen to still be hyperbolic but with its foci along an axis rotated by $\pi/2$ compared to the other case. This situation corresponds to an always phased-matched situation, and additionally exhibits a negative refraction that is illustrated in Fig. 5.21.

The proper characterization of a split-ring resonator in terms of constitutive tensors is therefore seen to be fundamentally important: failing to realize that a ring induces a bianisotropic behavior might lead to the expectation that the behavior will follow Fig. 5.20(a), whereas reality might be governed by Fig. 5.20(b).

In addition to the bianisotropy of Eqs. (5.46), chirality has been proposed as a way of achieving negative refraction (Pendry 2004a). A chiral medium is a medium that exhibits scalar bianisotropic constitutive parameters, so that the constitutive relations are written as

$$\mathbf{D} = \varepsilon\mathbf{E} + i\xi\mathbf{H}, \tag{5.48a}$$
$$\mathbf{B} = -i\xi\mathbf{E} + \mu\mathbf{H}. \tag{5.48b}$$

This medium supports two circularly polarized waves propagating at different velocities and can therefore induce the rotation of the wave-vector of an impinging linearly polarized wave. This rotation is referred to as optical activity, and is reciprocal (unlike the rotation induced by gyrotropic media (Agranovich et al. 2005), the Faraday rotation, which is non-reciprocal).

The dispersion relation for this material is given by

$$k = \pm\frac{\omega}{c}(\sqrt{\varepsilon\mu} \pm \xi). \tag{5.49}$$

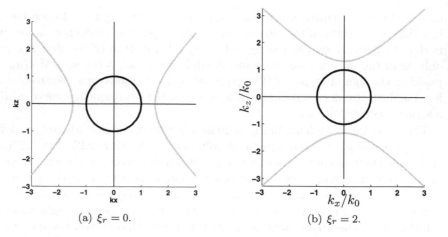

(a) $\xi_r = 0$.   (b) $\xi_r = 2$.

Figure 5.20 Iso-frequency contour of a medium governed by Eqs. (5.46) with relevant parameters being $\varepsilon_x = 1.5$, $\varepsilon_z = -1.5$, $\mu_y = -1.5$. The first case ($\xi = 0$) yields no phase matching along an $\hat{x}$-directed interface, whereas the second case ($\xi = 2$) yields an always phased-matched configuration that also exhibits negative refraction.

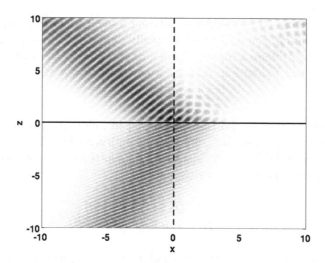

Figure 5.21 Illustration of the negative refraction of a Gaussian beam induced by a strong bianisotropy. The iso-frequency contour of the second medium is identical to Fig. 5.20(b).

For $\xi = 0$, the dispersion relation of course becomes the one of an isotropic medium so that the effect of $\xi \neq 0$ is to lift the degeneracy and split the modes. Fig. 3.27 on page 132 illustrates the band diagram (totally equivalent to the dispersion relation) of a medium for which the permittivity is frequency dispersive of the form

$$\varepsilon_r = 1 + \frac{\omega_0^2}{\omega_0^2 - \omega^2}. \tag{5.50}$$

When $\xi = 0$, the wave-number is real and propagating modes are sustained below the resonant frequency and above the plasma frequency (equal to $\sqrt{2}\omega_0$). When $\xi \neq 0$, however, the modes split and a region of negative group velocity (negative slope) and positive phase velocity appears, a signature of negative refraction. The point where $k = 0$ is particularly interesting on this diagram. Apart from the trivial solution $\omega = 0$, the point $k = 0$ corresponds to a situation where the phase velocity is infinite but the group velocity is finite, which is of particular interest for some antenna applications where multiple elements must be fed with a similar phase. Note that more details on chiral metamaterials can also be found in Section 3.4.

5.3.5 Flat lenses with anisotropic negative media

We conclude this section with a discussion on the possibility of realizing a flat lens using an anisotropic left-handed medium, whereas a more extensive discussion on the lensing effect using metamaterials is postponed until Chapter 8.

A lens is traditionally thought of as a dielectric device that refracts light and concentrates beams toward a focal point or plane, providing the well-known imaging capabilities. Lenses are usually made of glass or other dielectric materials, and can have various shapes such as biconvex and biconcave, plano-convex and plano-concave, convex-concave or meniscus. A common feature of all these configurations is that they all present at least one curved interface, whereas a lens with two flat interfaces (equivalent to a slab) would have no focusing effect on the propagating beam.

This was true until the advent of metamaterials and the possibility of achieving negative constitutive parameters. The negative refraction thus produced allows a lens to operate in two possible regimes. The first one is standard in which a beam parallel to the axis of the lens is focused to a point: negative refraction allows in this case for more design parameters but still requires at least one curved surface (Greegor et al. 2005). The second regime is proper to left-handed media lenses, in which a point source located at a distance ℓ from the first interface is refocused at a distance $\ell + 2d$, where d is the thickness of the flat lens. This lens was first proposed in Veselago (1968) where it was shown that its constitutive parameters have to be exactly anti-matched to those of the surrounding background in order for the refocusing to occur. Hence, if the left-handed medium lens is surrounded by free-space characterized by the permittivity ε_0 and the permeability μ_0, the lens must have a permittivity of $(-\varepsilon_0)$ and a permeability of $(-\mu_0)$.** Under this condition, and provided that the lens is thick enough (the thickness needs to be greater than the distance between the first interface and the source), all the rays emanating from the point source can be bent with the exact proper angles to form an image inside as well as an image outside the lens, as shown in Fig. 5.8. Note that it is assumed that the lens is of infinite extent in the lateral direction, an assumption that is examined in further detail in Section 8.3.2 on page 305.

The fundamental principle by which the lens proposed in Veselago (1968) can refocus a point source is the phenomenon of negative refraction. One is therefore left wondering if other media, which also exhibit a negative refraction, could be used to realize a flat lens. In particular, Tab. 5.1 has identified two other metamaterials, anisotropic, that exhibit a negative refraction: when $\mu_z < 0$ and all other parameters are positive, and when $\mu_x > 0$ and all other parameters are negative. The latter case is immediately excluded because it presents a cutoff angle below which rays are totally reflected. The former case, however, does not present such disadvantage and is examined in further detail subsequently.

The main difference between the isotropic and anisotropic $\mu_z < 0$ metamaterials is the dispersion relation: circular in the first case and obeying the standard Snell's law with a negative refractive index, it becomes hyperbolic in the second case and obeys the generalized Snell's laws derived in Section 5.3.2. As much as a circular iso-frequency contour yields exactly matching angles and refocuses all rays to a single point, a similar phenomenon does not happen with a hyperbolic contour so that it is expected that the image either does not exist or is smeared out. In fact, both phenomena can happen, as function of μ_z. If μ_z is too large, the hyperbola becomes too flat and the negative refraction of the power is not sufficient to fold back the rays onto themselves. This can be seen by comparing the ray tracing figures as well as the evolution of θ_s between Fig. 5.22 (for the case of $\mu_z = -1$ also examined in Smith et al. (2004b)) and Fig. 5.23 (for the case of $\mu_z = -2$). In the latter case, θ_s is smaller for a similar θ_i than in the former case and the rays indeed do not converge to form an image either inside the flat lens or outside. In the case of $\mu_z = -0.5$ shown in Fig. 5.24, however, the concavity of the hyperbola is much enhanced and a stronger negative refraction is obtained, clearly refocusing the rays both inside and outside the lens. Note, however, that this focusing is never perfect and the image formed is seriously aberrated: some

---

** Variations away from these ideal values directly impact the quality of the refocused image. In particular, it has been shown in Section 1.3.1 that the causality condition prevents the constitutive parameters to be negative and purely real, i.e., a left-handed medium is necessarily lossy. Consequently, a medium with constitutive parameters of $(-\varepsilon_0, -\mu_0)$ is not physical and the metamaterial lens must be intrinsically lossy. Discussion on the impact of mismatch parameters on the image quality is postponed to Chapter 8.

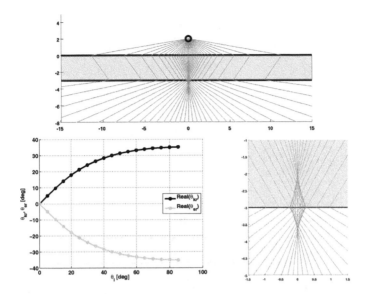

Figure 5.22 Top and bottom right: Ray tracing from a point source atop a flat lens. Bottom left: Relation between the incident angle θ_i and the refracted angles (θ_k, θ_s). Flat lens parameters: $(\varepsilon_{ry}, \mu_{rx}, \mu_{rz}) = (1, 1, -1)$. A small focused image close to the second boundary can be seen.

rays are either not refocused or are not in phase at the expected location of the image, and the resolution is worsened compared to the isotropic case with $(-\varepsilon_0, -\mu_0)$ (Dumelow et al. 2005, Grzegorczyk et al. 2005b).

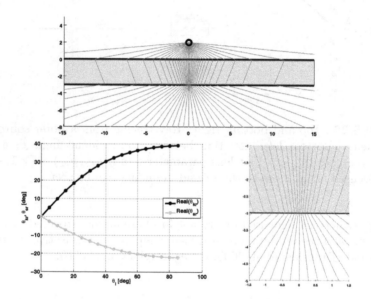

Figure 5.23 Top and bottom right: Ray tracing from a point source atop a flat lens. Bottom left: Relation between the incident angle θ_i and the refracted angles (θ_k, θ_s). Flat lens parameters: $(\varepsilon_{ry}, \mu_{rx}, \mu_{rz}) = (1, 1, -2)$. The rays are seen not to focus and the flat lens does not produce any image.

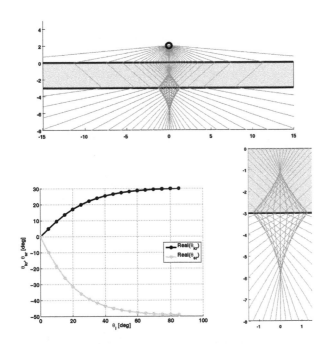

Figure 5.24 Top and bottom right: Ray tracing from a point source atop a flat lens. Bottom left: Relation between the incident angle θ_i and the refracted angles (θ_k, θ_s). Flat lens parameters: $(\varepsilon_{ry}, \mu_{rx}, \mu_{rz}) = (1, 1, -0.5)$. The rays are seen to focus farther away from the second boundary compared to the case of Fig. 5.22

6

Energy and momentum in negative refractive index materials

The previous chapter has extensively addressed properties of media exhibiting a negative permittivity and permeability, and has led us to revisit some of the most common effects of refraction, radiation, and propagation. In this chapter, we return to a somewhat more theoretical topic: the propagation of energy in left-handed media, with its immediate counterpart, the transfer of momentum from the propagating electromagnetic wave to matter and the traversal of pulses through such media.

The study of energy propagation is deeply rooted in the study of left-handed media since negative constitutive parameters can lead to paradoxical conclusions if not considered carefully. Questions such as *where does the energy go*; *in which direction does the phase propagate*; or *can a signal propagate in such media without violating fundamental physical laws* were therefore addressed within the framework of left-handed media and frequency dispersive media in general.

We start this chapter with a study of energy in frequency dispersive media, keeping causality in perspective. Upon including losses in the media, we emphasize that the energy has to be treated with care, since quantities such as the energy density are not thermodynamically well defined when radiation interacts with a lossy dissipative medium. The description of energy in such media therefore either necessitates the specification of the micro-structure of the medium which implies a model for dispersion of the material response functions, or the specification of the integrals over the time history of the interactions. Within this chapter, we specifically focus on the energy dynamics of radiation in a lossy medium idealized to be an assembly of independent oscillators, yielding a Lorentz type of dispersion for both the permittivity and the permeability.

In parallel, the transfer of momentum from an electromagnetic wave to matter is formulated within the classical framework of the Maxwell stress tensor and the Lorentz force. Two physical subsystems are considered: the electromagnetic subsystem constituted by the propagating electric and magnetic fields governed by the Maxwell equations, and the matter subsystem described by the type of material the wave is propagating through. We specifically identify the contribution of the electromagnetic wave and the contribution of matter to the total momentum, a distinction closely related to the Abraham-

Minkowski debate initiated in the 1920s (Minkowski 1908, Abraham 1909). We finally conclude this discussion by applying the theory again to a medium governed by a Lorentz dispersion relation.

When considering physical subsystems, one is often interested in knowing if energy or momentum is conserved. This is not to say, of course, that energy is not conserved in general, but that it may not be when one restricts the consideration to specific subsystems. Mathematically, the equations of conservation of energy and momentum are cast in the most general form as

$$\nabla \cdot \mathbf{S} + \frac{\partial W}{\partial t} = -\phi \,, \tag{6.1a}$$

$$\nabla \cdot \bar{\bar{T}} + \frac{\partial \mathbf{G}}{\partial t} = -\mathbf{f} \,, \tag{6.1b}$$

where $\mathbf{S}$ and $\bar{\bar{T}}$ denote the energy and momentum flow, respectively, while W and $\mathbf{G}$ denote the energy density and momentum density, respectively. The quantities ϕ and $\mathbf{f}$, if non-zero, imply that the subsystem considered is open, i.e., that energy and/or momentum is transferred to or from another subsystem that has not been considered. Hence, writing Eq. (6.1b) with $\mathbf{f} \neq \mathbf{0}$ in fact implies that the momentum is *not* conserved, but part of it is transmitted to matter in the form of a force able to do work (Gordon July 1973). This is in contrast with other subsystems for which the right-hand side of Eq. (6.1) is zero. An immediate example in electromagnetics is the well known charge conservation law

$$\nabla \cdot \mathbf{J} + \frac{\partial \rho}{\partial t} = 0 \,. \tag{6.2}$$

In the following sections, we study the propagation of energy in left-handed media and cast their equations to follow the form of Eqs. (6.1).

6.1 Causality and energy density in frequency dispersive media

6.1.1 Causality in left-handed media

Causality is a simple but very fundamental principle that states that the cause should precede the effect: a physical quantity may depend on other physical quantities evaluated in the past but not in the future. A restatement of this principle within the context of special relativity is that information cannot travel faster than the speed of light in vacuum (c).

This principle has deep consequences in electromagnetics: apart from the finiteness of the speed of light, it also implies restrictions on the functions that describe the frequency variations of the permittivity and the permeability. It is indeed well known that there is a non-local connection (in time) between the electric flux $\mathbf{D}(t)$ and the electric field $\mathbf{E}(t)$ (See Eq. (1.5)), where the Fourier components of these quantities are related by the frequency-dependent permittivity function (Landau et al. 1984, Jackson 1999). Similar arguments, of course, hold for the magnetic permeability that relates $\mathbf{B}(t)$ and $\mathbf{H}(t)$. As a consequence of this non-instantaneous response and of the causality principle, the real and the imaginary parts of the permittivity (and of the permeability) are related by the Kramers-Kronig relations (see also Eqs. (1.9a) on page 17)

$$\epsilon'(\omega)/\epsilon_0 = 1 + \frac{1}{\pi} PV \int_{-\infty}^{\infty} d\omega' \frac{\epsilon''(\omega')/\epsilon_0}{\omega - \omega'}, \tag{6.3a}$$

$$\epsilon''(\omega)/\epsilon_0 = -\frac{1}{\pi} PV \int_{-\infty}^{\infty} d\omega' \frac{\epsilon'(\omega')/\epsilon_0 - 1}{\omega - \omega'}, \tag{6.3b}$$

where PV denotes the principal value of the integral. Eqs. (6.3) are theoretical relations that have an immediate application in practice. For example, knowledge of the imaginary part of the permittivity from absorption measurements across frequencies allows for the determination of the real part via Eq. (6.3a), as illustrated in Fig. 1.8 on page 18.

Within the framework of left-handed media, two important consequences should be mentioned. First, Eqs. (6.3) impose restrictions on the models that are used for $\epsilon(\omega)$ and $\mu(\omega)$. As a matter of fact, one needs to derive or assume functions that do not violate causality in the relation between $\mathbf{D}(t)$ and $\mathbf{E}(t)$. The Lorentz model, commonly used to model left-handed media as we have seen in Chapter 1 and to which we shall return in greater length subsequently in this chapter, is one such example. The second important consequence of Eq. (6.3) is that lossless left-handed media are unphysical and are merely a theoretical idealization that should be understood as such (Eq. (6.3b) indicates that the imaginary part of $\epsilon(\omega)$ cannot be zero if the real part is non-zero). In particular, the assumption of lossless left-handed media breaks down at frequencies that are close to the resonance of the permittivity and of the permeability.

6.1.2 Causality and phase propagation

Let us consider a more intuitive aspect of the causality principle and its consequences on left-handed media: signals carried by an electromagnetic wave cannot travel faster than the speed of light. This principle was erroneously used to argue that negative refraction is impossible, with the following argument (Valanju et al. 2002): consider a plane wave impinging on a boundary between two semi-infinite media, one being free-space and the other one being case dependent. When the second medium is a standard dielectric, the phase fronts of the electric field look like those shown in Fig. 6.1(a). Consider that there is a time variation (e.g. one period) between points A_i and B_i, then again one period between B_i and C_i. At a given time step, A_1 and A_2 are on the same phase front, and become B_1 and B_2 at the next time step, then C_1 and C_2 at the following time step. Throughout this process, the phase front is simple to track and obeys the standard Snell's refraction law of positive refraction. The situation of a left-handed medium is somewhat different, schematically shown in Fig. 6.1(b). The points with subscripts 2 behave unambiguously like in the previous case: A_2 becomes B_2 which in turn becomes C_2 at the next time increment. The points with subscripts 1 are apparently more problematic: what happens after A_1 has become B_1? In order to be in phase with B_2 and then C_2, B_1 needs to become C_1', thus propagating along the path $B_1 C_1 B_1'$ at an infinite velocity. Hence, this argument (again, erroneous) apparently indicates that negative refraction violates causality and is therefore unphysical.

A more striking illustration of an apparent impossibility is provided by considering the effect of frequency dispersion inherent to left-handed media on the refraction of a signal composed of the superposition of two incident waves. Let us consider two $\hat{y}$ polarized (TE) incident waves of the type shown in Fig. 6.1, with similar incident angle but different frequencies. Upon transmitting into the left-handed medium, each wave is refracted with a different angle because of the different value of the permittivity at the two respective frequencies. The electromagnetic fields are simply written as (we assume a unit amplitude for the two waves):

$$E_{iy\ell} = e^{ik_{x\ell}x - ik_{iz\ell}z} + R_\ell e^{ik_{x\ell}x + ik_{iz\ell}z} , \qquad (6.4a)$$

$$H_{ix\ell} = \frac{k_{iz\ell}}{\omega_\ell \mu_0} \left(e^{ik_{x\ell}x - ik_{iz\ell}z} - R_\ell e^{ik_{x\ell}x + ik_{iz\ell}z} \right) , \qquad (6.4b)$$

$$H_{iz\ell} = \frac{k_{x\ell}}{\omega_\ell \mu_0} \left(e^{ik_{x\ell}x - ik_{iz\ell}z} + R_\ell e^{ik_{x\ell}x + ik_{iz\ell}z} \right) , \qquad (6.4c)$$

$$E_{ty\ell} = T_\ell e^{ik_{x\ell}x - ik_{tz\ell}z} , \qquad (6.4d)$$

$$H_{tx\ell} = \frac{k_{tz\ell}}{\omega_\ell \mu_{t\ell}} T_\ell e^{ik_{x\ell}x - ik_{tz\ell}z} , \qquad (6.4e)$$

$$H_{tz\ell} = \frac{k_{x\ell}}{\omega_\ell \mu_{t\ell}} T_\ell e^{ik_{x\ell}x - ik_{tz\ell}z} , \qquad (6.4f)$$

where the subscripts i and t denote the incident and transmitted media, respectively, $\ell = 1, 2$ distinguishes the two frequencies, and where R_ℓ and T_ℓ are the Fresnel reflection and transmission coefficients given by (as a special case of Eqs. (2.98) on page 63; see also Appendix A)

$$R_\ell = \frac{\mu_{t\ell} k_{iz\ell} - \mu_0 k_{tz\ell}}{\mu_{t\ell} k_{iz\ell} + \mu_0 k_{tz\ell}} , \qquad T_\ell = \frac{2k_{iz\ell} \mu_{t\ell}}{\mu_{t\ell} k_{iz\ell} + \mu_0 k_{tz\ell}} . \qquad (6.5)$$

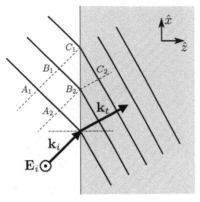

(a) Refraction between free-space and a standard dielectric.

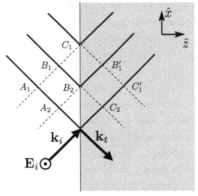

(b) Refraction between free-space and a medium exhibiting a negative refraction.

Figure 6.1 Wave-fronts illustrating the refraction of a plane wave from free-space to (a) a standard dielectric and (b) a left-handed medium. The points A_i, B_i, B_i', and C_i are defined for the purpose of discussion.

The pattern of the electric field is shown in Fig. 6.2. In both media one can see wave-fronts and an interference pattern. Although the wave-fronts are refracted negatively, the interference pattern is clearly "refracted" positively.*

We subsequently present two arguments to resolve these paradoxes: the first argument illustrates why causality is not violated in a negative refraction process, while the second proves mathematically the subtle fact that interference

---

*Note that we speak of refraction of an interference pattern, which is only a language simplification for the purpose of illustration. Rigorously, refraction only occurs for the phase and the energy.

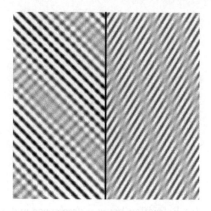

Figure 6.2 Electric field from two waves at two frequencies ($f_1 = 10$ GHz and $f_2 = 11$ GHz) impinging from free-space onto a left-handed medium. At the two frequencies, the relative parameters are $(\epsilon_1, \mu_1) = (-2.5, -1)$ and $(\epsilon_2, \mu_2) = (-1.5, -1)$. The interference pattern is positively refracted while the wave-fronts are negatively refracted.

pattern and Poynting vector are not perpendicular to each other.

The paradox presented in Fig. 6.1 is resolved by realizing that a time harmonic plane wave is a theoretical idealization: it exists everywhere and for all time in the past and the future. Hence, the point B_1 does not need to move infinitely fast to B_1': the point C_1 which is part of the same plane wave and of the same wave-front (one period away) is the one that propagates to B_1' at the speed of the electromagnetic wave, thus without violation of causality.[†] Consequently, what has gone astray in the original argument is the assimilation of phase motion with information transfer.

In addition, for a transient phenomenon, such as a sharp pulse of finite duration, Fig. 6.1 is not a valid illustration. When the pulse first impinges on the interface, the negative refraction does not happen instantaneously but takes time to build up. Eventually, the pulse is negatively refracted, but with a time delay compared to the same pulse refracted positively by a standard dielectric (Foteinopoulou et al. 2003, Alù et al. 2006b). Moreover, the problem of pulse propagation is intrinsically multi-frequency in nature and cannot be understood properly by considering a single frequency component. The time required for traversal of pulses is dealt with separately in Section 6.5.2.2 on page 247.

The seemingly positive refraction of Fig. 6.2 is clarified by realizing that

---

[†]One has to be cautious with the notion of causality and to what it is applied. As a matter of fact, special relativity only puts restrictions on information carrying signals travelling faster than the speed of light in vacuum. Hence the notion of signal itself needs to be properly defined.

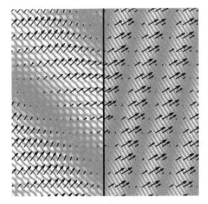

(a) Refraction between free-space and a standard dispersive dielectric.

(b) Refraction between free-space and a medium exhibiting negative refraction.

Figure 6.3 The background shows the $\hat{x}$ component (along the interface) of the Poynting vector at a given time from the same two waves as in Fig. 6.2. Arrows indicate the time average Poynting vector and therefore correspond to the direction of energy. Black arrows correspond to $< S_x >> 0$ while white arrows correspond to $< S_x >< 0$.

the interference pattern created in the left-handed medium is due to the two waves refracting at different angles, and is therefore not a wave-front for the energy of the total transmitted field (Pacheco, Jr. et al. 2002). In particular, the interference pattern is not perpendicular to the Poynting vector and therefore cannot be used to conclude on the positive refraction of the energy (note that a time animation of Fig. 6.2 does show the interference pattern seemingly propagating with a positive refraction – this is just a visual effect due to the infinite extent of the patterns). A more visual realization of this fact is proposed in Fig. 6.3: Fig. 6.3(a) shows the refraction of the signal composed of the two waves when the transmitted medium is a frequency dispersive dielectric with a positive permittivity, while Fig. 6.3(b) shows again the case when the transmitted medium is a left-handed medium. Being a refraction between two media with positive index of refraction, Fig. 6.3(a) should show a positive refraction while the interference pattern seems to indicate a negative refraction. This counter-example therefore proves that the interference pattern does not provide the necessary information on the refraction of energy. Conversely, Fig. 6.3(b) shows a positive refraction of the interference pattern, while the transmitted medium is a left-handed medium and should therefore exhibit a negative refraction. In order to unambiguously conclude on the propagation direction of the energy, the time averaged Poynting vectors need to be calculated. The results are shown in both cases as superposed arrows, and confirm the positive and negative refractions in Figs. 6.3(a) and 6.3(b), respectively. A closer examination of the sign of the $\hat{x}$ component of the Poynting vector

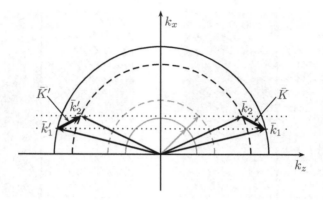

Figure 6.4 Spectral diagram of Fig. 6.2 to visualize the direction of the interference pattern.

(not shown) confirms these conclusions (Pacheco, Jr. et al. 2002).

A schematic argument based on the spectral representation of phase matching can also be used to understand the direction of the interference pattern. Taking first the case of refraction in a standard dielectric, we write the transmitted waves-vectors of the two waves as $\mathbf{k}_1 = \hat{x}k_{1x} + \hat{z}k_{1z}$ and $\mathbf{k}_2 = \hat{x}k_{2x} + \hat{z}k_{2z}$. Assuming that the transmission coefficient is identical for the two waves (this is a simplification without quantitative consequence on the final result if the transmission coefficient is relatively constant between the two frequencies), the total electric field can be written as

$$\mathbf{E}_{\text{tot}} \simeq e^{ik_{1x}x+ik_{1z}z} + e^{ik_{2x}x+ik_{2z}z}$$

$$= 2e^{i(k_{1x}+k_{2x})x/2}\, e^{i(k_{1z}+k_{2z})z/2} \cos\left[\frac{k_{1x}-k_{2x}}{2}x + \frac{k_{1z}-k_{2z}}{2}z\right] \quad (6.6)$$

The interference pattern is governed by the cosine term and therefore has an equivalent wave-vector of $\mathbf{K} = \hat{x}(k_{1x}+k_{2x})+\hat{z}(k_{1z}+k_{2z})$, shown in Fig. 6.4. As can be seen, $\mathbf{K}$ points downward (negative $\hat{x}$ component) and therefore gives the impression of a negative refraction. The group velocity was incorrectly defined by Valanju et al. (2002) as the reciprocal of the frequency derivative of this vector $\mathbf{K}$. However, as already pointed out, $\mathbf{K}$ has no relation to the power direction, which in this case points upward (positive $\hat{x}$ component).

The case of a refraction into a left-handed medium is very similar, and is also illustrated in Fig. 6.4. The wave-vectors are denoted with primes: $\mathbf{k}'_1 = \hat{x}k_{1x} - \hat{z}k_{1z}$ and $\mathbf{k}'_2 = \hat{x}k_{2x} - \hat{z}k_{2z}$. The vector $\mathbf{K}'$ related to the interference pattern has therefore the same $\hat{x}$ component as $\mathbf{K}$ but an opposite $\hat{z}$ component, as seen in Fig. 6.4. The interference pattern is therefore "refracted" positively (with a positive $\hat{x}$ component), but again does not correspond to the direction of the power. The use of a beam of light with finite transverse width is also very illustrative in this case: a simulation using two

Gaussian beams (Smith et al. 2002b) with slightly differing frequencies clearly shows that the beams experience negative refraction while the normal to the interference fronts point in a different direction that would correspond to a positive angle of refraction. Finally, note that this phenomenon is visible with any wave packet that is localized in the transverse direction (Lu et al. 2002; 2004) whereas this sideways motion of the interference fringes is not apparent in the case of plane waves due to their infinite transverse extent.

6.1.3   Energy in dispersive media

We proceed by studying the propagation of energy in frequency dispersive media from a more fundamental point of view. The permittivity $\epsilon(\omega)$ and permeability $\mu(\omega)$, yet to be related to the time domain electromagnetic fields, do not assume any specific form (such as Lorentz or Drude forms) but are only supposed to obey obvious conditions such as causality, reality of the fields, etc.

The propagation of the energy of an electromagnetic wave is given by the Poynting vector, directly obtained from the Maxwell equations. The latter are here cast in the Chu form, where the material contributions are clearly identified by the polarization currents $\mathbf{P} = \mathbf{D} - \epsilon_0 \mathbf{E}$ and $\mu_0 \mathbf{M} = \mathbf{B} - \mu_0 \mathbf{H}$:

$$\nabla \times \mathbf{H}(t) - \epsilon_0 \frac{\partial \mathbf{E}(t)}{\partial t} = \frac{\partial \mathbf{P}}{\partial t} \equiv \mathbf{J}_e(t) \,, \tag{6.7a}$$

$$\nabla \times \mathbf{E}(t) + \mu_0 \frac{\partial \mathbf{H}(t)}{\partial t} = -\mu_0 \frac{\partial \mathbf{M}}{\partial t} \equiv -\mathbf{J}_m(t) \,, \tag{6.7b}$$

$$\mu_0 \nabla \cdot \mathbf{H}(t) = -\nabla \cdot \mu_0 \mathbf{M} \equiv \rho_m(t) \,, \tag{6.7c}$$

$$\epsilon_0 \nabla \cdot \mathbf{E}(t) = -\nabla \cdot \mathbf{P} \equiv \rho_e(t) \,. \tag{6.7d}$$

In this form, the fundamental electromagnetic fields are $\mathbf{E}(t)$ and $\mathbf{H}(t)$ and the electric and magnetic sources in the medium $(\mathbf{J}_e, \rho_e)$, $(\mathbf{J}_m, \rho_m)$.[‡] It is well known that the electromagnetic energy and momentum quantities can be derived solely for the electromagnetic subsystem without assuming specific models for $\mathbf{P}$ and $\mathbf{M}$, as is the case for lossless frequency dispersive media. For lossy frequency dispersive media, however, this is not the case and failing to realize this may lead to unphysical conclusions on negative energy for example. We address this issue in Section 6.2, on page 230.

The Poynting vector is obtained by combining Eq. (6.7a) and Eq. (6.7b) dot multiplied by $\mathbf{E}(t)$ and $\mathbf{H}(t)$, respectively. After using a standard vector identity, the law of conservation of energy is cast in the form of Eq. (6.1a)

[‡] Note that the magnetic sources are added purely on a theoretical basis, as a dual of their electric counterparts.

where the following terms are defined:

$$\mathbf{S}_{eh} = \mathbf{E} \times \mathbf{H}, \tag{6.8a}$$

$$W_{eh} = \frac{\varepsilon_0}{2}\mathbf{E} \cdot \mathbf{E} + \frac{\mu_0}{2}\mathbf{H} \cdot \mathbf{H}, \tag{6.8b}$$

$$\phi_{eh} = \mathbf{J}_e \cdot \mathbf{E} + J_h \cdot \mathbf{H}, \tag{6.8c}$$

and where the subscript "eh" is used to specifically denote the electromagnetic subsystem. This is the well-known Poynting theorem for non-dispersive media.

Note that within this Chu formulation, only $\mathbf{E}$ and $\mathbf{H}$ are considered fundamental electromagnetic quantities, whereas the material is considered to contribute to sources (as seen from the definitions of the polarization currents $\mathbf{P}$ and $\mathbf{M}$). For this reason, the energy term of Eq. (6.8b) is only proportional to the free-space permittivity ε_0 and permeability μ_0, while the material contribution is included in ϕ_{eh}. Within the more standard Minkowski formulation where all the fields $\mathbf{E}$, $\mathbf{D}$, $\mathbf{H}$, and $\mathbf{B}$ are considered fundamental and where the sources are either conduction currents or induced currents at boundaries, the energy is proportional to the material constitutive parameters ε and μ. In this case, it is seen that the negative constitutive parameters of left-handed media would imply a negative stored energy, which is physically unacceptable. The correction to this paradox stems from recognizing the fundamental difference between a monochromatic wave and a quasi-monochromatic wave propagating through a dispersive medium such as a left-handed medium, and is considered next.

Let us return to the most general definition of the time domain electric field $\mathbf{E}(t)$ and flux $\mathbf{D}(t)$, expressed as a Fourier transformation of their spectral domain quantities (Landau et al. 1984, Jackson 1999) (see also Chapter 1). In order to express the electric field energy $W_e = \mathbf{E}(t) \cdot \partial \mathbf{D}(t)/\partial t$ (the magnetic field energy is obtained following similar steps), we write:

$$\mathbf{E}(t) = \int_{-\infty}^{+\infty} d\omega \mathbf{E}(\omega) e^{-i\omega t}, \tag{6.9a}$$

$$\mathbf{D}(t) = \int_{-\infty}^{+\infty} d\omega \mathbf{D}(\omega) e^{-i\omega t} \tag{6.9b}$$

with the constitutive relation $\mathbf{D}(\omega) = \varepsilon(\omega)\mathbf{E}(\omega)$. Let us assume that the electric field is dominated by a narrow range of frequencies, centered at ω_0, weighted by a slowly varying envelope $\mathbf{E}_\alpha(t)$ (in the limiting case of an infinitely narrow band and an infinitely slowly varying envelope, this assumption reduces to the common plane-wave):

$$\mathbf{E}(t) = \mathbf{E}_\alpha(t)\cos(\omega_0 t), \quad \text{where } \mathbf{E}_\alpha(t) = \mathbf{E}_0\cos(\alpha t), \tag{6.10a}$$

so that

$$\mathbf{E}(t) = \frac{\mathbf{E}_0(t)}{2}\Big(\cos[(\omega_0 + \alpha)t] + \cos[(\omega_0 - \alpha)t]\Big). \tag{6.10b}$$

Expressed in terms of its spectral components:

$$\mathbf{E}(t) = \int_{-\infty}^{+\infty} d\omega \frac{\mathbf{E}_0}{4} \left[\delta(\omega_0 + \alpha) + \delta(-\omega_0 - \alpha) + \delta(\omega_0 - \alpha) + \delta(-\omega_0 + \alpha) \right] e^{-i\omega t}.$$

(6.11)

The electric displacement is then given by

$$\mathbf{D}(t) = \int_{-\infty}^{+\infty} d\omega \varepsilon(\omega) \mathbf{E}(\omega) e^{-i\omega t}, \tag{6.12a}$$

$$= \frac{\mathbf{E}_0}{2} \left[\varepsilon(\omega_0 + \alpha) e^{-i(\omega_0 + \alpha)t} + \varepsilon(\omega_0 - \alpha) e^{-i(\omega_0 - \alpha)t} \right.$$

$$\left. + \varepsilon^*(\omega_0 + \alpha) e^{+i(\omega_0 + \alpha)t} + \varepsilon^*(\omega_0 - \alpha) e^{+i(\omega_0 - \alpha)t} \right], \tag{6.12b}$$

where we have used the condition $\varepsilon(-\omega) = \varepsilon^*(\omega)$ due to the reality of the fields (and the reality of ω since we have not deformed the integration path in Eq. (6.12a)). The permittivity is subsequently expanded in a Taylor series about ω_0:

$$\varepsilon(\omega_0 \pm \alpha) = \varepsilon(\omega_0) \pm \alpha \frac{\partial \varepsilon(\omega_0)}{\partial \omega} + \dots , \tag{6.13}$$

and similarly for $\varepsilon^*(\omega)$, where we neglect the higher order terms in α, and where the second term is a short-hand notation for $\frac{\partial \varepsilon(\omega)}{\partial \omega}\big|_{\omega=\omega_0}$. Introducing the permittivity expression:

$$\mathbf{D}(t) = \frac{\mathbf{E}_0}{4} \left[\varepsilon(\omega_0) e^{-i(\omega_0 + \alpha)t} + \varepsilon(\omega_0) e^{-i(\omega_0 - \alpha)t} \right.$$

$$\left. + \varepsilon^*(\omega_0) e^{i(\omega_0 + \alpha)t} + \varepsilon^*(\omega_0) e^{i(\omega_0 - \alpha)t} \right]$$

$$+ \frac{\mathbf{E}_0}{4} \alpha \left[\frac{\partial \varepsilon(\omega)}{\partial \omega} e^{-i(\omega_0 + \alpha)t} - \frac{\partial \varepsilon(\omega)}{\partial \omega} e^{-i(\omega_0 - \alpha)t} \right.$$

$$\left. + \frac{\partial \varepsilon^*(\omega)}{\partial \omega} e^{i(\omega_0 + \alpha)t} - \frac{\partial \varepsilon^*(\omega)}{\partial \omega} e^{i(\omega_0 - \alpha)t} \right],$$

$$= \varepsilon'(\omega_0) \mathbf{E}_\alpha(t) \cos \omega_0 t + \frac{\partial \varepsilon'(\omega_0)}{\partial \omega} \frac{\partial \mathbf{E}_\alpha(t)}{\partial t} \sin \omega_0 t$$

$$+ \varepsilon''(\omega_0) \mathbf{E}_\alpha(t) \sin \omega_0 t - \frac{\partial \varepsilon''(\omega_0)}{\partial \omega} \frac{\partial \mathbf{E}_\alpha(t)}{\partial t} \cos \omega_0 t, \tag{6.14}$$

where Eq. (6.10a) has been used as well as some simple trigonometric identities. In the previous equation, we remind the reader that the primed quantities denote the real parts whereas the double primed quantities denote the imaginary parts ($\varepsilon'(\omega) = \mathrm{Re}\{\varepsilon(\omega)\}$ and $\varepsilon''(\omega) = \mathrm{Im}\{\varepsilon(\omega)\}$, and similarly for μ). Taking the derivative with respect to time, multiplying by $\mathbf{E}(t)$, and taking the time average with respect to ω_0 yields

$$\left\langle \mathbf{E}(t) \cdot \frac{\partial \mathbf{D}(t)}{\partial t} \right\rangle_{\omega_0} = \frac{1}{4} \frac{\partial(\omega \varepsilon'(\omega))}{\partial \omega}\bigg|_{\omega=\omega_0} \frac{\partial |\mathbf{E}_\alpha(t)|^2}{\partial t} + \frac{1}{2} \omega_0 \varepsilon''(\omega_0) |\mathbf{E}_\alpha(t)|^2.$$

(6.15)

Adding the magnetic counterpart, we finally obtain

$$\langle \mathbf{E}(t) \cdot \frac{\partial \mathbf{D}(t)}{\partial t} + \mathbf{H}(t) \cdot \frac{\partial \mathbf{B}(t)}{\partial t} \rangle = \frac{\partial W}{\partial t} + \frac{1}{2}\omega\varepsilon''(\omega_0) < |\mathbf{E}(t)|^2 >$$
$$+ \frac{1}{2}\omega\mu''(\omega_0) < |\mathbf{H}(t)|^2 >, \qquad (6.16)$$

where we have omitted the subscript α and ω_0 for convenience and where

$$W = \frac{1}{4}\frac{\partial(\omega\varepsilon'(\omega))}{\partial\omega}\langle|\mathbf{E}(t)|^2\rangle + \frac{1}{4}\frac{\partial(\omega\mu'(\omega))}{\partial\omega}\langle|\mathbf{H}(t)|^2\rangle \qquad (6.17)$$

is the electromagnetic energy in a frequency dispersive medium. It can now be easily verified that even when $\varepsilon(\omega) < 0$ and $\mu(\omega) < 0$, the energy is positive provided that the frequency derivatives in the expression are positive.[§] The apparent paradox that a medium without dispersion and with $\varepsilon < 0$ and $\mu < 0$ would yield a negative energy is easily lifted: such medium cannot exist since it would violate the causality condition and would therefore not represent a physical medium.

As we shall see in the next section, however, Eq. (6.17) should be interpreted carefully within the frame of left-handed media: these media being necessarily lossy, W does not represent the total energy density. Failing to realize this leads to erroneous conclusions that the energy can still take negative values in some frequency ranges.

6.2 Electromagnetic energy in left-handed media

6.2.1 Erroneous concept of negative energy in lossy dispersive media

The causality conditions require that the constitutive parameters of left-handed media be both frequency dispersive and lossy. Although Eq. (6.17) has been derived for a frequency dispersive medium, it fails to give the stored energy when losses are not negligible. We shall illustrate this point by considering a medium that exhibits a Lorentz dispersion relation. In order to yield a positive energy, Eq. (6.17) requires that

$$\frac{\partial(\omega\varepsilon'(\omega))}{\partial\omega} > 0 \quad \text{and} \quad \frac{\partial(\omega\mu'(\omega))}{\partial\omega} > 0. \qquad (6.18)$$

Yet, for non-negligible losses, these relations are not verified around the resonance ω_{eo} (see Eqs. (6.26) subsequently): at a frequency slightly beyond

---
[§]This result was first obtained by Brillouin (1960).

resonance, the dispersion is anomalous and both the real part of the permittivity $\varepsilon'(\omega)$ and its slope $\partial \varepsilon'(\omega)/\partial \omega$ are negative, producing a negative energy as illustrated in Fig. 6.5.

The situation of lossy frequency dispersive media must therefore be considered with care since it is in general not possible to introduce the concept of electromagnetic energy density. Mathematically, this comes from the impossibility of uniquely dividing the Poynting theorem into parts corresponding to the change in energy and parts corresponding to dissipation: it can be found that both the conductivity (or the imaginary part of the permittivity) and the permittivity itself may contribute to the stored energy (Agranovich and Ginzburg 1966, Askne and Lind 1970, Ginzburg 1970). Physically, the impossibility of deriving a closed-form expression of the stored energy in arbitrary lossy media is due to the fact that two media with identical complex $\varepsilon(\omega)$ and $\mu(\omega)$ can store different amounts of energies. One can think of a direct analogy in circuit theory where a similar phenomenon occurs: it is impossible to uniquely determine the circuit layout from the sole knowledge of its input impedance. Indeed, two circuits with identical input impedances at all frequencies may store different energies. In other words, the knowledge of the input impedance does not provide the knowledge of the micro-structure of the circuit, from which the stored energy can be calculated (Tretyakov 2005). One can easily see this by looking at the example in Fig. 6.6 (Ginzburg 1961): when $R_1 = R_2 = R = \sqrt{L/C}$ the input impedance of the circuit is $Z_{in} = R$, which conveys no information concerning the energy stored in the circuit, which is obviously non-zero. The same situation occurs in materials: the sole knowledge of $\varepsilon(\omega)$ and $\mu(\omega)$ does not provide the knowledge of the micro-structure of the medium and therefore the stored energy cannot be predicted. The unambiguous determination of W in lossy frequency media is therefore possible only when the explicit functions of the permittivity and permeability are known as functions of the frequency, the collision frequency, the plasma frequency, etc. The implication is that the energy density needs to be solved for every material separately. Alternatively, if one knows the electromagnetic fields inside the medium at all previous times, one can express the energy density in terms of integrals over the field quantities at all past times (Glasgow et al. 2001).

6.2.2 Lossy Lorentz media

In the following, we assume that the lossy dispersive medium can be modeled as an idealized assembly of independent oscillators where the motion of the electrons is described by (see also Chapter 1)

$$m \left(\frac{d^2 \mathbf{r}_i}{dt^2} + \gamma_i \frac{d\mathbf{r}_i}{dt} + \omega_i^2 \mathbf{r}_i \right) = -e\mathbf{E}, \qquad (6.19)$$

where $\mathbf{r}_i$ is the displacement vector of the N electrons and m is their mass. Defining the polarization vector $\mathbf{P} = -N e \mathbf{r}$ where $-e$ is the charge of the

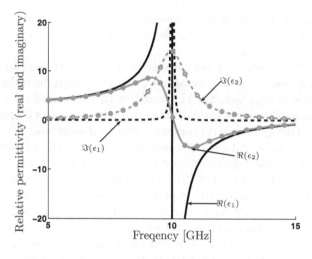

(a) Relative permittivity.

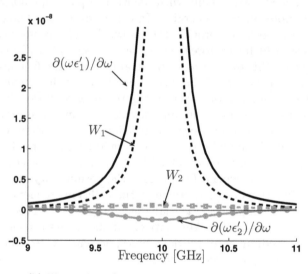

(b) Illustration of various energy and energy terms.

Figure 6.5 Lorentz-like relative permittivity and various energy definitions. It is seen that it is possible to produce $\partial(\omega\epsilon')/\partial\omega < 0$ with a very lossy Lorentz medium. The values of the Lorentz model in Eq. (6.26a) are $\omega_{eo} = 2\pi \times 10$ [rad/s], $\omega_{ep} = 2\pi \times 15$ [rad/s] and the two cases denoted by the subscripts "1" and "2" correspond to different loss amount: $\gamma_{e_1} = 10^8$ Hz and $\gamma_{e_2} = 10^{10}$ Hz. The value of the energy W in both cases is obtained from the second term of the right-hand side of Eq. (6.21b).

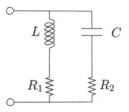

Figure 6.6 Illustration of the fact that the knowledge of the input impedance Z_{in} at all frequencies does not provide information on the stored energy in the electrical circuit: by taking $R_1 = R_2 = R = \sqrt{L/C}$ one finds $Z_{in} = R$ which conveys no information on the stored energy.

electrons, the equation of motion of the electrons under the action of an electric field can be rewritten as

$$\frac{\partial^2 \mathbf{P}}{\partial t^2} + \gamma_e \frac{\partial \mathbf{P}}{\partial t} + \omega_{eo}^2 \mathbf{P} = \varepsilon_0 \omega_{ep}^2 \mathbf{E}, \tag{6.20a}$$

where ω_{ep} and ω_{eo} are the electric plasma and resonant frequencies and where γ_e is the electric collision frequency responsible for losses in the medium. Similarly, the magnetization of the medium due to the series of split ring resonators introduced in Chapter 3 is governed by

$$\frac{\partial^2 \mathbf{M}}{\partial t^2} + \gamma_m \frac{\partial \mathbf{M}}{\partial t} + \omega_{mo}^2 \mathbf{M} = F \omega_{mp}^2 \mathbf{H}, \tag{6.20b}$$

where ω_{mp} and ω_{mo} are the magnetic plasma and resonant frequencies, γ_m is the magnetic collision frequency, and where F is the filling factor of the magnetic resonators.

In order to obtain the energy conservation law of the entire system composed of the electromagnetic wave and of the material, the material contribution needs to be added to the laws of Eqs. (6.8). This is done by dot multiplying Eq. (6.20a) by $\partial \mathbf{P}/\partial t$ and Eq. (6.20b) by $\partial(\mu_0 \mathbf{M})/\partial t$ which yields a set of two scalar equations, and adding them to Eq. (6.8a). After rearranging the terms, we obtain the time domain energy conservation which takes the form of Eq. (6.1a) with the following definitions:

$$\mathbf{S} = \mathbf{S}_{eh} = \mathbf{E} \times \mathbf{H}, \tag{6.21a}$$

$$W = \left[\frac{\varepsilon_0}{2} \mathbf{E} \cdot \mathbf{E} + \frac{\mu_0}{2} \mathbf{H} \cdot \mathbf{H} + \frac{1}{2\varepsilon_0 \omega_{ep}^2} \left(\frac{\partial \mathbf{P}}{\partial t} \cdot \frac{\partial \mathbf{P}}{\partial t} + \omega_{eo}^2 \mathbf{P} \cdot \mathbf{P} \right) \right.$$
$$\left. + \frac{\mu_0}{2F\omega_{mp}^2} \left(\frac{\partial \mathbf{M}}{\partial t} \cdot \frac{\partial \mathbf{M}}{\partial t} + \omega_{mo}^2 \mathbf{M} \cdot \mathbf{M} \right) \right], \tag{6.21b}$$

$$\phi = -\frac{\gamma_e}{2\varepsilon_0 \omega_{ep}^2} \frac{\partial \mathbf{P}}{\partial t} \cdot \frac{\partial \mathbf{P}}{\partial t} - \frac{\mu_0 \gamma_m}{2F\omega_{mp}^2} \frac{\partial \mathbf{M}}{\partial t} \cdot \frac{\partial \mathbf{M}}{\partial t}. \tag{6.21c}$$

As expected, the energy flow is entirely defined by the electromagnetic contribution since it is the only cause of energy propagation in the medium. The energy density W, however, contains the electromagnetic contribution of Eq. (6.8b) but also the kinetic and potential energy of the electric and magnetic dipoles. Under this form, Fig. 6.5 shows that the values of W are positive across the entire frequency range, and in particular at those frequencies where Eq. (6.17) produces negative values. Finally, the connection to other subsystems is provided by the losses in the medium, function of γ_e and γ_m. Hence, in the limit of a lossless medium, the system composed of the electromagnetic wave and matter is closed. For lossy media, however, the system is open and contributes for example to thermodynamic effects (heating).

6.3 Momentum transfer in media with negative material parameters

The topic of momentum density in left-handed media is related to an important part of electrodynamics (electro-magneto-dynamics in fact), which relates electromagnetic effects to forces and motion of matter. The relation between field and forces is, of course, expressed via the Lorentz force

$$\mathbf{f} = \rho_e(t)\mathbf{E}(t) + \rho_m(t)\mathbf{H}(t) + \mathbf{J}_e(t) \times \mu_0\mathbf{H}(t) + \mathbf{J}_m(t) \times \epsilon_0\mathbf{E}(t). \qquad (6.22)$$

This force, usually well known on free currents and charges, can be written in terms of the field quantities only by substituting the expressions of $\rho_e(t)$, $\rho_m(t)$, $\mathbf{J}_e(t)$, and $\mathbf{J}_m(t)$ from Eqs. (6.7). Upon doing so, one obtains an equation similar to Eq. (6.1b) where

$$\bar{\bar{T}}_{\text{eh}} = \frac{1}{2}\left(\varepsilon_0\mathbf{E}\cdot\mathbf{E} + \mu_0\mathbf{H}\cdot\mathbf{H}\right) - \varepsilon_0\mathbf{E}\mathbf{E} - \mu_0\mathbf{H}\mathbf{H}, \qquad (6.23a)$$

$$\mathbf{G}_{\text{eh}} = \varepsilon_0\mu_0\mathbf{E} \times \mathbf{H}, \qquad (6.23b)$$

$$\mathbf{f}_{\text{eh}} = \rho_e\mathbf{E} + \rho_h\mathbf{H} + \mathbf{J}_e \times \mu_0\mathbf{H} - \mathbf{J}_h \times \varepsilon_0\mathbf{E}, \qquad (6.23c)$$

where again we have used the subscript "eh" to denote quantities solely related to the electromagnetic subsystem. In Eqs. (6.23), $\bar{\bar{T}}_{\text{eh}}$ is the free-space Maxwell stress tensor in which the terms $\mathbf{E}\mathbf{E}$ and $\mathbf{H}\mathbf{H}$ are dyadic products, $\mathbf{G}_{\text{eh}}$ is the Abraham momentum density, and $\mathbf{f}_{eh}$ is the resulting force on the medium. Like already mentioned in the introduction, the fact of having a non-zero force implies that the system is open and one might think what other subsystems must be included (typically the kinematic subsystem) in order to close it.¶

¶For example, one can show that the kinetic and hydrodynamic subsystems are both open but that combined together, they form a closed system (Penfield and Haus 1967).

Like in the case of the energy, the material contribution must be added to Eqs. (6.23) in order to obtain the momentum conservation law for the entire system formed by the electromagnetic field and matter (Loudon et al. 1997, Kemp et al. 2007). This is done by dot-multiplying Eq. (6.20a) by $-\nabla\mathbf{P}$ and Eq. (6.20b) by $-\mu_0\nabla\mathbf{M}$ (which are dyads) to obtain

$$-\mathbf{E}\cdot\nabla\mathbf{P} = -\frac{\nabla\mathbf{P}}{\varepsilon_0\omega_{ep}^2}\cdot\left(\frac{\partial^2\mathbf{P}}{\partial t^2} + \gamma_e\frac{\partial\mathbf{P}}{\partial t} + \omega_{eo}^2\mathbf{P}\right), \tag{6.24a}$$

$$-\mu_0\mathbf{H}\cdot\nabla\mathbf{M} = -\frac{\mu_0\nabla\mathbf{M}\cdot}{F\omega_{mp}^2}\cdot\left(\frac{\partial^2\mathbf{M}}{\partial t^2} + \gamma_m\frac{\partial\mathbf{M}}{\partial t} + \omega_{mo}^2\mathbf{M}\right). \tag{6.24b}$$

Adding these two relations to the conservation law of Eq. (6.23) and performing some algebraic manipulations yield a momentum conservation equation as in Eq. (6.1b) where

$$\bar{\bar{T}} = \left[\frac{1}{2}(\mathbf{D}\cdot\mathbf{E}+\mathbf{B}\cdot\mathbf{H})\bar{\bar{I}} - \mathbf{D}\mathbf{E} - \mathbf{B}\mathbf{H}\right] + \left[\frac{1}{2}(\mathbf{P}\cdot\mathbf{E}+\mu_0\mathbf{M}\cdot\mathbf{H})\bar{\bar{I}}\right]$$
$$+ \left[\frac{1}{2\varepsilon_0\omega_{ep}^2}\left(\frac{\partial\mathbf{P}}{\partial t}\cdot\frac{\partial\mathbf{P}}{\partial t} - \omega_{eo}^2\mathbf{P}\cdot\mathbf{P}\right)\bar{\bar{I}}\right]$$
$$+ \left[\frac{\mu_0}{2F\omega_{mp}^2}\left(\frac{\partial\mathbf{M}}{\partial t}\cdot\frac{\partial\mathbf{M}}{\partial t} - \omega_{mo}^2\mathbf{M}\cdot\mathbf{M}\right)\bar{\bar{I}}\right], \tag{6.25a}$$

$$\mathbf{G} = \mathbf{D}\times\mathbf{B} - \frac{1}{\varepsilon_0\omega_{ep}^2}\nabla\mathbf{P}\cdot\frac{\partial\mathbf{P}}{\partial t} - \frac{\mu_0}{F\omega_{mp}^2}\nabla\mathbf{M}\cdot\frac{\partial\mathbf{M}}{\partial t}, \tag{6.25b}$$

$$\mathbf{f} = -\frac{\gamma_e}{\varepsilon_0\omega_{ep}^2}\nabla\mathbf{P}\cdot\frac{\partial\mathbf{P}}{\partial t} - \frac{\mu_0\gamma_m}{F\omega_{mp}^2}\nabla\mathbf{M}\cdot\frac{\partial\mathbf{M}}{\partial t}. \tag{6.25c}$$

Unlike the energy case in which the energy flow was unchanged and expressed by the Poynting vector, we see here that the medium contributes to the momentum flow with additive terms directly obtained from the polarization vectors $\mathbf{P}$ and $\mathbf{M}$. The momentum flow is therefore expressed as the sum of the momentum in non-dispersive media and a material contribution. Likewise, the momentum density $\mathbf{G}$ contains the Minkowski momentum $\mathbf{D}\times\mathbf{B}$ and terms due to the dispersion of the material. Finally, the force $\mathbf{f}$ is directly due to losses in the medium via γ_e and γ_m, and therefore represents dissipation. All these terms are expressed as function of the micro-structure of the medium, where the plasma frequency, the resonance frequency, and the damping rates appear explicitly.

6.4 Limit of plane wave and small losses

The previous sections did not suppose a specific form for the electromagnetic wave, but only a specific form of the medium micro-structure given by Eqs. (6.20). Here, we particularize the previous results to a monochromatic electromagnetic wave, where field quantities in the frequency domain are related to the field quantities in the time domain by $\mathbf{E} = \text{Re}\{\mathbf{E}(t)e^{-i\omega t}\}$, and similarly for all other quantities. In this regime, we can substitute the operators $\partial/\partial t$ by $-i\omega$ and ∇ by $i\mathbf{k}$ which yields the specific Lorentz model for the permittivity and permeability:

$$\varepsilon = \varepsilon_0 \left(1 - \frac{\omega_{ep}^2}{\omega^2 - \omega_{eo}^2 + i\omega\gamma_e} \right) , \tag{6.26a}$$

$$\mu = \mu_9 \left(1 - \frac{F\omega_{mp}^2}{\omega^2 - \omega_{mo}^2 + i\omega\gamma_m} \right) . \tag{6.26b}$$

Likewise, the electric and magnetic polarizabilities, as well as their derivatives, are expressed as

$$|\mathbf{P}|^2 = \frac{\varepsilon_0^2 \omega_{ep}^4}{(\omega^2 - \omega_{eo}^2)^2 + \omega^2\gamma_e^2} |\mathbf{E}|^2 , \tag{6.27a}$$

$$|\mathbf{M}|^2 = \frac{F\omega_{mp}^4}{(\omega^2 - \omega_{mo}^2)^2 + \omega\gamma_m^2} |\mathbf{H}|^2 , \tag{6.27b}$$

$$\left| \frac{\partial \mathbf{P}}{\partial t} \right| = \omega^2 |\mathbf{P}^2| , \tag{6.27c}$$

$$\left| \frac{\partial \mathbf{M}}{\partial t} \right| = \omega^2 |\mathbf{M}^2| . \tag{6.27d}$$

6.4.1 Energy

The previous expressions can be directly used in the energy relation Eq. (6.21b). Upon taking the time average, we obtain

$$\langle W \rangle = \frac{\varepsilon_0}{2} \left[1 + \frac{\omega_{ep}^2 (\omega^2 + \omega_{eo}^2)}{(\omega^2 - \omega_{eo}^2)^2 + \gamma_e^2\omega^2} \right] |\mathbf{E}|^2$$

$$+ \frac{\mu_0}{2} \left[1 + \frac{F\omega_{mp}^2 (\omega^2 + \omega_{mo}^2)}{(\omega^2 - \omega_{mo}^2)^2 + \gamma_e^2\omega^2} \right] |\mathbf{H}|^2 , \tag{6.28}$$

which again yields positive energy values across the entire frequency spectrum of the left-handed medium (Loudon 1970, Cui and Kong 2004, Boardman and Marinov 2006, Kemp et al. 2007). Although left-handed media cannot be lossless, we can still suppose that in some regions of the frequency dispersion the real parts of the permittivity and permeability are much larger than their imaginary counterparts. These regions correspond to transparent regions, where we can assume the losses to be negligible. Taking the limit of $\gamma_e \to 0$ and $\gamma_m \to 0$, Eq. (6.28) takes the known form of Eq. (6.17). Note also that the dissipation rate given by Eq. (6.21c) is directly related to the imaginary parts of the permittivity and permeability via

$$\phi = -\omega\epsilon''|\mathbf{E}|^2 - \omega\mu''|\mathbf{H}|^2 , \qquad (6.29)$$

where (from Eqs. (6.26))

$$\varepsilon'' = \varepsilon_0 \frac{\omega\gamma_e\omega_{ep}^2}{(\omega^2 - \omega_{eo}^2)^2 + \gamma_e\omega^2} , \qquad (6.30a)$$

$$\mu'' = \mu_0 \frac{\omega\gamma_m F\omega_{mp}^2}{(\omega^2 - \omega_{mo}^2)^2 + \gamma_m\omega^2} . \qquad (6.30b)$$

Since the time average of the rate of change of energy is zero, the energy conservation given by Eq. (6.1a) then reads

$$-\langle \nabla \cdot \mathbf{S} \rangle = \frac{1}{2} \left[\omega\epsilon''|\mathbf{E}|^2 + \omega\mu''|\mathbf{H}|^2 \right] , \qquad (6.31)$$

which is the complex Poynting theorem with $\langle S \rangle = \frac{1}{2}\mathrm{Re}\{\mathbf{E} \times \mathbf{H}^*\}$.

6.4.2   Momentum

Applying a similar procedure to the momentum density, the time average is obtained from Eq. (6.25b) as

$$\langle \mathbf{G} \rangle = \frac{1}{2}\mathrm{Re}\left[\mathbf{D} \times \mathbf{B}^* + \mathbf{k}\frac{\varepsilon_0\omega\omega_{ep}^2}{(\omega^2 - \omega_{eo}^2)^2 + \gamma_e^2\omega^2} |E|^2 \right.$$
$$\left. + \mathbf{k}\frac{\mu_\omega F\omega_{mp}^2}{(\omega^2 - \omega_{mo}^2)^2 + \gamma_m^2\omega^2} |H|^2 \right] . \qquad (6.32)$$

In the lossless case, this expression takes the known form (Veselago 1968)

$$\langle \mathbf{G} \rangle = \frac{1}{2}\mathrm{Re}\left[\mathbf{D} \times \mathbf{B}^* + \frac{\mathbf{k}}{2}\frac{\partial\varepsilon(\omega)}{\partial\omega}|E|^2 + \frac{\mathbf{k}}{2}\frac{\partial\mu(\omega)}{\partial\omega}|H|^2 \right] , \qquad (6.33)$$

where we have used

$$\frac{\partial\varepsilon(\omega)}{\partial\omega} = \varepsilon_0 \frac{2\omega\omega_{ep}^2}{(\omega^2 - \omega_{eo}^2)^2} , \qquad (6.34a)$$

$$\frac{\partial\mu(\omega)}{\partial\omega} = \mu_0 \frac{2\omega F\omega_{mp}^2}{(\omega^2 - \omega_{mo}^2)^2} , \qquad (6.34b)$$

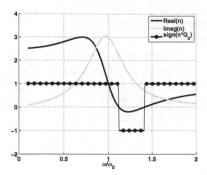

Figure 6.7 Index of refraction (real and imaginary part) and sign of $n' \cdot G_z$ for a $\hat{z}$ propagating plane wave in a Lorentz medium characterized by Eqs. (6.26) with $\omega_{eo} = \omega_{mo}$, $\omega_{ep} = \omega_{mp} = 1.5\omega_{ep}$, $\gamma_e = \gamma_m = 0.5\omega_{eo}$. It is seen that the momentum density is not always parallel to the wave-vector in the presence of losses.

from Eqs. (6.26). It is important to realize that relation (6.33) is shown here to be valid in the limit of a lossless medium, while the momentum in lossy media should be computed from the more general relation (6.32). This has an important consequence on the collinearity between the momentum density and the Poynting power flow.

As is apparent from Eq. (6.33), the momentum is obtained from the sum of three terms: $\mathbf{D} \times \mathbf{B}^*$ which is anti-parallel to the vector $\mathbf{E} \times \mathbf{H}^*$ and parallel to the vector $\mathbf{k}$, and two terms proportional to $\mathbf{k}$ and to the slope of the permittivity and of the permeability. In the limit of a lossless Lorentz medium, these two slopes are always positive and do not affect the direction of the second and third terms with the effect that the momentum is anti-parallel to the Poynting power. The lossy situation governed by Eq. (6.32), however, is different: there exist frequency regions where the momentum and the Poynting power are parallel to each other, as illustrated in Fig. 6.7. Consequently, the momentum flow may or may not be anti-parallel to the phase vector or to the energy flow, depending on the dissipation in the medium (Kemp et al. 2007).

The momentum flow reduces to

$$\langle \bar{\bar{T}} \rangle = \frac{1}{2}\text{Re}\left\{ \frac{1}{2}(\mathbf{D} \cdot \mathbf{E}^* + \mathbf{B} \cdot \mathbf{H}^*)\bar{\bar{I}} - \mathbf{D}\mathbf{E}^* - \mathbf{B}\mathbf{H}^* \right\} \qquad (6.35)$$

since the dispersive terms tend to zero upon time averaging. Similarly to the energy density, the time average of the rate of change of momentum density is zero, i.e., $\langle \partial \mathbf{G}/\partial t \rangle = 0$, so that

$$-\langle \nabla \cdot \bar{\bar{T}} \rangle = \frac{1}{2}\text{Re}\left\{ \omega\epsilon''\mathbf{E} \times \mathbf{B}^* - \omega\mu''\mathbf{H} \times \mathbf{D}^* \right\}, \qquad (6.36)$$

which is recognized to be the force density on free currents $\mathbf{J}_c = \omega\varepsilon''\mathbf{E}$ (and similarly for the magnetic part) (Loudon et al. 2005, Kemp et al. 2006).

6.5 Traversal of pulses in materials with negative material parameters

The phase velocity of waves in negative refractive index media is negative. "What then does it imply for the speed of light traveling through the medium?" is a question that one often encounters from an enthusiastic undergraduate student. A more experienced graduate student would immediately voice the often heard response that the phase velocity could be anything, but it is the group velocity that has information on the rate of motion of light through the medium. While this is correct to some extent, it does not encompass the whole truth as simply evidenced by media where the group velocity of light within some frequency bandwidth could also be negative. Moreover, the group velocity does not properly account for absorption or gain in a medium. The absorption or gain can drastically modify the spectrum of a pulse of light propagating in the medium and the pulse can be enormously reshaped. For example, a preferential absorption of the leading or trailing edge of a pulse can make it appear to have an ultra-slow or ultra-fast velocity. If the medium has rapidly varying material parameters over the bandwidth of the pulse, the effects on the pulse can be drastic, such as breaking the original pulse into a number of pulses. Hence a description by the group velocity alone might not suffice and one might need to study the energy flow in many such cases.

Metamaterials with negative material parameters usually have large dispersion in the effective medium parameters, reasonably large levels of dissipation, and additionally the energy flow represented by the Poynting vector is opposite in direction to the phase vector in isotropic negative refractive index media. Further, as shown in Section 3.3.2 (see Fig. 3.25), some metamaterials can also have a negative group velocity in addition to the negative phase velocity. There can only be evanescent waves inside media that have only one of ε or μ negative. The transport of evanescent pulses is interesting in its own right (Büttiker and Thomas 1998). Thus metamaterials have all the ingredients that can make the study of propagation of pulses through them very interesting.

In this section, we discuss the time taken by a pulse to traverse a medium – a quantity that encompasses several aspects of pulse propagation in the medium. The traversal time for a pulse through a dispersive medium is interesting and important both from a fundamental viewpoint, as well as for technological applications such as delay lines and systems with enhanced nonlinearities. We refer the reader to the many sources available in the literature, e.g. Landauer and Martin (1994) and Chiao and Steinberg (1997), for more detailed discussions of the fundamental importance of this quantity and various aspects of the traversal time. A related growing topic is the superluminal propagation of pulses through dispersive media, i.e., a pulse of light *apparently* traveling faster than the speed of light in vacuum (Milonni 2005). Although it is almost eighty years at the time of this writing since Brillouin and Sommerfeld showed that information transport represented by a sharp edge (irregularity) in the wave cannot be superluminal (Brillouin 1960), this topic has held the Physics community fascinated. The apparent superluminality for smooth pulses can arise either out of group velocity effects (Wang et al. 2000) or reshaping due to absorption or gain in the medium (Bolda et al. 1994). We reiterate that any pulse with a shape described by a holomorphic function of time (i.e., without any discontinuities or singularities) can exhibit superluminal propagation without violating causality or special relativity. This is because there is no information in the peak that is not already contained in the leading edge of the pulse. The peak can, in fact, be obtained by a Taylor series expansion about any point in the leading edge. Information can only be encoded by a meromorphic function through singularities in the function itself or in some higher order derivative of the function. These singularities, however, cannot propagate faster than the speed of light in vacuum.

In this section, we examine some of the aspects (Nanda et al. 2006, Nanda and Ramakrishna 2007) of pulse propagation through a medium with negative material parameters using the Wigner delay time (Wigner 1955), which is based on an extrapolation of the group velocity, and another quantity that is based on the flow of energy through a medium (Peatross et al. 2000). To model causal metamaterials with negative medium parameters, we assume that the dielectric permittivity is described by that of a plasma (a Drude model), while the magnetic permeability has a Lorentz-like dispersion around a resonance frequency (see Fig. 3.21 for the lossless case):

$$\varepsilon(\omega) = 1 - \frac{\omega_p^2}{\omega(\omega + i\gamma_p)}, \tag{6.37a}$$

$$\mu(\omega) = 1 + \frac{\omega_b^2}{\omega_{0m}^2 - \omega^2 - i\omega\gamma_m}. \tag{6.37b}$$

6.5.1   Wigner delay time for pulses in NRM

The Wigner (group) delay times (Wigner 1955) defined as

$$\tau_w = \partial\phi/\partial\omega, \tag{6.38}$$

where ϕ, the phase of the wave, has been a popular measure for the time delay of a pulse with a slowly varying envelope and a well-defined carrier wave frequency of ω. The Wigner delay time is based on tracking a fiducial point on a wavepacket that moves with the group velocity $v_g = \partial\omega/\partial k$, but has been generalized to include the phase shifts that arise due to scattering as well. The Wigner delay time can often unsurprisingly be superluminal or even negative as there is no causal relationship between the peaks of the input and output pulses (Martin and Landauer 1992). By a negative delay time, it is implied that the peak (for example) of the output pulse emerges out of the system under consideration even before the peak of the input pulse has entered the system. Surprising as this is, it should not be a cause for consternation as the two peaks are not causally connected. In a homogeneous medium, this arises from the superluminality or negativity of the group velocity (also see Section 5.3):

$$v_g = \left(\frac{\partial k}{\partial \omega} \bigg|_{\omega_c} \right)^{-1} = \frac{c}{n(\omega_c) + \omega \frac{\partial n}{\partial \omega} \big|_{\omega_c}}, \tag{6.39}$$

where n is the refractive index, ω_c is the carrier frequency of the pulse, and the derivatives are evaluated at the carrier frequency. It is clear that when $\frac{\partial n}{\partial \omega}\big|_{\omega_c}$ is large and negative as happens in the case of anomalous dispersion, the group velocity can be superluminal and even negative. On the other hand, where this quantity is a large and positive number, the group velocity can become very small and gives rise to ultra-slow light where the pulse travels at terrestrial speeds of a few meters per second. It can also be seen that a positive group velocity is possible for media with $n < 0$ only when the dispersion in the refractive index is large enough.

The fact that the Wigner delay time (or the group velocity) can be superluminal or negative has often been used to term the Wigner delay time (or the group velocity) as physically unimportant. But the group delay time, superluminal or otherwise, has been shown to well describe the arrival of electromagnetic pulses across absorptive media (Garrett and McCumber 1970) and amplifying media (Bolda et al. 1994), particularly for narrow bandwidths and short propagation distances. In such cases, the apparent superluminal propagation of the pulse can be explained by a preferential attenuation or amplification of the trailing or leading edge, respectively. However, the experiments of Wang et al. (2000) on superluminal pulse propagation in an almost non-absorptive or non-amplifying medium but with highly and anomalously dispersive refractive index have been a significant achievement in underlining the importance of the group velocity and the Wigner delay time in describing pulse propagation.

Let us calculate the Wigner delay time for the transmission of radiation, incident at some oblique angle, across a slab of material with some arbitrary material parameters (ε and μ). The transmission coefficient across the slab is given by

$$\mathbf{T}(\omega) = \frac{tt' e^{(ik_{z2}d)}}{1 - r'^2 e^{(2ik_{z2}d)}}, \tag{6.40}$$

where d represents the slab thickness and t, t', and r' represent the Fresnel coefficients of transmission and reflection relating the magnetic fields across the slab interfaces and are given by

$$t = \frac{2\frac{k_{z1}}{\varepsilon_1}}{\frac{k_{z1}}{\varepsilon_1} + \frac{k_{z2}}{\varepsilon_2}}, \quad t' = \frac{2\frac{k_{z2}}{\varepsilon_2}}{\frac{k_{z2}}{\varepsilon_2} + \frac{k_{z3}}{\varepsilon_3}}, \quad r' = \frac{\frac{k_{z2}}{\varepsilon_2} - \frac{k_{z3}}{\varepsilon_3}}{\frac{k_{z2}}{\varepsilon_2} + \frac{k_{z3}}{\varepsilon_3}}$$

for P-polarized light. Here the unprimed and primed coefficients stand for the first and the second boundaries, respectively. For S-polarized light, the ε are simply replaced by the corresponding μ in the expressions of the Fresnel coefficients. Also the Fresnel coefficients would then relate the electric fields across the interface rather than the magnetic fields. The suffix 2 represents the parameters inside the slab medium while the suffixes 1 and 3, respectively, represent the parameters on the incident and exit side of the slab, respectively. We take the media to be the same on both sides of the slab.

The Wigner delay time can be calculated using the phase of the transmission coefficient given above, and is given by

$$\tau = \frac{\partial \phi}{\partial \omega} = \frac{\frac{\partial p}{\partial \omega} \tan(k_{z2}d) + p \sec^2(k_{z2}d) \frac{\partial k_{z2}}{\partial \omega} d}{1 + p^2 \tan^2(k_{z2}d)}, \tag{6.41}$$

where

$$p = \frac{1}{2} \left(\frac{k_{z1}\varepsilon_2}{k_{z2}\varepsilon_1} + \frac{k_{z2}\varepsilon_1}{k_{z1}\varepsilon_2} \right) \tag{6.42}$$

for P-polarized radiation, and

$$p = \frac{1}{2} \left(\frac{k_{z1}\mu_2}{k_{z2}\mu_1} + \frac{k_{z2}\mu_1}{k_{z1}\mu_2} \right) \tag{6.43}$$

for S-polarized radiation. It is understood that all quantities in the above are evaluated at the carrier wave frequency of the pulse. Note that the expressions for P and S polarized light become identical at normal incidence.

Consider a slab of metamaterial with $\omega_{0m} = 0.3\omega_p$, $\omega_b = 0.5197\omega_p$, and $\gamma_m = \gamma_p$. Our choice of ω_{0m} and ω_b corresponds to a magnetic plasma frequency of $\omega_m = 0.6\omega_p$ when $\mu = 0$. We note that $\varepsilon < 0$ for $\omega < \omega_p$ and $\mu < 0$ for $\omega_{0m} < \omega < \omega_m$. In Fig. 6.8, we show the Wigner delay times obtained for transmission across the slab for normally incident radiation in different frequency bands when the medium has negative refractive index or positive refractive index, and for non-dissipative and slightly dissipative media. One can clearly see that the delay times can be large for pulse traversal in the negative index band ($\omega_{0m} < \omega < \omega_m$). This is actually due to the large values of μ (and correspondingly n) near the magnetic resonance frequency in this case. The delay time as a function of frequency shows a large number of peaks that arise due to the slab resonances which are determined by the poles of the transmission coefficient. In the negative index bandwidth, we appear to have two sets of resonances – one near the magnetic resonance frequency where the magnetic permeability has a large magnitude and another near the magnetic plasma frequency where the magnetic permeability has small magnitude and is negative. In both cases, the resonances affect the delay times significantly due to the large impedance mismatch with vacuum. When the dissipation is small to moderately large ($\gamma_p = \gamma_m \sim 10^{-2}\omega_p$ to $10^{-3}\omega_p$), one can have still large values of the delay times although much of the resonant features

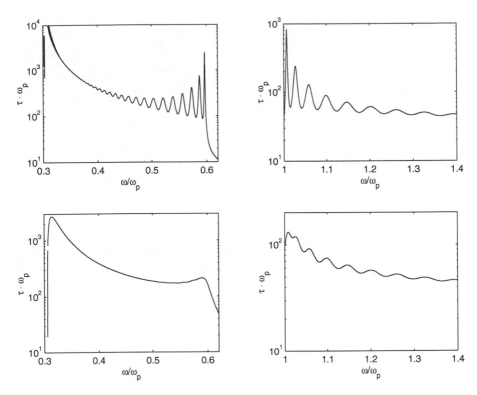

Figure 6.8 Wigner delay times for pulse propagation across a slab for radiation incident at normal incidence. Top left: For a non-dissipative negative refractive index medium ($\gamma_m = \gamma_p = 0$). Top right: For a non-dissipative positive refractive index medium. Bottom left: Same as for top left, but with moderate dissipation $\gamma_p = \gamma_m = 0.02$. Bottom right: Same as for top right, but with moderate dissipation $\gamma_p = \gamma_m = 0.02$.

due to the slab resonances become negligible. The pole structure for the positive index frequency band is different. The resonant features that appear are those due to the slab resonances caused by a small value for ε at frequencies beyond and near the plasma frequency. Similar results have been shown for metamaterials in the zero dissipation limit in Duttagupta et al. (2004), who also confirmed these results by numerically propagating pulses with narrow frequency bandwidths. We conclude that negative index media could have large group indices due to the resonant nature of the material parameters and this is not drastically modified by moderate levels of dissipation in the medium. This has much potential for applications where solid-state large group index media are required. The only restriction is that the frequency bandwidth of the pulse should be narrow enough compared to the bandwidth of the metamaterial resonance.

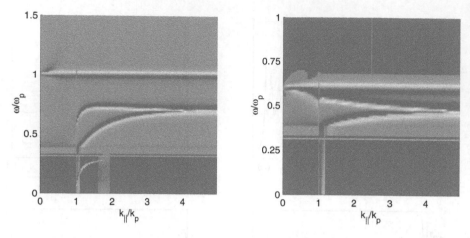

Figure 6.9 Wigner delay times for pulse propagation across a slab of thickness ($\Delta r = \lambda_p/5$) with moderate dissipation $\gamma_p = \gamma_m = 0.01\omega_p$ and embedded in vacuum as a function of frequency and parallel wave-vector magnitude. Left: The delay times for P-polarized light. Right: The delay times for S-polarized light. The grayscale maps predominantly trace out the conditions for the existence of surface plasmon polariton modes where the delay times are very large in magnitude.

We next consider the Wigner delay times for the case of oblique incidence of waves and for evanescent incident waves. We plot the Wigner delay time vs. both the frequency and the parallel wave-vector in a moderately dissipative slab ($\gamma_p = \gamma_m = 0.01\omega_p$). In a comparatively uniform landscape of delay times, the resonant conditions for the slab surface plasmon polaritons (SPPs) stand out in stark contrast where the magnitude of the delay times are very large. The entire dispersion of the SPPs can be traced out by the regions of large magnitude of the Wigner delay times as can be seen in Fig. 6.9. Both the antisymmetric and symmetric SPP modes of electric nature manifest for the P-polarized light, while similar SPP modes of magnetic nature manifest for S-polarized light. More discussion of the nature of these modes and their dispersion can be found in Chapter 7. The other notable features are the large negative delay times that occur near the electric (ω_p) and magnetic (ω_m) plasma frequencies for the P- and S-polarized light, respectively. The other very large feature in the plots occurs for both polarizations at the magnetic resonance frequency which is characterized by extremely large magnitudes for τ_w. We conclude that delay times of very large magnitudes are possible for evanescent waves interacting with the slab due to their coupling with the slab plasmon polariton modes. This can be an important and interesting manner of providing for delay times in plasmonic systems and future plasmonic devices.

6.5.2 Traversal times based on the flow of radiative energy

In this section we discuss a time scale, due to Peatross et al. (2000), that directly relates to the average flow of electromagnetic energy through the medium. For pulses, and particularly broadband pulses, the arrival time of a pulse at a point $\mathbf{r}$ can be well described by a time average over the component of the Poynting vector $\mathbf{S}$ normal to a (detector) surface at $\mathbf{r}$ as

$$\langle t \rangle_{\mathbf{r}} = \frac{\mathbf{u} \cdot \int_{-\infty}^{\infty} t\mathbf{S}(\mathbf{r}, t) dt}{\mathbf{u} \cdot \int_{-\infty}^{\infty} \mathbf{S}(\mathbf{r}, t) dt}. \tag{6.44}$$

Here $\mathbf{u}$ is taken to be the unit vector along the normal to the given surface. The time for traverse between two points $(\mathbf{r}_i, \mathbf{r}_f)$ is equal to the difference of the arrival times at the two points, and we call this quantity the *traversal time*. This traversal time can be shown analytically (Peatross et al. 2000) to consist of two parts: a contribution by the spectrally weighted average group delay at the final point $\mathbf{r}_f$,

$$\Delta t_G = \frac{\mathbf{u} \cdot \int_{-\infty}^{\infty} \mathbf{S}(\mathbf{r}_f, \omega) \left[(\partial \, \mathrm{Re} k / \partial \omega) \cdot \Delta \mathbf{r} \right] d\omega}{\mathbf{u} \cdot \int_{-\infty}^{\infty} \mathbf{S}(\mathbf{r}_f, \omega) d\omega}, \tag{6.45}$$

and a contribution that can be ascribed to the reshaping of the pulse by dissipation,

$$\Delta t_R = \mathcal{T} \left[\exp(-\mathrm{Im} \mathbf{k} \cdot \Delta \mathbf{r}) \mathbf{E}(\mathbf{r}_i, \omega) \right] - \mathcal{T} \left[\mathbf{E}(\mathbf{r}_i, \omega) \right], \tag{6.46}$$

which is calculated with the frequency spectrum of the pulse at the initial point $\mathbf{r}_i$. Here the operator $\mathcal{T}$ is

$$\mathcal{T} \left[\mathbf{E}(\mathbf{r}, \omega) \right] = \frac{\mathbf{u} \cdot \int_{-\infty}^{\infty} \mathrm{Re} \left[-i \frac{\partial \mathbf{E}(\mathbf{r}, \omega)}{\partial \omega} \times \mathbf{H}^*(\mathbf{r}, \omega) \right] d\omega}{\mathbf{u} \cdot \int_{-\infty}^{\infty} \mathbf{S}(\mathbf{r}, \omega) d\omega}, \tag{6.47}$$

which represents the arrival time of a pulse at a point $\mathbf{r}$ in terms of the spectral fields and $\mathbf{S}(\mathbf{r}, \omega) \equiv \mathrm{Re} \left[\mathbf{E}(\mathbf{r}, \omega) \times \mathbf{H}^*(\mathbf{r}, \omega) \right]$ represents the Poynting vector. Here we take the real parts of the quadratic terms since we use complex representation for the fields, i.e., $e^{i(\mathbf{k} \cdot \mathbf{r} - \omega t)}$ for a plane wave. The total traversal time has been shown (Peatross et al. 2000) to remain subluminal for broadband pulses for traversal across a medium with Lorentz dispersion for the dielectric permittivity. The most significant aspect of this proposal is that unlike any description based on the group velocity such as the Wigner delay time, it does not involve any perturbative expansion of the wave number around the carrier frequency. Also note that while there are significant difficulties related to the definition of an energy density in a dispersive medium as discussed earlier in this chapter, there is no such ambiguity about the Poynting vector or the power flux. This underlines the importance of this traversal time which does not depend on the specific details of the model for dispersion. The arrival time based on the average of the Poynting vector has also been shown to be equivalent to the arrival times measured by the average rate of absorption in an ideal impedance matched detector (Nanda et al. 2006). In fact, these times have also been experimentally measured for ultrashort pulses (Talukder et al. 2005) propagating through dispersive media as well as angularly dispersive systems (Talukder et al. 2007) underlining their importance as an important quantity to characterize the time spent in a medium by pulses. Even if the output pulses are badly deformed these delay times have been shown to experimentally well describe the motion of the centroid of the pulse (Talukder et al. 2007).

6.5.2.1   Traversal times through negative refractive index media

We discuss subsequently the nature of traversal times for a pulse traversing across a slab of negative refractive index medium as a function of the carrier frequency for a pulse with broad bandwidths. We show in Fig. 6.10 the delay times obtained by Eq. (6.45) and Eq. (6.46) for pulse traversal across a slab whose material parameters are described by a plasma form for the dielectric permittivity and a Lorentz form for the magnetic permeability, respectively. The relevant frequencies have been chosen to be $\omega_{p,b,0} = 2\pi f_{p,b,0}$ with f_p, f_b and f_0 taken to be 12 GHz, 6 GHz and 4 GHz, respectively. First, we note that for ultrashort pulses with large frequency bandwidths compared to the resonance bandwidths, the resonant features in the graph of the traversal time such as due to the slab resonances tend to get averaged out. In the presence of dissipation in the medium (finite width of the resonance), the Wigner delay time can become superluminal or even negative at frequencies near the resonance frequency due to the anomalous nature of dispersion. However, if the bandwidth for the pulse is made sufficiently large, the traversal times based on the energy flow become positive and eventually subluminal (See Fig. 6.10). Hence the problem of super-luminality is not present with this definition. The nature of large delay times near the magnetic resonance frequency outside the region of anomalous dispersion is still retained, although the corresponding delay times are not as large as in the limit of zero dissipation.

Another important issue is the relative contribution of the group delay time and the reshaping delay time to the total delay time. In an experiment, one usually measures the total traversal time. To avoid any resonant features arising from the slab resonances here, we place the source inside the negative index medium assumed to be semi-infinitely extended and a distance d away from the boundary with vacuum. A detector is assumed to be placed immediately outside the boundary as depicted schematically in Fig. 6.10. For a plane wave pulse propagating normal to the interface, the delay times based on the energy flow are shown in Fig. 6.10(c). We can immediately conclude that there is appreciable contribution to the total traversal time from both the group delay time as well as the reshaping delay time. Particularly, the group delay time dominates due to large dispersion near the resonance frequency, while away from the region of anomalous dispersion near the resonance, the reshaping delay times contribute very significantly. Hence in the interpretation of any experiment involving the delay times of pulse through negative index media such as the one reported by Dolling et al. (2006b), one would have to carefully sort out the relative contribution of the group delay times in order to conclude anything about the dispersion in the medium and the refractive index. As an aside we also note here an unusual result that the reshaping delay time for a medium with $\varepsilon(\omega) = \mu(\omega)$ is identically zero (Nanda et al. 2006) and only group delay effects are present in such a medium.

6.5.2.2   Traversal times for evanescent waves

We consider in this section the traversal times for pulses composed entirely of evanescent waves. This is analogous to quantum mechanical tunneling of a particle under a potential barrier. Such situations arise directly in the transport of radiation across a metal slab or under conditions of total internal reflection when the parallel component of the wave-vector becomes larger than the wave number in the medium. Since the phase vectors for the evanescent waves are imaginary, interesting questions arise regarding their traversal times (Landauer and Martin 1994). One of the most paradoxical aspects is the Hartman effect (Hartman 1962), which is the saturation of the Wigner delay time with the barrier thickness, i.e., the delay time stops increasing with the thickness of the barrier beyond a certain point. Hence the delay time of evanescent pulses through sufficiently thick barriers always appears superluminal. We study here the traversal time associated with the energy transport for a pulse composed predominantly of evanescent modes. It is actually difficult to envisage the motion of such a pulse. At any point away from the source, the intensity of such a pulse would simply appear to increase and then decrease in time. If the source were located in an infinitely extended medium (of spatial extent large compared to the wavelength of light), then the maximum of the pulse would always appear to be concentrated at the point of the source! But that is the nature of the non-radiative near field of a source. The problems that we consider arise in the detection of such near fields either by a local absorbing detector or by coupling to propagating modes.

Consider the complex wave-vector in a medium,

$$k^2 = \varepsilon\mu\frac{\omega^2}{c^2}. \tag{6.48}$$

In the limit of small imaginary parts of ε and μ, one can write for the wave-vector

$$k_r = \text{Re}(k) \simeq \frac{\omega}{c}\sqrt{\varepsilon_r\mu_r - \varepsilon_i\mu_i}, \tag{6.49}$$

$$k_i = \text{Im}(k) \simeq \frac{\omega}{c}\frac{\varepsilon_r\mu_i + \varepsilon_i\mu_r}{2\sqrt{\varepsilon_r\mu_r - \varepsilon_i\mu_i}}. \tag{6.50}$$

For propagating waves, the real part of the wave-vector depends primarily on ε_r and μ_r, while the imaginary part is directly proportional to ε_i and μ_i or the dissipation. This, however, becomes the other way around for evanescent waves. To make the discussion for evanescent waves clear, we consider an absorbing electric plasma with $\varepsilon_r < 0$, $\varepsilon_i > 0$, $\mu = \mu_r$, and then

$$k_r \simeq \frac{\omega}{c} \frac{1}{2} \sqrt{\frac{\mu_r}{|\varepsilon_r|}} \varepsilon_i, \tag{6.51}$$

$$k_i \simeq \frac{\omega}{c} \sqrt{|\varepsilon_r| \mu_r}. \tag{6.52}$$

Thus the real part of the wave-vector depends on the levels of dissipation in the medium (ε_i), while the imaginary part of the wave-vector which determines the decay of the wave depends on $|\varepsilon_r|$. This implies, in turn, that the definitions of the group delay time and the reshaping delay time given by Eq. (6.45) and Eq. (6.46), respectively, get interchanged for evanescent waves. This is an important difference for the arrival times of evanescent waves from that of propagating waves (Ramakrishna and Kumar 2002). This behavior can also be analytically continued for larger values of the imaginary parts. This difference arises because the quintessential decay length for evanescent waves is determined by ε_r and not ε_i.

We consider, first, the time for traversal of an evanescent pulse through a plasma. There is an inherent negativity associated with the the time of transport for an evanescent wave in an infinitely extended dissipative medium. The energy transport for such a wave happens through the phase shift caused by the imaginary part of the dielectric permittivity (Ramakrishna and Armour 2003). At any point in the medium, the energy flow away from the source is proportional to the energy dissipated in the entire region of space beyond that point all the way up to infinity. Thus, it is almost as if the source has to correctly anticipate the dissipation at regions far away and radiate accordingly. This is confirmed by calculating the traversal time using the energy flow at a point in the plasma, a distance d away from the source placed inside the plasma medium. The total traversal time comes out to be negative (see Fig. 6.11(a)) and dominated by the reshaping delay time indicating the dominance of dissipation in the energy transport in this configuration. This behavior is rather independent of the bandwidth of the pulse. This situation is, however, unnatural as the detector would have to be placed inside the plasma medium.

It is better to consider a semi-infinitely extended plasma medium with the source placed inside the plasma a distance d away from the boundary and the detector to be in vacuum outside the plasma. A corresponding physical situation would be an atom or an antenna located within the plasma and emitting radiation, whose leakage is detected outside the plasma or the metamaterial with $\varepsilon < 0$. For simplicity, consider only waves with zero parallel wave-vector. This results in the wave not coupling to any surface plasmon modes of the plasma-vacuum interface (see Chapter 7). Thus our source would be an infinite plane of current parallel to the interface. In this situation and for small distances, $d \sim c/100\gamma$ and $\simeq c/10\gamma$, the reshaping delay time is actually negligibly small compared to the group delay time even for $\omega < \omega_p$ and the group delay dominates the total delay time (see Fig. 6.11(b) and (c)). This important difference from the case of an unbounded plasma results because of the presence of a reflected evanescent wave from the boundary. Now energy transport is primarily determined by the phase difference of the incident evanescent wave and the reflected wave, and not merely by the dissipation in the medium. Thus the group delay time plays the determining role for small distances when the effect of the reflected wave can be considerably well felt. For large distances $d \sim c/\gamma$, however, the reshaping delay time ($\bar{\omega} < \omega_p$) tends to that of the behavior in an infinite plasma, while the group delay time strongly moderates this contribution to the total delay, and the total delay time is positive everywhere, at least asymptotically with increase in the frequency bandwidth of the pulse.

Another interesting aspect is the presence of a Hartman effect even for the energy transport of a pulse through the plasma medium. We show here the traversal time calculated for the detection of an evanescent pulse emitted by a source placed inside the plasma. The schematic diagram for the arrangement is shown in Fig. 6.10. Figure 6.11 (d) shows the traversal time for the source distances of $c/1000\gamma_p$, $c/10\gamma_p$ and for c/γ. We find that over frequency band of the plasma behavior between $2\gamma_p$ and $8\gamma_p$, there is hardly any appreciable change in the traversal both with distance and frequency. In fact, hardly any change in the delay time can be seen when the distance is changed from $d = c/(1000\gamma_p)$ to $c/(10\gamma_p)$. For large thicknesses, however, the dissipation in the medium begins to play a greater role and the Hartman effect is lost. The new feature here is the presence of the Hartman effect for broad-band pulses as well.

It should be re-emphasized that even superluminal energy transport by a pulse described by a holomorphic function does not actually violate causality or special relativity. But the traversal times in such cases reassuringly turn positive and even sub-luminal whenever the bandwidth of the pulse is made large enough! Finally, it should be noted that in some of the cases that we have discussed the amount of energy that is actually transmitted across the plasma medium via the evanescent waves may be minuscule although large enough to be detected. But the time at which the pulse arrives and the amplitude with which the pulse arrives are entirely different matters.

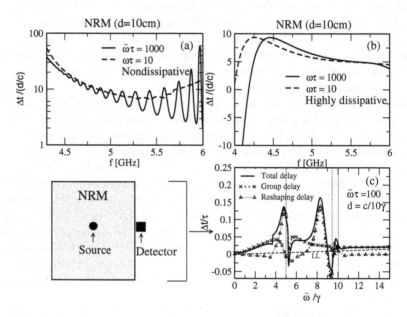

Figure 6.10 (a) Traversal times for pulse propagation across an NRM slab of thickness ($d = \lambda_p/5$) for broadband pulses as a function of the carrier frequency. (b) When there is large dissipation in the medium, $\gamma_p = 0.1\omega_p$. (c) Pulse traversal time in a negative index medium showing the relative contributions of the group delay time and the reshaping delay time in an NRM. The geometry of the source and detector for case (c) is shown. In all cases here, τ is the pulse width in time and $\bar{\omega}$ is the carrier frequency.

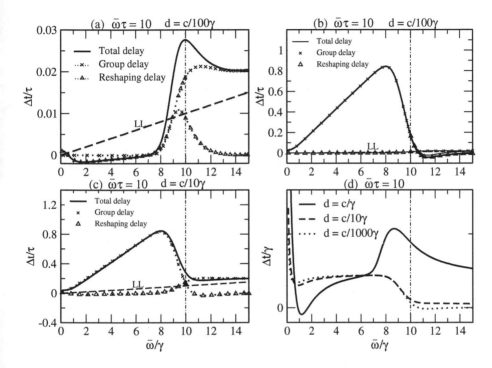

Figure 6.11 (a) Traversal time in an infinitely extended dissipative plasma when the radiation is both emitted and detected a distance d away inside the plasma. The reshaping delay time is negative and determines the total traversal time as well. (b) and (c) Traversal times for pulse propagation across a plasma medium composed of a plasma of thickness for broadband pulses. (d) Pulse traversal time in a plasma medium showing the Hartman effect for the traversal time. The traversal time does not change appreciably in magnitude for over two orders of magnitude change in the distance. In all cases the traversal time is plotted as a function of the carrier frequency and the plasma frequency is taken to be $\omega_p = 10\gamma_p$.

Figure ...

7

Plasmonics of media with negative material parameters

An important and unique feature of materials with negative material parameters is their ability to support a variety of surface electromagnetic modes. These surface modes have electromagnetic fields that have maximum amplitude at the surface of the medium with negative material parameters and the fields decay exponentially inside both the bulk of the medium and in vacuum (or the positive medium) outside. This feature is well known and extensively studied in the case of metals or plasmas, which have negative dielectric coefficients, and the surface modes are called surface plasmon (Ritchie 1957, Raether 1986). On a metal surface, the surface plasmons are essentially collective excitations of electrons with the charges displaced parallel to the (real part of the) wave vector on the surface of the metal. The interior of the metal is, however, shielded from these electromagnetic fields and the wave amplitude decays into the bulk of the metal as in a regular conducting medium. These surface plasmon modes have also been termed Zenneck waves or Sommerfeld waves in the context of the ionosphere. Fig. 7.1 qualitatively represents a surface plasmon mode on the surface of a metal. These charge density waves flow on the surface, they scatter off obstacles on the plane, reflect and refract off interfaces between two surfaces: thus they can literally be considered two-dimensional entities that exist on the surface. Surface plasmons on a plane surface cannot directly interact with propagating radiation in vacuum and are coupled mainly through scattering events (surface roughness, Bragg scattering in the case of periodic scatterers such as a diffraction grating, etc.). There are also localized surface plasmons that can be confined to the surface of a sphere, a cylinder or, in general, any scatterer made of a negative dielectric medium. These plasmons can be directly excited by an incident plane wave of light.

The surface plasmons become approximately degenerate for larger wavevectors on a metal surface. This is the essential reason why modeling the interaction of radiation with metallic structures of complex shape is so difficult – the radiation couples (through scatterers) to the surface plasmons at all lengthscales near the resonant frequencies. The surface plasmons have a P-polarized nature (TM polarization) on a metal surface. Apart from sensor applications, the main attraction for studying the properties of propagation and confinement of plasmons on metallic surfaces is the promise that

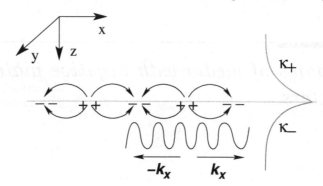

Figure 7.1 Schematic representation of a surface plasmon at an interface between a negative and a positive dielectric medium, and the associated charge density fluctuations. The exponential decay of the fields normal to the surface and the propagating nature along the surface are depicted schematically. (Reproduced with permission from (Ramakrishna 2005). © 2005, Institute of Physics Publishing, U.K.)

plasmons hold for miniaturization of all-optical circuits and communication devices. Photonic components due to photonic bandgap materials or conventional waveguides are just too bulky for integration with nonsecular electronic components. With surface plasmons, however, the wavelength on the surface can be much smaller than the wavelength of free radiation. This has implications for the possible miniaturization of the waveguides, the circuit channels, etc. required for optical switches and digital logic circuits. There can also be large local field enhancements due to the excitation of localized resonances, which implies that nonlinear effects can be made large, thus enabling easy switching of optical pulses. This has been the main driving force behind basic research on the control of surface plasmons on a metal surface. For more detailed discussions of surface plasmon modes on metallic surfaces, we refer the reader to Maier (2007).

The study of surface plasmons has regained interest and importance with the realization of metamaterials, in particular due to the possibility of reaching negative permeabilities. Until now, only P-polarized (TM-modes) surface plasmons could be excited on a metallic surface due to the necessity of having a normal component of the electric field. The possibility of having negative magnetic permeability gives rise to a electromagnetic modes of an alternate magnetic nature for the S-polarized waves (TE-modes) as well. This can give rise to new polarization-dependent phenomena in metamaterials shaped into complex geometrical shapes. In this chapter, we discuss, mainly in the context of metamaterials, some of the basic aspects of surface plasmon modes, their dispersion and interaction with radiation, and negative refraction at the interface between two surfaces. We also explore some sub-wavelength structured conducting films that show plasmonic responses. Surface modes on nonlinear metamaterials are briefly touched upon toward the end of the chapter.

7.1 Surface electromagnetic modes in negative refractive materials

In this section, we primarily focus on the conditions for the existence of surface plasmon modes and their dispersion in a variety of geometries and materials. The plasmon modes turn out to be crucial for the performance of a super-lens that can image the evanescent near-field modes of a source (see Chapter 8). An understanding of the conditions under which they exist and can be excited is important. The dispersive properties of the surface plasmons are discussed assuming a *local* response for the media concerned whereby the material parameters such as the dielectric permittivity (ε) and the magnetic permeability (μ) depend only on the frequency ω and are not spatially dispersive.

7.1.1 Surface plasmon modes on a plane interface

The dispersive properties of surface plasmon modes on the surface of a semi-infinitely extended negative refractive index medium have been considered in Ruppin (2000b). Let us consider the interface ($z = 0$ plane) between two semi-infinitely extended media whose material parameters are given by (ε_1, μ_1) and (ε_2, μ_2). Consider the time harmonic fields at frequency ω in the two media for P-polarized light (magnetic field along the y axis):

$$\mathbf{H}(x, y, z) = \begin{cases} \hat{y} H_2 \exp[i(k_x x + k_y y - \omega t) - \kappa_{z2} z], \; \forall \; z > 0, \\ \hat{y} H_1 \exp[i(k_x x + k_y y - \omega t) + \kappa_{z1} z], \; \forall \; z < 0, \end{cases} \quad (7.1)$$

where $k_x^2 + k_y^2 - \kappa_{z1}^2 = \varepsilon_1 \mu_1 \omega^2 / c^2$ and $k_x^2 + k_y^2 - \kappa_{z2}^2 = \varepsilon_2 \mu_2 \omega^2 / c^2$. Note that the fields decay exponentially away from the interface into the bulk of the media in either side. The transverse components of both fields are the same due to the translation invariance along the transverse directions (necessity of phase matching). The coefficients H_1 and H_2 in the above equations are solved for by enforcing the conditions of continuity of the tangential magnetic field and the electric field across the interface:

$$H_2 - H_1 = 0, \quad \frac{\kappa_{z2}}{\omega \varepsilon_2} H_2 + \frac{\kappa_{z1}}{\omega \varepsilon_1} H_1 = 0. \quad (7.2)$$

This system of homogeneous equations has a non-trivial solution when the determinant is zero:

$$\frac{\kappa_{z1}}{\varepsilon_1} + \frac{\kappa_{z2}}{\varepsilon_2} = 0, \quad (7.3)$$

which is the condition for the existence of the surface modes. It is obvious that solutions to this equation require that the permittivities of the two media be of opposite signs.

¿From the above condition, the dispersion for these surface modes is obtained as

$$k_x = \left[\frac{\varepsilon_1 \varepsilon_2 (\varepsilon_2 \mu_1 - \varepsilon_1 \mu_2)}{\varepsilon_2^2 - \varepsilon_1^2} \right]^{1/2} \frac{\omega}{c}. \tag{7.4}$$

A similar condition for S-polarized light (electric field along the $\hat{y}$ direction) can be obtained as

$$\frac{\kappa_{z1}}{\mu_1} + \frac{\kappa_{z2}}{\mu_2} = 0, \tag{7.5}$$

which obviously requires only one of the μ involved be negative. We can similarly obtain the dispersion for these surface modes as

$$k_x = \left[\frac{\mu_1 \mu_2 (\mu_2 \varepsilon_1 - \mu_1 \varepsilon_2)}{\mu_2^2 - \mu_1^2} \right]^{1/2} \frac{\omega}{c}. \tag{7.6}$$

Note that the P-polarized and S-polarized modes do not coexist at any frequency except where $\varepsilon_2 \mu_2 = 1$.

In order to understand the dispersion of these surface modes, it is important to consider causal, frequency dispersive models for the negative permittivity and permeability. Let us, as usual, consider a plasma model for the dielectric permittivity (see Section 3.1),

$$\varepsilon_2 = 1 - \frac{\omega_p^2}{\omega(\omega + i\gamma_p)}, \tag{7.7}$$

where the plasma frequency is ω_p and a magnetic resonance model for the magnetic permeability (see Section 3.2) is:

$$\mu_2 = 1 + \frac{f\omega^2}{\omega_{0m}^2 - \omega^2 - i\omega\gamma_m}, \tag{7.8}$$

where the magnetic plasma frequency (when $\mu_2 = 0$) is given by $\omega_{mp}^2 = \omega_{0m}^2/(1-f)$. While investigating the dispersion of these modes, let us assume the limit of zero damping $\gamma_p \to 0$ and $\gamma_m \to 0$. Let us take the positive media to have $\varepsilon_1 = 1$, $\mu_1 = 1$.

The dispersion for the cases of the P-polarized and the S-polarized surface modes are shown in Fig. 7.2. Note that the surface modes exist only for frequencies when the waves are evanescent in both media. This requires $k_x^2 > \omega^2/c^2$ and $k_x^2 > \varepsilon_2 \mu_2 \omega^2/c^2$. It can be seen from the figure that the P- and S-polarized modes become degenerate for large wave-vectors k_x at the frequencies $\omega = \omega_{0m}$, $\omega_p/\sqrt{2}$ and $\omega = \omega_{ms} = \omega_{0m}/\sqrt{1 - f/2}$, respectively. For the parameters associated with the graph on the top panel of Fig. 7.2, there are two branches of the P-polarized modes: one at low frequencies for

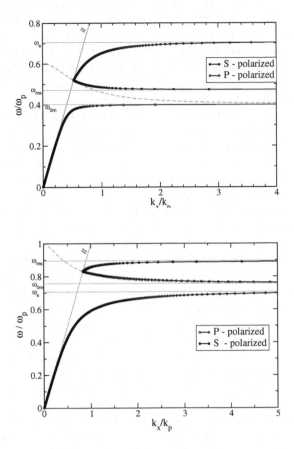

Figure 7.2 Top panel: Dispersion of the surface plasmon modes for a medium with the material parameters given by Eqs. (7.7) and (7.8). There are two branches for the P-polarized modes (with positive slope) and one for the S-polarized mode (negative slope). The assumed parameters are $\omega_{0m} = 0.4\omega_p$, $f = 0.56$. Bottom panel: Dispersion of the surface plasmon modes for a medium with $\omega_{0m} = 0.75\omega_p$, $f = 0.56$. Note that when $\omega_{mp} > \omega_p$ the dispersion curves for the S- and P-polarized modes change the sign of their slopes. In both panels, the light line ($\omega = ck_x$) indicated as "ll" and the curve $k_x = \varepsilon(\omega)\mu(\omega)\omega^2/c^2$ (dashed curve) are indicated in grey. The surface modes can occur only to the right of these curves. The dielectric surface plasmon frequency ω_s when $\varepsilon_2 = -1$, the magnetic surface mode frequency ω_{ms} when $\mu_2 = -1$, and the magnetic resonance frequency ω_{0m} are indicated by grey lines in both panels.

$\omega < \omega_{0m}$ where $\varepsilon_2 < 0$ while $\mu_2 > 0$ and is large near the resonance, and another one for $\omega < \omega_p$. The low frequency P-polarized branch is the only branch present for an interface between vacuum and a metal with $\varepsilon_2 < 0$. Part of the second branch lies in the spectral region where both $\mu_2 < 0$ and $\varepsilon_2 < 0$, thus in a region of negative refractive index. This branch essentially originates on the intersection of the light line in medium 1 ($k_x = \sqrt{\varepsilon_1 \mu_1} \omega / c$) and the curve $k_x > \sqrt{\varepsilon_2 \mu_2} \omega / c$. There is, by comparison, only one branch for the S-polarized mode that lies entirely within the left-handed regime where $\varepsilon_2 < 0$ and $\mu_2 < 0$. Note that the locations of the high frequency branch for the P-polarized mode and the S-polarized mode are interchanged if we choose $\omega_{mp} > \omega_p$. This is shown in the bottom panel of Fig. 7.2 where $\omega_{0m} = 0.75 \omega_p$ and $f = 0.56$. It can be seen that when the magnetic plasma frequency $\omega_{mp} > \omega_p$ the S-polarized mode has a positive slope while the P-polarized mode has negative slope. In this case, we also have that both the high frequency S and P modes lie entirely within the region of negative refractive index, i.e., when $\varepsilon_2 < 0$ and $\mu_2 < 0$. Note that our treatment includes the surface modes at the interface between the two media if $\varepsilon_1 < 0$, $\mu_1 > 0$ and $\varepsilon_2 > 0$, $\mu_2 < 0$, in which case, there are no propagating modes in either of the two media as $k_x^2 > \varepsilon_1 \mu_1 \omega^2 / c^2$ and $k_x^2 > \varepsilon_2 \mu_2 \omega^2 / c^2$ for all k_x, however small.

In the previous discussion, the dissipation in the medium was neglected while examining the dispersion of the surface plasmon modes. When a finite imaginary part is introduced into the negative medium parameters, the transverse wave-vector becomes complex. The imaginary part corresponds to the inverse propagation length on the surface. The surface plasmon amplitude decays to $1/e$ of its initial values while propagating over this length. Due to the algebraic complexity of the expressions, it is fruitful to examine them separately for different cases.

(*i*) *A negative dielectric metal*: For case of a metal with only $\varepsilon_2 < 0$ and $\mu_2 = \mu_1 = 1$, we can simplify the dispersion equation for the P-polarized modes to

$$k_x = \left(\frac{\varepsilon_1 \varepsilon_2}{\varepsilon_1 + \varepsilon_2} \right)^{1/2} \frac{\omega}{c}, \qquad (7.9)$$

Noting that $\varepsilon_1 = \varepsilon_2' + i \varepsilon_2''$, in the limit $\varepsilon_2'' \ll \varepsilon_2'$ we obtain

$$k_x \approx \left(\frac{\varepsilon_2'}{\varepsilon_1 + \varepsilon_2'} \right)^{1/2} \frac{\omega}{c} + i \left(\frac{\varepsilon_2'}{\varepsilon_1 + \varepsilon_2'} \right)^{3/2} \frac{\varepsilon_2''}{2(\varepsilon_2')^2} \frac{\omega}{c}. \qquad (7.10)$$

In order to satisfy the conditions for a surface plasmon mode, we require $\varepsilon_2 < 0$, and for real k_x' an additional requirement is $|\varepsilon_2'| > \varepsilon_1$. The propagation length for the surface plasmon is

$$\ell_{sp} = \{2\text{Im}(k_x)\}^{-1}. \qquad (7.11)$$

For surface plasmons on silver at about 500 nm wavelength radiation, the propagation length evaluates to about $\ell_{sp} \simeq 20\ \mu$m. Basically as $|\varepsilon_2'|$ increases, the fields penetrate lesser into the metal and the consequent losses are lesser.

At about 1200 nm wavelength, the surface plasmon propagation length in comparison comes out to be about 1 mm. To measure ℓ_{sp}, we would need to inject energy in the surface modes at one point and measure the fields of the mode some distance away, which can be accomplished by near-field microscopy techniques (Paesler and Moyer 1996). Another related quantity that is of interest is the decay time of the surface plasmon which can be probed by a pulse of light. Upon considering k_x to be a real quantity (excitation by a plane wave), the real and imaginary parts of the complex frequency are given by

$$\omega' - i\omega'' = (k_x c)\left(\frac{\varepsilon_1 + \varepsilon_2'}{\varepsilon_1 \varepsilon_2'}\right)^{1/2} - i(k_x c)\frac{\varepsilon_2''}{2\varepsilon_2'^2}\left(\frac{\varepsilon_1 \varepsilon_2'}{\varepsilon_1 + \varepsilon_2'}\right)^{1/2}. \tag{7.12}$$

The imaginary part of the frequency is negative and represents the rate of dissipation in the medium, whereas the decay time is inversely proportional to this and $t_{sp} = (2\pi)/\omega''$. The spatial decay length on the surface and the decay time are related by $\ell_{sp} = v_g t_{sp}$, where v_g is the group velocity on the surface.

(*ii*) *Interface between vacuum and a negative refractive index medium:* Let us consider the case when $\varepsilon_1 = 1$ and $\mu_1 = 1$. Then the dispersion reduces to

$$k_x = \left[\frac{\varepsilon_2(\varepsilon_2 - \varepsilon_1)}{\varepsilon_2^2 - 1}\right]^{1/2}\frac{\omega}{c}. \tag{7.13}$$

We can separate the real and imaginary parts of the right-hand side in the limit of $\varepsilon_2'' \ll \varepsilon_2'$ and $\mu_2'' \ll \mu_2'$ in a straightforward but slightly tedious calculation and obtain

$$k_x' \approx \left[\frac{\varepsilon_2'(\varepsilon_2' - \mu_2')}{\varepsilon_2'^2 - 1}\right]^{1/2}\frac{\omega}{c}, \tag{7.14}$$

$$k_x'' \approx \left[\varepsilon_2''\frac{(\varepsilon_2'\mu_2' + 2)(\varepsilon_2'^2 - \mu_2'\varepsilon_2')^{1/2}}{(\varepsilon_2'^2 - 1)^{3/2}} - \mu_2''\left(\frac{\varepsilon_2'}{(\varepsilon_2'^2 - 1)(\varepsilon_2' - \mu_2')}\right)^{1/2}\right]\frac{\omega}{2c}. \tag{7.15}$$

It is clear that the additional possibility of having a negative magnetic permeability increases the number of possibilities. Given that $\varepsilon_2 < 0$, we require for real k_x' that

$$\frac{(\varepsilon_2' - \mu_2')}{\varepsilon_2'^2 - 1} > 0.$$

This implies that the imaginary part of the magnetic permeability can actually increase the propagation length via a phase shift in the fields compared to the case when $\mu_2'' = 0$. Surprising as this result might seem, one should

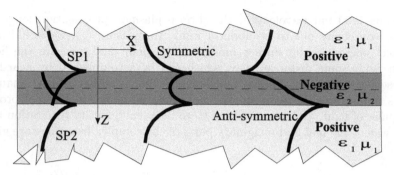

Figure 7.3 Schematic representation of the geometry of the slab of a material with negative ε or μ. The surface plasmons on two interfaces are no longer independent and get coupled to give rise to coupled slab plasmon polariton excitations. The slab modes can have field distributions that are either symmetric or anti-symmetric with respect to the center of the slab.

remember that the total dissipation is due to both the electric and magnetic susceptibilities. We discuss the full implications for the real part of k'_x, and the possibility of negative refraction for the surface plasmon in Section 7.3.

The most important aspect about all these modes is that all of them lie to the right of the dispersion of free light (shown by the dotted curve in the figures). This implies that propagating modes of light incident on the surface can never couple to these surface plasmon modes as the phase matching condition can never be satisfied. The surface plasmon modes can, however, be excited through a periodic structure on the surface where the Bragg scattering for propagating modes can give the extra wave-vector (momentum) along the transverse direction. As another example, surface roughness can also accomplish the same effect (the surface roughness can be considered to be a "white" distribution of different frequencies). Alternatively, they can be excited by the near-field evanescent modes of a source placed very close by. These aspects have been very thoroughly discussed in Raether (1986) for the case of surface plasmon modes on a metal surface.

7.1.2 Surface plasmon polariton modes of a slab

The surface plasmon modes in a slab of a medium with negative material parameters and a finite thickness d are interesting since a slab has two interfaces where degenerate surface plasmon modes can be supported (see Fig. 7.3). Each interface plasmon feels the fields of the surface plasmon on the other interface, and they hybridize to form a pair of non-degenerate slab modes. The problem is completely analogous to the problem of quantum levels in two identical potential wells separated by some distance.

Consider that the slab has a width d, material parameters (ε_2, μ_2) given by Eqs. (7.7) and (7.8), and is embedded in a medium whose material parameters are (ε_1, μ_1). The conditions for the existence of independent slab modes can be obtained in a similar manner by matching the tangential components of the electric and magnetic fields across the two interfaces. Note that the fields inside the slab can be written as a superposition of an exponentially decaying and an exponentially amplifying wave along the thickness of the slab. The conditions for the P-polarized modes come out to be

$$\tanh(\kappa_{z2}d/2) = -\frac{\varepsilon_2\kappa_{z1}}{\varepsilon_1\kappa_{z2}}, \tag{7.16a}$$

$$\coth(\kappa_{z2}d/2) = -\frac{\varepsilon_2\kappa_{z1}}{\varepsilon_1\kappa_{z2}}, \tag{7.16b}$$

where $\kappa_{zj} = \sqrt{k_x^2 - \varepsilon_j\mu_j\omega^2/c^2}$. Each of these conditions corresponds to a slab mode. The two modes have different symmetries of the fields with respect to the center of the slab: one has a symmetrical field distribution $(\cosh[\kappa_z(z - d/2)])$ while the other has an anti-symmetric field distribution $(\sinh[\kappa_z(z - d/2)])$ corresponding to the first and the second conditions of Eq. (7.16), respectively. Thus, the slab develops a gross polarization and hence these modes are called surface plasmon polariton (SPP) modes. Similarly the conditions for the S-polarized modes are

$$\tanh(\kappa_{z2}d/2) = -\frac{\mu_2\kappa_{z1}}{\mu_1\kappa_{z2}}, \tag{7.17a}$$

$$\coth(\kappa_{z2}d/2) = -\frac{\mu_2\kappa_{z1}}{\mu_1\kappa_{z2}}. \tag{7.17b}$$

The dispersions for the slab plasmon polaritons have to be obtained as the solution to the above transcendental equations. These dispersion relations are plotted in Fig. 7.4. In fact, if we calculate the transmission or reflection coefficients for light incident on the slab, the conditions for the SPP modes correspond to the poles of the transmission or reflection coefficient. This is a typical effect whenever a system is excited at resonance – the scattering coefficients (transmission and reflection coefficients in this case) diverge. We see that there are six branches in total: four for the P-polarized modes and two for the S-polarized modes. Basically each surface plasmon branch for the semi-infinite media splits into two separate branches. The separation of the branches that have hybridized reduces as the thickness of the slab increases and the branches merge asymptotically as $d \to \infty$. Once again the SPP can only exist in regions in the (k_x, ω) plane where $k_x^2 > \varepsilon_j\mu_j\omega^2/c^2$ in either of the two media. The lower two P-polarized branches for $\omega < \omega_{0m}$ occur in the region $\varepsilon_2 < 0$, $\mu_2 > 0$; the two S-polarized branches occur entirely within the negative refractive index band when $\varepsilon < 0$, $\mu_2 < 0$. The higher frequency P-polarized modes cross over to the region of $\varepsilon_2 < 0$, $\mu_2 > 0$ from the region of negative refractive index. For large wave-vectors, the slab polariton modes (for all k_x) become degenerate at ω_{0m} for the low frequency P-polarized modes,

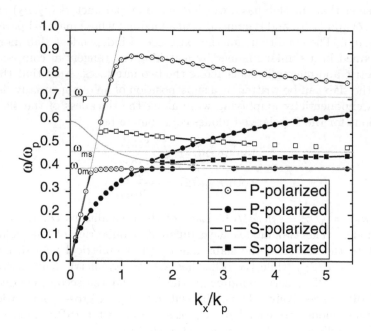

Figure 7.4 Dispersion of the slab plasmon polariton modes for a medium with the material parameters given by Eqs. (7.7) and (7.8). The surface plasmon branches essentially split into two modes: one symmetric and another anti-symmetric. The assumed parameters are $\omega_{0m} = 0.4\omega_p$, $f = 0.56$, and $k_p d = 0.3$. The light line ($\omega = ck_x$) and the curve $k_x = \varepsilon(\omega)\mu(\omega)\omega^2/c^2$ (dashed curve) are indicated in grey. The surface modes can occur only to the right of these curves. The dielectric surface plasmon frequency ω_s when $\varepsilon_2 = -1$, the magnetic surface mode frequency ω_{ms} when $\mu_2 = -1$, and the magnetic resonance frequency ω_{0m} are indicated by grey horizontal lines.

at $\omega = \omega_p/\sqrt{2}$ for the high frequency P-polarized modes, and $\omega = \omega_{ms}$ where the $\mu = -1$ for the S-polarized modes.

We also note that, for $\omega > \omega_{0m}$, the branches of the anti-symmetric P-polarized and S-polarized modes smoothly cross over to the dispersion of the first-order slab waveguide mode in the negative refractive index medium when the light is propagating inside the negative refractive index slab ($k_x^2 < \varepsilon_2\mu_2\omega_2/c^2$) and evanescent outside ($k_x^2 > \varepsilon_1\mu_1\omega^2/c^2$). We discuss waveguide modes separately in the next section.

In the quasi-static limit of $k_x \gg \omega/c$, we can easily obtain an analytic expression for the dispersion of the SPP modes. In this limit, $\kappa_{zi} \to k_x$ and the P-polarized modes becomes independent of μ while the S-polarized modes become independent of ε. Note that this does not hold for the P-polarized modes when $\mu_2 \to \infty$ as $\omega \to \omega_{0m}$. One can analytically obtain the dispersion for the P-polarized modes as

$$\omega_\pm = \frac{\omega_p}{2}[1 \pm \exp(-k_x d)]^{1/2}, \tag{7.18}$$

where the $\pm$ sign indicates the frequencies of the two non-degenerate modes. For the S-polarized modes, with the magnetic resonance model we obtain the dispersion as

$$\omega_\pm = \frac{\omega_{0m}}{1 - f/2[1 \pm \exp(-k_x d)]^{1/2}}. \tag{7.19}$$

Note that had we used a Lorentz model for the dispersion of μ_2, we would instead obtain

$$\omega_\pm = \left[\omega_{0m}^2 + \frac{\omega_b^2}{2}(1 \pm e^{-k_x d})\right]^{1/2}. \tag{7.20}$$

These equations illustrate the splitting of the modes and that the symmetric mode ω_+ has a higher frequency.

One should note that these results in the non-retarded regime should be treated as only approximate, particularly very close to the surface plasmon frequency. For example, the dispersions of both the symmetric and the anti-symmetric modes, at large slab thickness or large k_x, have to tend to the uncoupled plasmon dispersion for a single surface as the two plasmons are essentially uncoupled. Thus, even the slopes of the dispersion curves should be the same for large k_x. The non-retarded approximation predicts opposite signs for the group velocity on the surface, $(\partial\omega)/(\partial k_x)$. In reality, however, the two curves actually tend to the surface plasmon frequency either both from below (for $\omega_{mp} < \omega_p$) or both from above (for $\omega_{mp} > \omega_p$). In the former case, the upper dispersion curve crosses over to below ω_s at some point (Ramakrishna et al. 2002). This can have important physical implications: in the case of the silver lens discussed in Section 8.2.2, this would imply that the anti-symmetric mode is always excited.

Consider next the dissipation and the propagation of SPP modes in very thin slabs of the negative material. For concreteness, we consider the P-polarized modes only and the case of a metal ($\mu_2 = 1$, $\varepsilon_2 < 0$). In this case, several researchers (Kovacs and Scott 1977, Quail et al. 1981, Sarid 1981) reported both theoretic and experimental results that the damping for the symmetric mode reduces as the thickness of the metallic film decreases as much by an order of magnitude while the damping for the anti-symmetric mode increases by modest amounts. We can easily estimate the thickness dependence of the imaginary part of k_x for very thin films: using $\tanh(x) \sim x$ for $x \ll 1$ in Eq. (7.16), and for the symmetric mode $k_x = k_1 + \Delta$ where $k_1 = \varepsilon_1 \omega^2/c^2$ and $\Delta \ll k_0$, we obtain that

$$\frac{\Delta}{\omega/c} \simeq \frac{\varepsilon_1(\varepsilon_1 - \varepsilon_2)^2(k_0 d/2)^2}{2[\varepsilon_2^2 - \varepsilon_1^2(\varepsilon_2 - \varepsilon_2)d^2 k_0^2]}, \tag{7.21}$$

from which we obtain in the limit of small $k_1 d$, the imaginary part of the wave-vector as

$$\frac{k_x''}{\omega/c} = \varepsilon_1(\varepsilon_2' - \varepsilon_1)\frac{\varepsilon_2''}{\varepsilon_2'^2}\left(\frac{k_1 d}{2}\right)^2, \tag{7.22}$$

which clearly demonstrates that $\ell_{sp} \sim d^{-2}$, i.e. the damping reduces with reducing thickness of the slab. This property is due to the field structure that is more spread out in the surrounding medium than inside the slab where dissipation takes place. It can be similarly shown that the damping actually increases but more moderately for the anti-symmetric SPP mode. More exact estimates can of course be obtained by solving Eq. (7.16) numerically for the real and imaginary parts of the wave-vector ($k_x = k_x' + ik_x''$) and such analyses indicate that a reduction in k_x'' by one order of magnitude is possible when the film thickness becomes of the order of 20 nm (Sarid 1981, Raether 1986). Such SPP modes would have an enhanced range of propagation on the surface and they are, therefore, referred to as *long range plasmons* (Sarid 1981). These long range plasmons penetrate highly into the surrounding media, and their potential for sensor applications at almost single molecule sensitivity and enhanced nonlinear phenomena has been demonstrated.

Finally, we consider the case when the slab has an asymmetric environment. This is typical of many applications, say, a silver film (ε_2) deposited on a quartz substrate (ε_1) with vacuum/air ($\varepsilon_0 = 1$) on the other side. The dispersion of the SPP modes can be determined by locating the poles of the transmission or the reflection coefficients (see Appendix C) in the (k_x, ω) plane. The condition for P-polarized light is

$$\left(\frac{\kappa_{z1}}{\varepsilon_1} + \frac{\kappa_{z2}}{\varepsilon_2}\right)\left(\frac{\kappa_{z1}}{\varepsilon_1} + \frac{\kappa_{z0}}{\varepsilon_0}\right) + \left(\frac{\kappa_{z1}}{\varepsilon_1} - \frac{\kappa_{z2}}{\varepsilon_2}\right)\left(\frac{\kappa_{z1}}{\varepsilon_1} - \frac{\kappa_{z0}}{\varepsilon_0}\right)e^{i\kappa_{z2}d} = 0. \tag{7.23}$$

For complex ε_1 and ε_2, one needs to determine the complex roots k_x of this equation. One tractable way to determine the roots is to scan the (k_x, ω) plane and monitor the derivative of the phase of the transmission coefficient. The phase changes rapidly in the vicinity of a resonance and the derivative shows a peak at the resonance. The main effects of the asymmetry are to give rise to a shift in the real part of k_x. In addition, when the wave can propagate in the substrate medium, it also gives a contribution to the imaginary part of k_x which corresponds to the energy radiated by the surface plasmon into the substrate medium. Fig. 7.4 showing the dispersion of the slab plasmon polariton modes, albeit for a symmetric slab, should be compared to Fig. 6.9 where similar information about the dispersion can be obtained from the frequency derivative of the phase of the transmission coefficient.

7.2 Waveguides made of negative index materials

The guidance condition listed in Eqs. (7.16) and (7.17) can be studied in a more systematic manner in order to visualize the evolution of the associated modes with some typical parameters, such as the frequency and the thickness of the slab. In order to render the equations symmetric and with only dimensionless quantities, Eqs. (7.17) are rewritten as

$$\kappa_{z1}d = -\frac{\mu_1}{\mu_2}(\kappa_{z2}d)\tanh\left(\frac{\kappa_{z2}d}{2}\right),\qquad(7.24a)$$

$$\kappa_{z1}d = -\frac{\mu_1}{\mu_2}(\kappa_{z2}d)\coth\left(\frac{\kappa_{z2}d}{2}\right),\qquad(7.24b)$$

which correspond to the symmetric and asymmetric modes shown in Fig. 7.3. In addition, combining the dispersion relations in both regions yields the condition

$$(\kappa_{z1}d)^2 - (\kappa_{z2}d)^2 = (k_2^2 - k_1^2)d^2.\qquad(7.25)$$

A mode that simultaneously satisfies Eqs. (7.24) and (7.25) is therefore a mode supported by the structure. An intuitive graphical representation of the solution can be obtained by looking at Fig. 7.5, where the equations have been represented as a set of curves depending on the ratio $(-\mu_1/\mu_2)$. The graphical representation of these modes, in Fig. 7.3, illustrates that one is symmetric whereas the other one is asymmetric, which refers to the profile of the electric field. By analogy to regular media where the electric field follows sine and cosine functions, the electric field in this case follows hyperbolic sine and cosine functions,* suggesting to term these modes "cosh" and "sinh," for the symmetric and asymmetric modes, respectively (Wu et al. 2003).

The existence and characteristics of a mode can be directly obtained from Fig. 7.5 once the configuration is determined, i.e. once the frequency of operation is known, the slab thickness d, as well as the material parameters (ε_1, μ_1) for the medium surrounding the slab and (ε_2, μ_2) for the slab. These parameters define a value for $(k_2^2 - k_1^2)d^2$ that maps onto a horizontal line in both panels. Note that an increase in frequency or in thickness simply corresponds to a translation of this line away from the $(k_2^2 - k_1^2)d^2 = 0$ line, in either direction depending on the material parameters. The intersection of the $(k_2^2 - k_1^2)d^2$ line with the guidance condition curves corresponding to the $(-\mu_1/\mu_2)$ ratio of interest yields the solution(s), if any, that are supported by the given configuration. It can then immediately be seen for example that when supported, the cosh modes do not have a cutoff frequency whereas the sinh modes do. As another example, the perfect lens situation (see Chapter 8)

---

*The demonstration is straightforward and is left as an exercise to the reader.

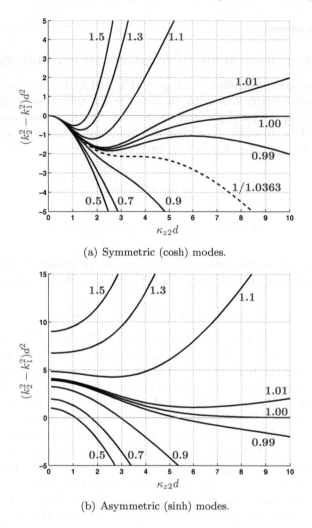

(a) Symmetric (cosh) modes.

(b) Asymmetric (sinh) modes.

Figure 7.5 Graphical representation of Eqs. (7.24) and (7.25) as function of $\kappa_{2z}d$ for various values of $(-\mu_1/\mu_2)$ (indicated as numbers next to the curves).

corresponds to $k_2^2 - k_1^2 = 0$ and $-\mu_1/\mu_2 = 1.0$, for which it is seen that the intersecting point occurs at infinity.

The region $(k_2^2 - k_1^2)d^2 < 0$ can be accessed in two situations. The first corresponds to $k_2^2 < k_1^2$, i.e. a slab less dense than the background medium, supposing that all wavenumbers are real. The second situation, more unusual, occurs when the slab is made of a plasma medium, in which case k_2 is imaginary and $k_2^2 < 0$. The existence of hyperbolic modes in this case is well known (Oliner and Tamir 1962) and it is seen that these modes are identical to those supported by a slab of left-handed medium (for which k_2 is negative) when $|k_2| < |k_1|$. In particular, the specific value of $(-\mu_1/\mu_2 = 1/1.0363)$ was pointed out as being a limit above which a single mode propagation occurs and below which multi-modes are sustained. This conclusion is confirmed here graphically in Fig. 7.5(a).

The complex modes supported by the slab of left-handed media are therefore additional modes, which need to be accounted for. In particular, they coexist with the regular modes obtained when the transverse wavenumber in the slab is real. The guidance condition for these modes can be written as

$$\alpha_{z1}d = \frac{\mu_1}{\mu_2}\,(k_{z2}d)\,\tan\,(k_{z2}d/2)\;, \qquad (7.26a)$$

$$\alpha_{z1}d = \frac{\mu_1}{\mu_2}\,(k_{z2}d)\,\cot\,(k_{z2}d/2)\;, \qquad (7.26b)$$

$$(k_{2z}d)^2 + (\alpha_{z1}d)^2 = (k_2^2 - k_1^2)d^2 \qquad (7.26c)$$

and is illustrated in Fig. 7.6. Since these modes have been extensively studied in the literature, for example in Collin (1990), we shall not study them in more detail here.

7.3 Negative refraction of surface plasmons

In analogy with the negative refraction of transverse electromagnetic waves at the interface between a positive index medium and a negative index medium, one can wonder whether it is possible to find interfaces between two surfaces where a surface plasmon wave refracts negatively. We show a schematic picture of such a process at the interface between the surfaces of two different media in Fig. 7.7. Since the surface modes are not purely transverse waves (there is a component of the fields along the wave-vector on the surface), we cannot examine if the waves are left-handed as for transverse waves in a three-dimensional negative refractive index medium. This question has been addressed by Kats et al. (2007) through an analysis of the relative directions of the energy flow based on the Poynting vector and the phase-vector $\mathbf{k}_\parallel = \hat{x}k_x + \hat{y}k_y$ on the surface.

For simplicity of calculations, we assume the zero dissipation limit where the imaginary parts of the material parameters are δ functions at the specified resonance frequencies, whereas at other frequencies, the media are assumed non-dissipative. Consider the dispersion for the P-polarized surface plasmon modes on a surface given by Eq. (7.4)

$$k_\parallel = \left[\frac{\varepsilon_1\varepsilon_2(\varepsilon_2\mu_1 - \varepsilon_1\mu_2)}{\varepsilon_2^2 - \varepsilon_1^2}\right]^{1/2}\frac{\omega}{c}.$$

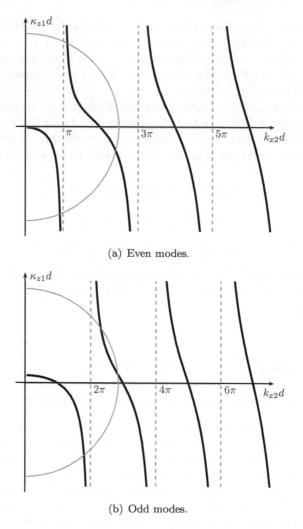

(a) Even modes.

(b) Odd modes.

Figure 7.6 Illustration of the guidance condition in a regular waveguide configuration when the transverse wavenumber inside the slab is positive. In both cases, $-\mu_1/\mu_2 = 0.2$.

An essential condition for the existence of such modes is $\varepsilon_1\varepsilon_2 < 0$. The time averaged Poynting vector associated with such a mode given by $\mathbf{S} = (1/2)\mathrm{Re}(\mathbf{E}\times\mathbf{H}^*)$ is oppositely directed in the media on either side[†] and decays

---

[†]The demonstration is straightforward.

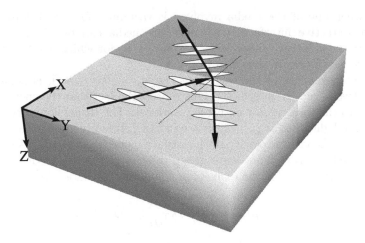

Figure 7.7 A schematic picture of surface plasmon modes at the surfaces of two media (shown by different shades of gray). The surface waves can undergo negative refraction at the interface between the two surfaces (media) as shown depending on the material parameters of the two media.

exponentially with distance from the interface

$$\mathbf{S} = \begin{cases} \frac{\mathbf{k}_\parallel}{\varepsilon_0 \varepsilon_1 \omega} |\mathbf{H}|^2 \exp(\kappa_{z1} z) & \forall \ z < 0, \\ \frac{\mathbf{k}_\parallel}{\varepsilon_0 \varepsilon_2 \omega} |\mathbf{H}|^2 \exp(-\kappa_{z2} z) & \forall \ z > 0. \end{cases} \tag{7.27}$$

One has to determine the energy flow along the interface associated with this surface mode. This can be accomplished by integrating the Poynting vector with respect to the normal z direction. We obtain for the integrated energy flow

$$\mathbf{W} = \int_{-\infty}^{\infty} \mathbf{S} \, dz = \frac{\mathbf{k}_\parallel}{\varepsilon_0 \omega \kappa_{z1} \varepsilon_1} \left[1 - \frac{\varepsilon_1^2}{\varepsilon_2^2} \right] |\mathbf{H}|^2. \tag{7.28}$$

Noting that $k_\parallel > 0$, $\varepsilon_1 > 0$, the condition for the wave-vector and the energy flow at the surface to be anti-parallel is

$$\varepsilon_1^2 > \varepsilon_2^2, \tag{7.29}$$

in which case such interfaces can be called *negative refractive interfaces*.

Note that the interface of an ordinary metal with $\varepsilon_2 < 0$, $\mu_2 = 1$, and vacuum cannot satisfy the above condition as no surface plasmon modes exist for $\varepsilon_1 > |\varepsilon_2|$. Consider, however, an interface between vacuum and a medium with $-1 < \varepsilon_2 < 0$, $\mu_2 < 0$: a surface plasmon mode on the interface with oppositely directed wave-vector and energy flow can be excited if $\varepsilon_2 - \mu_2 < 0$. This is the only kind of negative refractive interface possible for P-polarized modes when one of the media involved is vacuum. The conditions for the negative refractive interfaces with different media can be divided into six cases and the different ranges for $\varepsilon_1/\varepsilon_2$ and μ_1/μ_2 for which one has negative refractive interfaces are shown in Fig. 7.8.

The group velocity of the surface waves, $v_g = (\partial\omega)/(\partial k_\parallel)$, is negative on the negative refractive interfaces. This reflects the fact that the energy flow is opposite to the wave-vector on these interfaces. The dispersion curve for the P-polarized light when $\omega_{mp} > \omega_p$ shown in Fig. 7.2 clearly has a negative slope. For $k_\parallel \sim k_0$, the group velocity of the surface mode ($v_g = (\partial\omega)/(\partial k_\parallel)|_{\omega_0}$) can be expressed in terms of the wave group velocity in the two media. Consider the Taylor expansion for the dispersion equation $\varepsilon(\omega)\mu(\omega)\omega^2/c^2$:

$$\omega^2/c^2\varepsilon(\omega)\mu(\omega) = k_0^2 + 2k_0\frac{\partial k}{\partial\omega}|_{\omega_0}\delta\omega + \mathcal{O}(\delta\omega)^2, \qquad (7.30)$$

where we retain only terms linear in the frequency difference, and the dispersion equation for the waves in the two media

$$\kappa_{z1}^2 = k_\parallel^2 - \varepsilon_1(\omega)\mu_1(\omega)\omega^2/c^2,$$
$$\kappa_{z2}^2 = k_\parallel^2 - \varepsilon_2(\omega)\mu_2(\omega)\omega^2/c^2.$$

For $k_\parallel \sim k_0$, substituting the expansion above into the right-hand side of the equation, we obtain

$$v_g = \frac{v_1 v_2(\kappa_{z1}^2 - \kappa_{z2}^2)}{v_1\kappa_{z1}^2 - v_2\kappa_{z1}^2}, \qquad (7.31)$$

i.e. one can understand the group velocity of the surface mode at an interface as a weighted average of the group velocities of waves in the two media. The meaning of this becomes very clear in the case when one medium has a negative refractive index and the other has a positive refractive index. A particularly interesting case occurs when $\varepsilon_1 = \mu_1 = +1$ and $\varepsilon_2 = \mu_2 = -1$, which results in a zero group velocity for all surface modes. The surface modes are completely degenerate at the frequency $\omega = \omega_s = \omega_{ms}$. Note that the case $|\varepsilon_1/\varepsilon_2| = 1$ and $|\mu_1/\mu_2| = 1$ corresponds to this special point in the case of (c) and (f) in Fig. 7.8 where there is no net energy carried by the surface wave in any direction: as $\kappa_{z1} = \kappa_{z2}$ in the two media, the oppositely oriented Poynting vectors on either side of the media cause the net energy flow along the surface to become zero. A more detailed discussion of this effect is offered in Chapter 8. In other media that do not support propagating modes at all (for example, metals or purely negative magnetic permeability media), the meaning of this average for the group velocity becomes less clear and it is then more meaningful to talk of the energy associated with the surface mode rather than the group velocities within the individual media.

Consider Fig. 7.7 where we depict the interface (along the $y - z$ plane) between the two media with different properties which support surface states in the xy plane. If one surface supports modes with a positive wave-vector

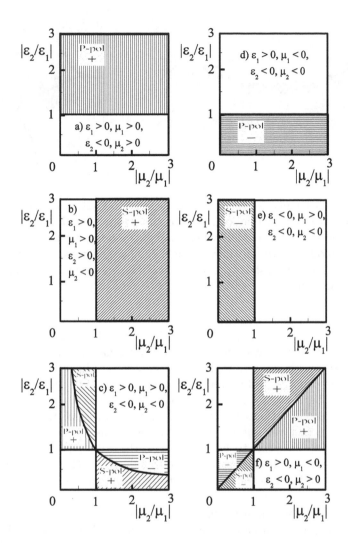

Figure 7.8 The six cases of negative refractive interfaces that are possible between media with different ε and μ. The parameter ranges when the interfaces support surface modes and have negative refractive interface (shown by a $-$ sign) or a positive refractive nature (shown by a $+$ sign) are marked by different gray shades for P-polarized and S-polarized surface modes. No surface modes are possible in the white regions. Redrawn based on the data of Kats et al. (2007).

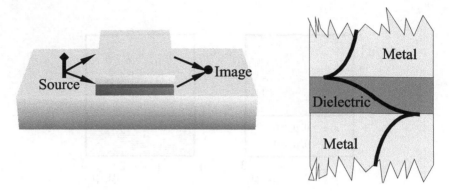

Figure 7.9 A schematic picture of the negative refraction of surface plasmon modes on a metal surface by coupling to the slab modes of a heterostructure. Negative refraction is enabled when the slab modes, particularly the anti-symmetric modes, have anti-parallel energy flow and wave-vector (shown on the right). The heterostructure acts as a flat lens focusing a point source placed on one side.

while the other surface supports modes with a negative wave-vector, it is clear that an incident surface wave is going to refract negatively across the interface between the two surfaces as depicted in the figure. This is in complete analogy with the negative refraction of transverse electromagnetic waves across the interface between two media: one positively refracting and one negative refracting. Note that there is a reflected wave from the interface as well. The solid arrows depict the energy flow. On one of the two surfaces the wave-vector on the surface has to be opposite to the energy flow. The reflection and transmission coefficients can also be calculated for these surface modes if one includes the fields of the resonant states at the interface between the two surfaces.

Another way to accomplish negative refraction of surface plasmon modes is to use a waveguide structure as shown in Shin and Fan (2006). The central idea is to couple the surface plasmon modes propagating on the surface into a waveguide-like structure where the guided modes have an energy flow opposite in direction to the wave-vector. This can be done as shown in Fig. 7.9 by using a metal-dielectric-metal structure where the metal is assumed to have a negative dielectric permittivity. It can be easily seen that the positive dielectric layer enclosed by the negative dielectric metal supports symmetric and anti-symmetric combinations of the surface plasmon modes as discussed earlier. Noting that the Poynting vector is oriented in opposite directions on either side of the metal-dielectric interface (due to change in the sign of the permittivity as discussed before), the flow of energy associated with the waveguide mode can be either parallel or antiparallel depending on whether the field is more concentrated in the dielectric or extended out to the metallic regions. The anti-symmetric mode is seen to extend out in the metallic regions under the appropriate conditions and can have a negative energy flow associated with it. Hence coupling the surface plasmon waves on the metal surface to the slab waveguide modes as shown in Fig. 7.9 results in negative refraction of the energy. The slab waveguide of finite width acts as a flat slab for surface plasmon waves. The exact conditions for an imaging geometry of a slab lens can be found in Shin and Fan (2006). Thus, not only surface modes, but slab modes can also be utilized for such purposes.

7.4 Plasmonic properties of structured metallic surfaces

In recent years, structured metallic surfaces and films have been shown to exhibit a wide variety of interesting plasmonic properties and novel phenomena [see Maier (2007) for a detailed exposition on this topic]. The prime among these effects is the extra-ordinary transmission of light through an array of subwavelength sized holes in a thick metallic film first described in Ebbesen et al. (1998). The problem of transmission of light through a small hole in a perfectly conducting surface is one of the few exactly solvable problems in rigorous diffraction theory (Bethe 1944) and predicts a transmitted intensity that is proportional to $(a/\lambda)^4$ (note that the similarity to the result from Rayleigh theory). In the experiments of Ebbesen et al. (1998), it was found that the transmittance through the array of subwavelength sized holes could be very high (∼90% of the incident light at certain resonant frequencies). This intriguing result of high transmittance was later shown to hold even for single holes provided the region around the hole had been suitably patterned, for example, by having ridges of the metal in concentric circles (Barnes et al. 2003). The resonant tunneling through single subwavelength holes can be controlled by any embedded non-linear medium and has also been used to demonstrate a "photon blockade" effect (Smolyaninov et al. 2002) akin to the Coulomb blockade effect in mesoscopic quantum dots. The surface plasmon modes on the metal and the modes in the holes appear to play a crucial role in these phenomena.

In this section, we discuss a general issue related to structured metallic surfaces and films including structured perfectly conducting surfaces: i.e. their ability to support surface modes (Pendry et al. 2004). Note that a plain flat perfect conductor cannot support any surface modes, so that these plasmon modes were termed as "spoof" surface plasmon modes in Pendry et al. (2004). The structuring of the surface, however, appears to bring in an ability to support surface modes. It turns out that a distinction between these structure-induced surface modes and true surface plasmon modes on the structured surface of a metal with negative dielectric permittivity might not even be

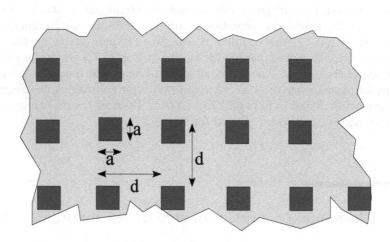

Figure 7.10 A film of a perfect conductor with an array of periodically placed square holes behaves as a plasma medium and can support surface plasmons.

possible as the excitations blend into each other. We discuss here only this general property as proposed in Pendry et al. (2004) while avoiding extended discussions of the large number of properties of subwavelength structured metallic films as well any controversies about their explanations in terms of the surface modes (Lezec and Thio 2004).

Consider a film of a perfectly conducting material containing periodically placed small holes (assumed square for simplicity) that go through the thickness of the film as shown in Fig. 7.10. Let the thickness of the film be t and the holes have a side of a placed on a square lattice of period d. Assume that the conditions for a homogeneous description of the surface are satisfied, i.e. $a < d \ll \lambda$ and consider radiation (with the electric field **E** along the y axis) to be incident on the film. The fields inside the perfectly conducting regions are zero and the fields exist only inside the holes. The incident radiation would primarily couple to the fundamental waveguide mode in the holes as higher modes decay faster with distance inside the hole. Consequently we can write the fields inside the holes as

$$\mathbf{E} = \hat{y}E_0 \sin(\pi x/a) \exp[i(k_z z - \omega t)], \tag{7.32}$$

where $k_z = \sqrt{\varepsilon_i \mu_i \omega^2/c^2 - (\pi/a)^2}$ and the subscript i refers to the medium in the interior of the holes. We would like to replace the structured film by a film of some equivalent material where we have the averaged macroscopic field

$$\mathbf{E}_{av} = \hat{y}E_0' \exp[i(k_x x + k_z z - \omega t)]. \tag{7.33}$$

Due to the symmetry in the xy plane, we have $\varepsilon_x = \varepsilon_y$ and $\mu_x = \mu_y$. For fields applied parallel to the axis of the holes (z direction), from the continuity of the E_z and H_z fields along the axis of the holes, we obtain that the material parameters should be the volume weighted average of the permittivities and permeabilities of the material in the hole and the perfect conductor. Hence we obtain

$$\varepsilon_z \rightarrow \infty, \qquad \mu_z \rightarrow \infty. \tag{7.34}$$

This implies that the dispersion in the homogenized film is

$$\frac{k_x^2}{\mu_z} + \frac{k_z^2}{\mu_x} = \varepsilon_y \frac{\omega^2}{c^2} \quad \Rightarrow \quad k_z = \pm \sqrt{\mu_x \varepsilon_y} \frac{\omega}{c}. \tag{7.35}$$

There is no dispersion of the modes with k_x or k_y for either polarization.

The average fields in the region of the film can be obtained as a volume average:

$$E_0' = \frac{E_0}{d^2} \int_0^a dy \int_0^a dx \ \sin(\pi x/a) = \frac{2a^2}{\pi d^2} E_0. \tag{7.36}$$

Another consistency condition would be that the component of the Poynting vector normal to the film has to be the same: whether calculated from the microscopic fields

$$(\mathbf{E} \times \mathbf{H})_z = \frac{-k_z E_0^2 \int_0^a dy \int_0^a \sin^2(\pi x/a)}{\mu_0 \mu_i \omega} \frac{}{d^2} = \frac{-k_z E_0^2}{\mu_0 \mu_i \omega} \frac{a^2}{2d^2}, \tag{7.37}$$

or the macroscopic averaged fields

$$(\mathbf{E}_{av} \times \mathbf{H}_{av})_z = \frac{-k_z E_0'^2}{\mu_0 \mu_x \omega}. \tag{7.38}$$

Although there are large inhomogeneous microscopic fields at the edges of the holes, we neglect these fields as they are localized (evanescent) and do not contribute to the energy flow. From the above two equations, we obtain

$$\mu_x = \frac{8a^2}{\pi d^2} \mu_i. \tag{7.39}$$

Noting now that k_z has to be the same in both cases, we obtain

$$\varepsilon_y = \frac{\pi^2 d^2}{8a^2} \left(1 - \frac{\pi^2 c^2}{\varepsilon_i \mu_i a^2 \omega^2} \right), \tag{7.40}$$

which is similar to the dielectric function of a plasma with a plasma frequency $\omega_p = (\pi c)/(a\sqrt{\varepsilon_i \mu_i})$. This is actually the value for the lower cutoff frequency of the waveguide mode in the square hole configuration.

The structured metallic film is equivalent to a homogeneous plasma-like medium with the specified anisotropic material parameters. This film can essentially support slab plasmon polariton modes much like a plasma medium.

In addition, the structure can enable weak coupling to propagating radiation. In general, the structures on the conducting film need not be holes, but could be bumps or any other structure that generates scattering modes that can be localized on the surface. However, the advantage of the hole-based geometry is to support a strong resonance, yielding a large response.

The homogenization view point is particularly useful for a dense set of small sized holes. Note that a discussion of the strong localized evanescent fields that would be present near the edges of the holes would be necessary for completeness of the description. These evanescent waves would be excited and would be required to account for the continuity of the fields at the surface of the conducting film. However, these evanescent modes do not contribute to the energy flow and can be neglected in the absence of dissipation in the film.

7.5 Surface waves at the interfaces of nonlinear media

We have seen in Section 3.6 that metamaterials can have nonlinear polarizability and magnetization. Thus, an interesting question concerns what happens to the surface modes at an interface if one or both media in question have nonlinear material parameters. This question assumes larger importance in view of the large local fields due to the surface states that enhance the nonlinear effects and drive the system into the nonlinear regime. Although nonlinear polarizabilities can be resonantly enhanced, it is advisable to remain reasonably away from the resonance as the unavoidable dissipation near the resonance itself effectively prevents both large local field enhancements and the expression of the nonlinear behavior. We essentially follow the approach of Shadrivov et al. (2004) here and present the elementary solutions for nonlinear surface modes on the interface of nonlinear metamaterials.

We specifically consider the case of third-order nonlinearities: $\chi_e^{(3)}$ and $\chi_m^{(3)}$ for both the electric polarization and the magnetization. For example, the SRR medium with Kerr nonlinear dielectric in the capacitive gaps has a Kerr-type nonlinear response at frequencies well away from the resonances where the system is bistable. In this case, the field at one frequency can act on itself through the nonlinearity (Boyd 2003) and the equation for the magnetic field of the P-polarized mode ($\mathbf{H}(x, z, \omega)$ is along $\hat{y}$ and $\mathbf{E}(x, z, \omega)$ lies in the xz plane) becomes a nonlinear Schrödinger equation:

$$\frac{\partial^2 H_y}{\partial x^2} + \frac{\partial^2 H_y}{\partial z^2} + \frac{\omega^2}{c^2} \left[\varepsilon(\omega)\mu(\omega) + \chi_m^{(3)}\varepsilon(\omega)|H_y|^2 \right] H_y = 0. \tag{7.41}$$

Here the nonlinear polarization in the medium is taken to be

$$M_{NL} = \mu_0 \chi_m^{(3)} |H(x, z, \omega)|^2 H_y(x, z, \omega).$$

There is a similar equation for the S-polarized mode, where the nonlinear electric polarization and the nonlinearity couple to the dispersion in μ. It is straightforward to see that an effective negative permittivity reverses the effect of self-focusing nonlinearity $\chi_m^{(3)} > 0$ to that of a non-focusing nonlinearity in a positive medium. Plane wave solutions in the nonlinear media can be written down by the ansatz:

$$H(x, z) = \xi(z) \exp[ik_x x], \tag{7.42}$$

where for localized solutions we have

$$\xi(z) = A \operatorname{sech}[\kappa_z(z - z')], \tag{7.43}$$

and z' is a variable to be fixed by the boundary conditions. Implementing this into the nonlinear Schrödinger equation, we have that

$$\kappa_z^2 = k_x^2 - \varepsilon\mu\omega^2/c^2, \tag{7.44}$$

$$A = \kappa_z \left[\frac{2}{\chi_m^{(3)}\varepsilon} \right]^{1/2}. \tag{7.45}$$

Thus, we can have solitonic solutions for the surface modes and z' turns out to be the center of the soliton.

We can have interfaces between two nonlinear media or between one linear medium and a nonlinear medium. These have been dealt with comprehensively in Shadrivov et al. (2004). The solutions in different cases are pictorially shown in Fig. 7.11. Note that nonlinearity can induce surface modes even in cases where there exist no surface modes in the linear limit: for example, S-polarized surface modes on the interface between a negative dielectric and positive dielectric medium. Here we only consider the interface between vacuum ($\varepsilon = 1$, $\mu = 1$) for $z < 0$, and a medium with $\varepsilon_2 < 0$ and $\mu_2 < 0$ and nonlinear Kerr-type polarization ($\chi_e^{(3)}$) and magnetization ($\chi_m^{(3)}$) for $z > 0$. Then we have the magnetic fields of the P-polarized modes as:

$$H_y(x, z) = \begin{cases} H_0 \, \exp[ik_x x] \, \exp[\kappa_{z1} z] & \forall \ z < 0, \\ \kappa_{z2} \left[\frac{2}{\chi_m^{(3)}\varepsilon_2} \right]^{1/2} \exp[ik_x x] \operatorname{sech}[\kappa_{z2}(z - z')] & \forall \ z > 0. \end{cases} \tag{7.46}$$

Enforcing the conditions on continuity of the tangential components of the fields at the interface, we get the conditions

$$H_0 = \kappa_{z2} \left[\frac{2}{\chi_m^{(3)}\varepsilon_2} \right]^{1/2} \operatorname{sech}[\kappa_{z2}(z')], \tag{7.47a}$$

$$-\frac{\kappa_{z1}\varepsilon_2}{\kappa_{z2}} = \tanh[\kappa_{z2}z'], \tag{7.47b}$$

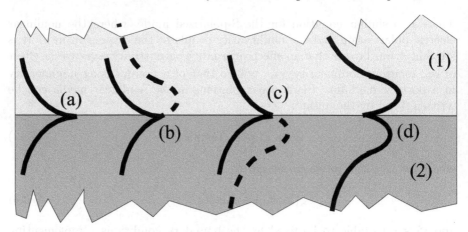

Figure 7.11 An interface between two media (1) and (2) can support a variety of surface waves, which are shown schematically: (a) Both media are linear. (b) Medium (1) is linear while medium (2) is nonlinear. (c) Medium (2) is linear while medium (1) is nonlinear. (d) Both media are nonlinear. The dashed line corresponds to the continuation of the sech solution into the other medium and peaks at the location of the displaced center of the solitonic solution (z').

from which we obtain the equation of dispersion for the nonlinear surface modes

$$\kappa_{z1} + \frac{\kappa_{z2}}{\varepsilon_2}\left(1 - \frac{\chi_m^{(3)}\varepsilon_2 H_0^2}{\kappa_{z2}^2}\right)^{1/2} = 0. \qquad (7.48)$$

A similar equation can be obtained for the S-polarized modes as

$$\kappa_{z1} + \frac{\kappa_{z2}}{\mu_2}\left(1 - \frac{\chi_e^{(3)}\mu_2 E_0^2}{\kappa_{z2}^2}\right)^{1/2} = 0. \qquad (7.49)$$

Note that the nonlinear contribution to the dispersion is essentially the nonlinear change of the dielectric permittivity or the magnetic permeability arising due to the particular field strengths. Further, the effect of the nonlinearity is more effective for smaller wave-vectors κ_{z2}. Typical numbers for these changes can be obtained by noting that some of the largest nonlinear coefficients for non-resonant Kerr processes in conventional nonlinear media are $\chi_e^{(3)} \sim 10^{-18}$ SI units. Hence even with surface electric fields of $E \sim 10^5$ V/m which occur in structured metallic surfaces with large local field enhancements, these corrections are of the order of 10^{-8}. However, for resonant processes or photorefractive processes, the nonlinear coefficients can be as large as $\chi_e^{(3)} \sim 10^{-12}$ SI units. In these cases, the corrections become of the order of 10^{-2} to 10^{-1}

and nonlinearity can have large effects, particularly for small wave-vectors, and can even actually switch the dispersion of the surface modes from a negative group velocity to a positive group velocity as detailed in Shadrivov et al. (2004) if the non-linear terms are large enough. However, the large nonlinearities come at the cost of large time response implying that the large changes in the surface mode dispersions obtained with such enhanced nonlinear processes are probably of scant importance for plasmonic applications where fast switching speeds are required. However, they may well be important for many other applications.

8

Veselago's lens is a perfect lens

It was mentioned in Chapter 5 that a slab of negative refractive index material with $n = -1$ can have a lens-like action: this slab (of infinite transverse width) can form the image of a source located on one its side at two locations, one within itself and another on the opposite side of the source. We call this flat lens a *Veselago lens* after its original proposer (Veselago 1968). Its imaging action arises as a direct consequence of the negative refraction of a ray across a planar interface between positive and negative index media. An additional condition for a real image to be formed is that the sum of the distances from the source to the slab (d_1) and the slab to the external image plane (d_2) in the positive medium equals the thickness of the negative index slab $(d = d_1 + d_2)$ as shown in Fig. 8.1. All this can be deduced with a simple ray analysis. The Veselago lens is a remarkable device: it maps each point on the object plane onto a point in the image plane and thus suffers from no geometrical aberrations. This lens is an example of an *Absolute Instrument* in geometric optics, preserves distances and angles in the image, and the imaging is projective (Caratheodory 1937). The image does, however, suffer from chromatic aberrations, given that media with negative refractive index are necessarily dispersive. The lens is also short-sighted and can only form images of objects placed within a distance d of the slab. The Veselago lens also accomplishes an image transfer while preserving the transverse translational symmetry, in contrast to conventional lenses, which have curved surfaces that enable them to image.

The Veselago lens is, however, much more than just a flat lens. In 2000, a full wave analysis of the flat lens revealed that, in principle, the image produced by the slab with $\varepsilon = -1$ and $\mu = -1$ had an infinite spatial resolution (Pendry 2000). That is to say, the lens can resolve geometrical details in the source that are much smaller than the imaging wavelength, theoretically without any limit. This capability of the Veselago lens to give image resolution beyond the so-called *diffraction limit* in conventional optics actually derives from the capacity of negative index materials to support surface states (discussed in Chapter 7). These surface states interact with and involve the non-radiative near-field modes of the source in the image formation process. These non-radiative fields are associated with the spatial features of the source at a subwavelength length scale and are confined to the immediate vicinity of the source. Hence their involvement enables image resolution beyond conventional optics.

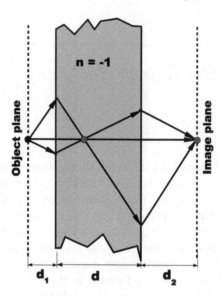

Figure 8.1 Drawing showing the focusing of rays emitted by a point source by a slab with refractive index of $n = -1$. One image is formed on the other side of the slab while another image is formed inside the slab. The condition for forming a real image is $d = d_1 + d_2$.

In practice, however, no imaging system can have infinite resolution and the resolution of the Veselago lens is also limited by other factors such as the impossibility of having a purely real, negative value for refractive index and the inherent spatial lengthscales of the metamaterial making up the slab. It turns out that the restoration of the evanescent components is quite sensitive to imperfections in the negative refractive index material (NRM) and this does somewhat curtail the subwavelength focusing capabilities of the lens. Nonetheless, subwavelength image resolution is possible to quite some extent even with these imperfect lenses, which we henceforth refer to as *super-lenses*.

The *perfect lens* effect (Pendry 2000) has been one of the most celebrated consequences of media with negative constitutive parameters, and its concept has been generalized to several configurations of spatially inhomogeneous negative refractive index media as well (Pendry and Ramakrishna 2003). These ideas have also given rise to another kind of a hyper-lens (Jacob et al. 2006) whereby the evanescent non-radiative modes are coupled into propagating modes in an indefinite medium (see Chapter 5) to be subsequently imaged at the far end of the medium.

8.1 Near-field information and diffraction limit

The diffraction limit imposed on the image generated by an optical system is one of the classic results in Optics: it relates the smallest discernible spatial features on the image to the wavelength of the illuminating or emitted light. Let us first re-examine it here. Consider the electromagnetic fields in the xy plane (called the object plane hereafter), typically emitted by some extended source. By imaging we usually imply that the intensity in this plane is re-produced on another parallel plane (called the image plane) at some distance along the (normal) z axis (called the optical axis hereafter). Conventional lenses or imaging devices collect radiation from the object plane and repro-duce them in the image plane. Consider the imaging action of a conventional convex lens: the lens has more material in its center than at its extremities in order to compensate for the longer path-lengths in the air of the rays prop-agating at larger angles. Thus all the waves propagating at different angles arrive at the image plane with the same phase shift.

Mathematically, the field emitted by the source can be Fourier decomposed in terms of plane waves propagating in different directions. Hence, for the electric field $\mathbf{E}(x, y, 0)$ at the object plane, the Fourier integral can be written as

$$\mathbf{E}(x, y, 0; t = 0) = \left(\frac{1}{2\pi}\right)^2 \int_{k_x} \int_{k_y} dk_x\, dk_y\, \tilde{\mathbf{E}}(k_x, k_y) \exp\left[i(k_x x + k_y y)\right],$$

$$(8.1)$$

where the amplitude of the Fourier components $\tilde{\mathbf{E}}(k_x, k_y)$ is given by the Fourier transform:

$$\tilde{\mathbf{E}}(k_x, k_y) = \int_x \int_y dx\, dy\, \mathbf{E}(x, y, 0) \exp[-i(k_x x + k_y y)]. \qquad (8.2)$$

Note that k_x and k_y (spatial frequencies) represent the Fourier components of the spatial variation in the source in the x and y directions, respectively. The spatial variation of the fields on the object plane can be thought of as arising from the superposition of periodic functions with different periodicities. Each spatial frequency $(k_{x,y})$ represents a periodicity $\Delta_{x,y} = 2\pi/k_{x,y}$ of variation of the electromagnetic fields on the source plane. In general, each set of spatial frequencies (k_x, k_y) on the source acts as an infinite sheet source emitting a plane wave along (k_x, k_y, k_z). The electromagnetic fields at any point for $z > 0$ can be written as a superposition of these plane waves

$$\mathbf{E}(x, y, z; t) = \left(\frac{1}{2\pi}\right)^2 \int_{k_x} \int_{k_y} dk_x\, dk_y\, \tilde{\mathbf{E}}(k_x, k_y) \exp\left[i(k_x x + k_y y + k_z z - \omega t)\right],$$

$$(8.3)$$

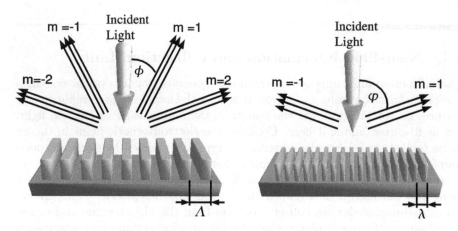

Figure 8.2 Pictorial representation of diffraction from periodic objects such as a grating. The object on the left has greater spatial period and the diffracted beams emerge at smaller angles. The diffracted beam from the more fine grating on the right has a much larger angle for the same diffracted order. Higher orders in this case have become evanescent and do not propagate.

where ω is the frequency of the radiation. The Maxwell equations impose that

$$k_x^2 + k_y^2 + k_z^2 = \frac{\omega^2}{c^2} = k_0^2 \qquad (8.4)$$

in free space.

A periodic planar object of period Δ_x has a periodic variation of the electromagnetic fields and acts as a sheet source emitting a plane wave along $k_x = 2\pi/\Delta_x$. Since we can decompose the fields of an arbitrary source into periodic variations on the object plane, it is sufficient to consider what happens to light emitted or scattered by a periodic object. Consider the diffraction of light from such periodic object (for example, a grating that is periodic along one dimension) as shown in Fig. 8.2. We assume that there are many periods involved in the scattering of the plane wave incident at normal incidence. The parallel component of the wave-vector can change upon scattering by a Bragg vector $G = \text{m}.2\pi/L$ along the direction of the periodicity (the $\hat{x}$ direction for example), where m is an integer and the diffracted beams emerge in a number of directions (diffraction orders) for the different values of m. If the periodicity of the object is large ($L > l$), the diffracted beams scatter through smaller angles ($\phi < \varphi$). For some maximum value of m, the parallel component of the wave-vector of the diffracted beam given by $k_x = \text{m}G$ becomes larger than the magnitude of the wave-vector ($k_0 = \omega/c$) in the medium (assumed to be air), yielding an imaginary longitudinal component k_z (because of the dispersion relation $k_x^2 + k_z^2 = k_0^2$) and waves evanescent along the $\hat{z}$ direction. These waves remain in the vicinity of the source, are non-propagating, and have appreciable amplitudes only over distances lesser than a wavelength near the source. Consequently, a grating of smaller periodicity shows a smaller number of diffracted beams as the beams for larger m are all evanescent. For a periodicity $L_x < 2\pi/k_0$, the beams are evanescent for any non-zero value of m and no diffracted waves other than the zeroth order wave propagate out.

Returning to the original problem of light emitted or scattered by an arbitrary object, the electromagnetic fields vary over multiple lengthscales. The sources can have intensity or field variations even over arbitrarily small distances. As an example, consider the case of isolated molecules on a surface emitting radiation in which case the $\Delta_{x,y}$ can literally be on atomic lengthscales. The corresponding transverse wave-vectors k_x and k_y are very large, and the waves are highly evanescent. Thus the waves with large $k_x > k_0$ and $k_y > k_0$ are evanescent and decay in amplitude exponentially away from the source $z = 0$ plane. These near-field modes have a decay length of $1/k_{x,y} < \lambda/2\pi$. In conventional imaging systems, these evanescent modes are never detected because of their minuscule amplitudes at the detectors, typically located at large distances (on the scale of a wavelength). Hence the corresponding source information about the fast varying features of the source is entirely missing. For these reasons, it is often assumed that the largest propagating wave-vector that contributes to the image is k_0 (assuming a unit numerical aperture), yielding the corresponding spatial frequency of the wavelength $\lambda = 2\pi/k_0$ to be the minimum limit on the spatial resolution in the image. Thus, even the best made lenses are limited in their resolution of small details on a lengthscale finer than the wavelength of light at the operating frequency.

It is worth noting that an image resolution of almost $\lambda/2$ can nevertheless be obtained, if one considers light to be incident at almost grazing incidence with a parallel wave-vector with $k_x \simeq \pm k_0$. In this case, light can scatter via two Bragg wave-vectors and emerge out with a parallel wave-vector of $k_x \simeq \mp k_0$. This is actually used in confocal imaging where one uses a very large numerical aperture lens to focus light on the source and to collect the scattered light over a broad cone of angles. Because of this possibility, the diffraction limit is stated to be $\lambda/2$ in many references.

In order to obtain an image resolution beyond the diffraction limit, one has therefore to measure the near-field radiation of a source and use the information it contains to reconstruct the image. This is the central idea of optical near-field microscopy where there are a variety of techniques that have been developed to measure the near-field and image subwavelength features on the source.* Near-field imaging is predominantly done by either feeding

---

*This does not violate any fundamental principle such as the Heisenberg uncertainty principle. Consider for example the identity for a wave $k_x \cdot \Delta_x \simeq 1$. In order to resolve features with $\Delta_x = \lambda$, we can count on $k_x = 2\pi/\lambda$ at most. For smaller Δ_x resolution, larger k_x are

energy into the near-field modes or by coupling out the energy in the near-field modes via a fine fiber tip tapered down to a few nanometers. The tapered fiber tip couples the near-field modes of the sample under investigation to the propagating modes in the fiber thus enabling excitation or detection of the near-field modes. The drawback is, of course, that the coupling efficiencies can be very small, typically of the order of 10^{-8}, which limits the measurements of fine effects. The reader is referred to more comprehensive treatises devoted to this subject (Paesler and Moyer 1996, Kaupp 2006) for the technical details of near-field imaging.

8.2   Mathematical demonstration of the perfect lens

Consider a lens as proposed by Veselago (1968) consisting of a slab of NRM of thickness d with $\varepsilon = -1$ and $\mu = -1$ and surrounded by a positive dielectric medium as shown in Fig. 8.3. Let ε_+ and μ_+ be the relative dielectric permittivity and the relative magnetic permeability of the positive medium outside. Similarly, let ε_- and μ_- be the relative dielectric permittivity and the relative magnetic permeability of the NRM. Consider that we need to calculate the fields at $z = 2d$ (the image plane) when a source is placed at $z = 0$ (the object plane) in the positive medium for which the transmission coefficient of the slab is required. The latter, along with the reflection coefficient for completeness, is (see Appendix C)

$$T = \frac{4\eta_1 \eta_2 \exp(ik_{z2}d)}{(\eta_1 - \eta_2)^2 - (\eta_1 + \eta_2)^2 \exp(2ik_{z2}d)}, \qquad (8.5)$$

$$R = \frac{(\eta_1^2 - \eta_2^2) \exp(i2k_{z2}d)}{(\eta_1 - \eta_2)^2 - (\eta_1 + \eta_2)^2 \exp(2ik_{z2}d)}, \qquad (8.6)$$

where $\eta_1 = k_{z1}/\varepsilon_+$ and $\eta_2 = k_{z2}/\varepsilon_-$ for P-polarized incident light and $\eta_1 = k_{z1}/\mu_+$ and $\eta_2 = k_{z2}/\mu_-$ for S-polarized incident light. Here $k_{z1} = \sqrt{\varepsilon_+ \mu_+ k_0^2 - k_x^2 - k_y^2}$ and $k_{z2} = \pm\sqrt{\varepsilon_- \mu_- k_0^2 - k_x^2 - k_y^2}$, where k_x and k_y are the parallel components of the wave-vector for the incident light. The sign of k_{z2} should be taken in accordance with the discussion in Section 5.2 for consistency. However, note that the transmission and reflection coefficients given above are invariant with respect to a sign change of k_{z2}.

---

needed ($k_x > k_0$), i.e. evanescent waves. For a perfect resolution, $\Delta_x \to 0$ and $k_x \to \infty$: *all* the evanescent waves are needed. Although the transverse image resolution might be very small the corresponding transverse component of the wave-vector is arbitrarily large, thus satisfying the wave property $k_x \cdot \Delta_x \sim 1$. The essential higher ($> 1$) dimensional nature of the problem makes this possible.

When the NRM has, for example, $\varepsilon_- = -\varepsilon_+ = -1$ and $\mu_- = -\mu_+ = -1$ (say), we obtain trivially for propagating waves that $k_{z2} = -k_{z1}$, $\eta_1 = \eta_2$, and

$$T = e^{-ik_{z1}d}, \qquad R = 0. \qquad (8.7)$$

Hence we have

$$E(z = 2d) = \exp(ik_{z1}d/2) \ \exp(-ik_{z1}d) \ \exp(ik_{z1}d/2) \ E(z = 0),$$
$$= E(z = 0), \qquad (8.8)$$

which clearly shows that the total phase change for propagation from the object plane (at $z = 0$) to the image plane (at $z = 2d$) is zero, thus yielding the same result as that obtained through an elementary ray analysis.

For evanescent waves with $k_{z1} = i\sqrt{k_x^2 + k_y^2 - \varepsilon_+\mu_+\omega^2/c^2} = i\kappa_z$, then we have $k_{z2} = k_{z1}$ and $\eta_1 = -\eta_2$. The transmission and reflection coefficients for the slab come out to be

$$T = e^{+\kappa_z d}, \qquad R = 0, \qquad (8.9)$$

i.e. the slab actually increases exponentially the amplitude of the evanescent waves at the same rate at which they decay in free space. We similarly have

$$E(z = 2d) = \exp(-\kappa_z d/2) \ \exp(\kappa_z d) \ \exp(-\kappa_z d/2) \ E(z = 0),$$
$$= E(z = 0). \qquad (8.10)$$

There is no net change in the amplitude of the evanescent waves at the image plane (at $z = 2d$) from the object plane at $z = 0$ and we have an exact compensation for the decay of the evanescent wave in free space (see Fig. 8.3). Thus, not only does the Veselago lens cancel the phase accumulation for the propagating waves, but it also restores the amplitudes of the evanescent components, bringing them both to a focus at the image plane. Further the reflection is identically zero for all incident plane waves due to the impedance matching ($\eta_1 = \eta_2$). The information carried by the evanescent waves, usually lost because of their exponential decay, is here totally transmitted to the image plane. As a consequence, all the fine features of the source can be reproduced and the lens has an infinite resolution in principle. This lens goes beyond the diffraction limit as it includes all the components of the near-field, too, and hence it is a *perfect lens*.

This result was first demonstrated in Pendry (2000) using the transmission and reflection coefficients of the slab calculated by a multiple scattering technique. Both methods of calculation obviously yield the same results, but the multiple scattering technique gives some more insight into the functioning of the perfect lens and some fundamental aspects of the physics involved in the problem. Hence we now examine the salient aspects of that solution. In the multiple scattering method, one decomposes the total transmission and reflection from the slab into the partial reflections from the interfaces and

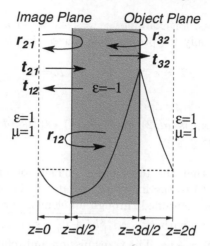

Figure 8.3 The perfect lens system consisting of a slab of NRM. The object is at a distance $d/2$ from the surface of the NRM and is focused on the other side of the slab. The restoration of the amplitude of an evanescent component is depicted schematically. The partial reflection/transmission coefficients across the boundaries are shown. (Reproduced with permission from Ramakrishna (2005). © 2005 Institute of Physics Publishing, U.K.)

sums the contribution from all such infinite number of partial waves as (see Appendix C)

$$T = \frac{t_{21}t_{32}e^{ik_{z2}d}}{1 - r_{12}r_{32}e^{2ik_{z2}d}}, \tag{8.11}$$

$$R = \frac{r_{21} + r_{32}e^{2ik_{z2}d}}{1 - r_{12}r_{32}e^{2ik_{z2}d}}, \tag{8.12}$$

where t_{jk} and r_{jk} are the Fresnel transmission and reflection coefficients across the interfaces between the different media (see Appendix A). For the configuration when $\varepsilon_- = -\varepsilon_+ = -1$ and $\mu_- = \mu_+ = -1$, we obtain trivially for propagating waves that $k_{z2} = -k_{z1}$, $t_{jk} = 1$, and $r_{jk} = 0$ due to the perfectly matched impedance across the interfaces. Hence we have $T = \exp(-ik_z d)$ and $R = 0$ as in the other mode of calculation. Now consider the evanescent waves with $k_{z1} = i\sqrt{k_x^2 + k_y^2 - \varepsilon_+\mu_+\omega^2/c^2} = i\kappa_z$. The proper choice of the wave-vector for evanescent waves in absorbing media implies that $k_{z2} = k_{z1}$, and when $\varepsilon_- = -\varepsilon_+$ and $\mu_- = -\mu_+$ the partial coefficients t_{jk} and r_{jk} actually diverge. However, the total transmission and reflection coefficients for the slab are still well defined in the limit:

$$\lim_{\substack{\varepsilon_- \to -1 \\ \mu_- \to -1}} T = e^{+\kappa_z d}, \qquad \lim_{\substack{\varepsilon_- \to -1 \\ \mu_- \to -1}} R = 0. \tag{8.13}$$

This is due to the fact that the partial transmission and reflection Fresnel coefficients diverge in the same manner (their denominators are identical). An identical result has been obtained with the previous method, confirming that the amplitudes of the evanescent waves increase exponentially upon traversal across the slab.

Note that the total transmission and reflection coefficients are well defined in spite of being the sum of a geometric series of the partial waves whose magnitude can be greater than unity, and are actually divergent in the case of evanescent waves. This is well known and justified on grounds of analytic continuity and convergence of infinite series: the validity of the sum of an infinite series transcends the divergence of the individual terms themselves provided that the sum was carried out to include all such infinite terms and provided that the series is not truncated at any point.[†] In fact, we could have obtained the solution for the evanescent waves directly from that of the propagating waves by a simple analytic continuation $k_z \to i\kappa_z$.

The process of restoring the amplitudes of the evanescent components, henceforth called amplification in this book,[‡] does not really involve any energy transport, as the Poynting vector associated with purely evanescent waves in a lossless medium is strictly zero (Ramakrishna and Armour 2003). These are the steady-state solutions of the Maxwell equations and give the field distributions essentially at infinite time. The large energy density associated with the exponentially enhanced fields in the NRM is obtained from the source itself and is built up over some time. One can see that in the region $d/2 < z < 3d/2$ the individual evanescent waves are required to be over-amplified in order to compensate for the decay of the fields outside in $3d/2 < z < 2d$. In the limit of large wave-vectors ($k_x \to \infty$) the amplification of these evanescent waves leads to an exponential mathematical divergence in the electromagnetic energy (Garcia and Nieto-Vesperinas 2002, Gomez-Santos 2003). These over-amplified fields do not correctly represent the fields in the region before the object plane ($z < 0$) and represent the price that has to be paid in order to project out a point source into free space. This is a manifestation of the pathological nature of the perfect lens and arises primarily from the lack of a large momentum (wave-vector) cutoff (Haldane 2002) in the problem. In principle, however, with all metamaterials there is always actually a cutoff for the parallel wave-vector that corresponds to the inverse

---

[†]A simple example is provided by the series expansion for

$$e^{-t} = 1 - \frac{t}{1!} + \frac{t^2}{2!} - \frac{t^3}{3!} + \cdots \quad ,$$

for $t > 1$. As $t \to \infty$, each term of the series individually diverges, while the sum monotonically and identically tends to zero. In fact, the sum cannot be obtained by truncating the series anywhere.

[‡]Note that the solution amplifies with distance and not with time. This amplification is distinct from the more usual amplification related to laser gain or amplifiers in electronics where there is an increase of energy with time.

of the lengthscale of the structure of the metamaterial $k_c \sim 1/a$. Some of the effects of non-locality are discussed in Section 8.6. The ideal lens conditions $\varepsilon_- = -1$ and $\mu = -1$ are very singular. All calculations should be carried out by adding an infinitesimal imaginary quantity, $i\delta$, to the frequency ω and the limit $\delta \to 0$ should be taken to recover the causal solutions. Since ε and μ are dispersive, this implies that calculations must always be made with a small absorptive imaginary part to ε and μ. Thus, the unphysical divergences disappear when one has even a finite level of absorption in the NRM, however small. Any finite amount of dissipation $\varepsilon_- = -1 + i\delta$ and $\mu_- = -1 + i\delta$ essentially sets its own large momentum cutoff. This is investigated in more detail in Section 8.3. In fact, the absorption in the material limits the ability to amplify the evanescent waves with a finite flux of energy available from the source. This limitation becomes particularly acute for the fields associated with large transverse wave-vectors, and hence affects the high resolution capabilities of the perfect lens.

The solution for evanescent waves is actually non-trivial and one is puzzled about the mechanism of their amplification. In the partial wave picture where the wave amplitude is decomposed, the proper choice of the sign of the wave-vector gives a deeper understanding of the process. The divergence of the scattering coefficients is a generic property of a linear system when the excitation frequency corresponds to some eigen frequency of the linear system. The divergence of the partial scattering coefficients for the evanescent waves therefore indicates the essential role of some resonances in the process of restoring the amplitudes for the evanescent waves. We investigate the nature of the resonance which is due to the surface plasmons in the next section.

The first demonstration of focusing of evanescent waves and subwavelength imaging was made in Grbic and Eleftheriades (2004), Iyer et al. (2003) using a slab of two-dimensional transmission line metamaterials (shown in Fig. 1.6) at microwave frequencies. The measured full width at half maximum (FWHM) for the image was about $\lambda/5$, clearly demonstrating the subwavelength image resolution and a transversally confined image was measured inside the slab as well.

8.2.1   Role of surface plasmons

In the proposal for the perfect lens (Pendry 2000), it was hinted that the surface plasmon excitations played a significant role. In the quasi-static limit of large wave-vectors ($k_x \to \infty$), $\varepsilon_- = -\varepsilon_+$ are exactly the conditions for the excitation of the surface plasmons on a metal surface for P-polarized light. Similarly, $\mu_- = -\mu_+$ are the conditions for the S-polarized surface modes of magnetic nature on the surface of a semi-infinite medium. These surface electromagnetic modes have been studied in Chapter 7. In the special case when both these conditions are satisfied at a single frequency ω, the conditions for the surface modes on a semi-infinite NRM,

$$\frac{\sqrt{k_x^2 - \varepsilon_+\mu_+\omega^2/c^2}}{\varepsilon_+} + \frac{\sqrt{k_x^2 - \varepsilon_-\mu_-\omega^2/c^2}}{\varepsilon_-} = 0, \tag{8.14}$$

$$\frac{\sqrt{k_x^2 - \varepsilon_+\mu_+\omega^2/c^2}}{\mu_+} + \frac{\sqrt{k_x^2 - \varepsilon_-\mu_-\omega^2/c^2}}{\mu_-} = 0, \tag{8.15}$$

hold for all k_x and all the surface plasmon modes (of both polarizations) become degenerate at ω. These dispersionless modes have a zero group velocity along the interface $v = (\partial\omega/\partial k_{||}) = 0$ (see Chapter 7).

For a slab of negative index material, the presence of the other interface detunes the surface plasmon resonance on each interface. The slab modes have resonance frequencies which are detuned by an exponentially small amount $\exp(-k_{||}d)$ for large wave-vectors. The dispersion of the slab polaritons for this case looks alike that obtained in the quasi-static limit of $\omega \to 0$. The condition for resonant excitation of a slab plasmon polariton at the frequency when $\varepsilon_- = -\varepsilon_+$ and $\mu_- = -\mu_+$ becomes $k_{||} \to \infty$. Thus, the action of the perfect lens crucially depends on not resonantly exciting any slab polariton resonance. When excited by evanescent waves from a source, these slab plasmon modes make their contribution to the total electromagnetic field off-resonantly. There is an increasing contribution from SPP modes with larger parallel wave-vector because the detuning from the frequency of the exciting radiation becomes exponentially small. Thus, one can produce exponentially large fields for the large wave-vectors and the predicted amplification of the electromagnetic fields results. The perfect lens condition of $\varepsilon_- = -\varepsilon_+$ and $\mu_- = -\mu_+$ is very special because it enables the excitation of the surface plasmon modes to just the right degree. However, as pointed out before, there is always some large wave-vector cutoff that prevents physical divergences of the fields or energy densities. This implies that for such large wave-vectors, our assumption of local response functions breaks down and the complete degeneracy of the surface plasmon modes also does not hold. These aspects are investigated in more detail in later sections of this chapter.

When the perfect lens conditions are satisfied, the total field can be written as a sum of the fields due to the exciting source, and the fields of the surface mode excitations on the two surfaces for any wave vector $i\kappa_z$ as

$$E(z) = Ae^{-\kappa_z|z|} + Be^{-\kappa_z|z-d_1|} + Ce^{-\kappa_z|z-d_2|}, \tag{8.16}$$

where $A = 1$ is the evanescent field of the source at $z = 0$, $B = -e^{-\kappa_z d_1}$ is the contribution from the surface plasmon at the left surface $z = d_1$, and $C = e^{-\kappa_z(d_2-2d)}$ is the field of the surface plasmon at the right surface $z = d_2$.[§] The field of the surface plasmon on the first interface exactly cancels the incident field for $z > d_1$ leaving behind only the fields of the second surface

---

[§] $d_1 = d/2$ and $d_2 = 3d/2$ in our case. But it is true as long as $d_1 + d_2 = 2d$.

plasmon in this region. Note that the amplitude C of this plasmon becomes exponentially large with $k_{||}$. On the left side of the slab $z < d_1$, the fields of the two surface plasmons exactly cancel (for zero reflectivity) and only the fields of the exciting source can be felt in this region. Thus, we are able to understand the exponentially large fields on the far-interface (from the source) as shown in Fig. 8.3. Contrastingly there are small fields on the near interface and it appears as if an unphysical *anti-surface plasmon* mode, with a field structure $\exp(+\kappa_z|z - d_1|)$ that appears to be exponentially growing with distance on both sides, has been excited at the first interface. This coherent interaction of the surface plasmons is responsible for the perfect lens action and renders the image resolution beyond the diffraction limit possible.

Actually the amplification of the evanescent waves is a more general property of the Helmholtz equation. For example, we can consider the one-dimensional time-independent Schrödinger equation (which is a Helmholtz equation) for a quantum mechanical particle and assume two identical negative δ potentials or two rectangular potential wells separated by some distance d. It is straightforward to show that if the incident energy of an incident quantum particle corresponds to the energy of the bound states in the potential wells, then the wave increases exponentially in the region of space between the two potential wells. The solutions are very similar in structure to the ones presented above. Thus, the crucial element is the presence of a bound state. The negative refractive index media can support surface plasmon states at the interfaces by themselves so that we do not require more structure like the rectangular potential well considered in our previous example. However, it should be noted that in the case of the quantum wells, we do not have a degeneracy of bound states for all incident wave-vectors in higher dimensions, which we do for negative index media. The modes disperse rapidly for larger wave-vectors which makes it difficult to obtain focusing using electronic waves. A similar discussion holds for acoustic waves where evanescent waves can be amplified in a similar manner using two resonant acoustic waveguides in place of the potential wells.

8.2.2   Quasi-static limit and silver lens

We note that the perfect lens works when the conditions on both the permittivity and the permeability, $\varepsilon_- = -\varepsilon_+$ and $\mu_- = -\mu_+$, are satisfied. From our discussion in Chapter 3, it is clear that generating media with both $\varepsilon < 0$ and $\mu < 0$ at the same frequency involves complex designs. Particularly at optical and ultraviolet frequencies, this problem is more acute due to the lack of magnetic materials with small enough unit cells. It is at the optical and ultraviolet frequencies that the prospect of imaging the near-field modes becomes most exciting, and there is enormous potential requirement for sub-wavelength imaging techniques. Thus, we would be severely handicapped by the impossibility of achieving a perfect lens at those frequencies due to the lack of proper materials.

An important simplification can be achieved (Pendry 2000) upon considering the situation when all the lengthscales in the problem such as the sizes of the objects involved, the distances involved, etc. are much smaller than the wavelength at the operating frequency. This is the *extreme near-field limit* or the *quasi-static limit*. All the wave-vectors involved are very large and we have $k_{x,y} \gg k_0 = \omega/c$ and $k_z \simeq ik_x$ in both the positive and negative media. In this limit, the Fresnel coefficients for the S-polarization across an interface between media j and k become approximately

$$r_{jk} \simeq \frac{\mu_k - \mu_j}{\mu_j + \mu_k}, \qquad t_{jk} \simeq \frac{2\mu_k}{\mu_j + \mu_k}. \qquad (8.17)$$

The coefficients are seen to be independent of ε of the two media involved. Similarly the Fresnel coefficients across an interface between media j and k for P-polarized light become approximately

$$r_{jk} \simeq \frac{\varepsilon_j - \varepsilon_k}{\varepsilon_j + \varepsilon_k}, \qquad t_{jk} \simeq \frac{2\varepsilon_k}{\varepsilon_j + \varepsilon_k}, \qquad (8.18)$$

which are independent of the magnetic permeabilities of the two media. In this extreme near-field limit, the P- and S-polarizations are therefore totally decoupled and appear like electrostatic and magnetostatic fields, respectively. Hence a slab with negative dielectric permittivity ($\varepsilon_- = -\varepsilon_+$) alone can act as a near-field lens for the P-polarized light, while a slab with negative magnetic permeability ($\mu_- = -\mu+$) alone can act as a near-field lens for the S-polarized light.

As emphasized in Chapter 3, some metals behave as good plasmas at optical and ultraviolet frequencies and have negative dielectric permittivities. A simple slab of a metal like silver or potassium has the potential to work as a near-field lens for imaging with P-polarized light at a frequency where $\varepsilon_- = -1$. Noting that dissipation affects the capability to resolve large wave-vectors, a highly conducting metal such as silver or potassium should be chosen. Alkali metals may be difficult to use due to their high reactivity and a noble metal like silver or gold would be preferable for direct use in such imaging applications. In Fig. 8.4, we show the intensities corresponding to the image of two slits separated by a small distance imaged by a slab of silver. These calculations have been carried in the near-field approximation and the subwavelength resolution capabilities are clearly visible. The resolution is, however, limited by the large levels of dissipation in silver. It actually turns out that the image intensities are also influenced by retardation effects, which are discussed in Section 8.3.

It is important to emphasize that these theoretical considerations have been verified experimentally, and in particular the silver lens has been implemented in practice. Fig. 1.5 shows its subwavelength imaging capability where subwavelength features on a nanosized object prepared by focused ion beam etching were reproduced in the near-field image produced by the silver lens. This

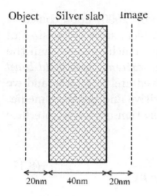

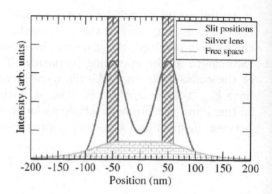

Figure 8.4 A thin film of silver can act as a near-field lens. The configuration used is shown on the left-hand side. The two slits, placed apart by 100 nm (subwavelength distance), are clearly resolved by the lens while they cannot be resolved when at the same distance in vacuum. The calculations have been carried out in the electrostatic limit and a value of $\varepsilon_- = -1 + i0.4$ has been used for silver. (Reproduced with permission from Ramakrishna (2005). © 2005 Institute of Physics Publishing, U.K.)

image was then used to lithographically etch the nanosized object in a pho-topolymer demonstrating the potential of this process for subwavelength photo lithography (Fang et al. 2005). Spatial features as small as 40 nm are clearly resolved. The silver lens has also been used by Melville and Blaikie (2005) for nano-lithography, and lithographic etching of periodic features of about $\lambda/4$ has been demonstrated.

8.2.3 "Near-perfect" lens with an asymmetric slab

The "perfect" lens requires the conditions of

$$\varepsilon_- = -\varepsilon_+, \qquad \mu_- = -\mu_+ \qquad (8.19)$$

to be satisfied at both the interfaces of the slab to enable the amplification of evanescent waves. Nonetheless the restoration of the amplitude for evanese-cent waves can be accomplished even if these conditions are satisfied at merely one interface only (Ramakrishna et al. 2002). This affords a considerable flex-ibility in the design of these lenses for practical applications where it may be desirable to have different media on either side of the flat lens. For example, a thin silver film deposited on a dielectric surface like GaAs or silica is much more mechanically robust than a free-standing thin film of silver in air.

Consider the geometry shown in Fig. 8.5 with the negative medium slab of thickness d placed between between two positive media of different permit-tivity and permeability. The transmission coefficient for the P-polarization is

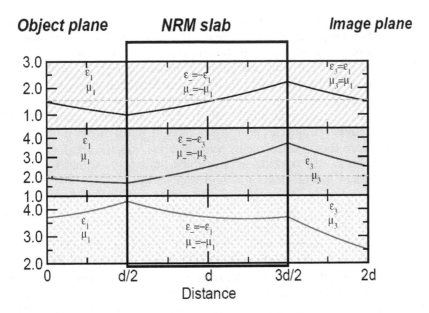

Figure 8.5 The asymmetric lens system consists of a slab of NRM with dielectric media with different ε on either side. The figure shows the three cases of (a) $\varepsilon_1 = \varepsilon_3 = -\varepsilon_2$ – the perfect lens, (b) Matching the perfect lens conditions on the far-side $\varepsilon_3 = -\varepsilon_2$ only and (c) Matching the perfect lens conditions on the near-side $\varepsilon_1 = -\varepsilon_2$ only. (Reproduced with permission from Ramakrishna (2005). © 2005 Institute of Physics Publishing, U.K.)

given by

$$T_p(k_x) = \frac{4\frac{k_{z1}}{\varepsilon_1}\frac{k_{z2}}{\varepsilon_2}\exp(ik_{z2}d)}{\left(\frac{k_{z1}}{\varepsilon_1} + \frac{k_{z2}}{\varepsilon_2}\right)\left(\frac{k_{z2}}{\varepsilon_2} + \frac{k_{z3}}{\varepsilon_3}\right) - \left(\frac{k_{z1}}{\varepsilon_1} - \frac{k_{z2}}{\varepsilon_2}\right)\left(\frac{k_{z2}}{\varepsilon_2} - \frac{k_{z3}}{\varepsilon_3}\right)\exp(2ik_{z2}d)}.$$
$$(8.20)$$

It is sufficient for the amplification of evanescent waves if the plasmon conditions on both ε and μ are satisfied at either one of the interfaces, i.e. for an incident evanescent wave if we have

$$\frac{k_{z2}}{\varepsilon_2} + \frac{k_{z3}}{\varepsilon_3} = 0$$

on the far surface from the source, or

$$\frac{k_{z1}}{\varepsilon_1} + \frac{k_{z2}}{\varepsilon_2} = 0$$

on the near surface. In either case, the transmission coefficient reduces to

$$T_p(k_x) = \frac{2(k_{z1}/\varepsilon_1)}{(k_{z3}/\varepsilon_3) + (k_{z1}/\varepsilon_1)}\exp(-ik_{z2}d),$$

which clearly shows the amplification for the evanescent modes by letting $k_z \to i\kappa_z$. For propagating waves (real k_z), we would have to choose the negative sign of the wave vector in the negative index medium. Under the condition of perfect impedance matching at any one interface, we can again obtain the same result as above which demonstrates the phase reversal for propagating waves.

Although we can obtain the amplification of the evanescent waves with this asymmetric configuration, the field strength of a plane wave at the image plane differs from the object plane by a factor $2(k_{z1}/\varepsilon_1)/[(k_{z3}/\varepsilon_3) + (k_{z1}/\varepsilon_1)]$ that depends on k_x. In the extreme near-field limit ($\omega \to 0$ and $k_z \to ik_x$), this term becomes independent of k_x and is given by

$$T_p(k_x) = \left(\frac{2\varepsilon_3}{\varepsilon_1 + \varepsilon_3}\right) \exp(-k_x d).$$

Thus only the image strength would be changed from that of the source in the near-field limit. It is also easily verified that wave-vectors with different k_x refocus at slightly different positions except in the quasi-static limit $k_x \to \infty$. There is no unique image plane and the focus of this lens is aberrated. There is also an impedance mismatch at one interface and the reflection coefficient is clearly non-zero. Hence, this asymmetric slab has been termed *a near-perfect lens* (Ramakrishna et al. 2002). We note that while the transmission is the same in both cases, the spatial variation of the field is completely different, depending on which interface the perfect lens conditions are satisfied on, as shown in Fig. 8.5 for an incident evanescent wave from a source placed in the object plane. Depending on which surface the perfect lens are satisfied on, resonant surface plasmons are excited on that surface and these resonances can contribute to the enhancement of the amplitude of the evanescent waves at the image plane. The reflection coefficient of the slab in the case of $\varepsilon_2 = -\varepsilon_1$ is also highly amplified. This is due to the resonant excitation of a surface plasmon field at the near interface. In the case $\varepsilon_2 = -\varepsilon_3$, we have a field variation over the distance from the object plane to the image plane in a manner that is more akin to the case when the perfect lens conditions are satisfied on both interfaces. In fact, this configuration turns out to give better image resolution in the presence of dissipation in the negative medium as discussed in Section 8.3, and is preferable to the first case.

A similar result holds for the S-polarized wave incident on the slab. The presence of a large reflectivity has serious consequences for the use of a lens for near-field imaging applications as it would disturb the object field. This aspect of the reflectivity deteriorating the image formation has been investigated by some researchers. In spite of these disadvantages, the possibility of using asymmetric slab structures simplifies the technology of making the lenses by deposition of thin films, a well-established technology. In some lithographic situations, it is also imperative to have different materials such as a mask layer so that the possibility of having asymmetric structures can facilitate the lithographic processes.

8.3 Limitations due to real materials and imperfect NRMs

As we have pointed out in our discussions, the perfect lens with infinite image resolution would be possible, in principle, only if we could have a medium with exactly $\varepsilon = \mu = -1$ and no dissipation. Further, it should be possible to regard the medium as homogeneous at all lengthscales, however small. In this section, we consider the effects of the underlying structure and dissipation in real materials which limits the image resolution to some finite levels. These limitations, which always arise in real materials, prevent as an exaggerated example, the possibility of imaging of atoms at lengthscales of 10^{-10}m using radio-frequency waves with wavelengths of over 10 m.

It has been emphasized that the infinite image resolution in the perfect lens arises from the absence of a large momentum cutoff. With all real materials, however, there exists an inherent large momentum (wave-vector) cutoff and this infinite resolution is actually not possible. In metamaterials, it is clear that once the radiation can probe the structural details of the metamaterial, effective medium parameters make no sense anymore (see Chapter 2). This already imposes a natural cutoff on the transverse wave-vector of $k_x < k_c = 2\pi/a$, where a is the lengthscale of the periodicity of the metamaterial structures. In the case of metamaterials, the boundaries of the slab are also indeterminate on the lengthscale of a unit cell, which can also lead to some extra dispersion on the surface modes thereby causing some degradation of the image (Feise et al. 2002). Even with metals, the dielectric constant becomes dependent on the wave-vector for large wave-vectors and is spatially dispersive. For example, the Lindhard form (Ashcroft and Mermin 1976) for the dielectric permittivity of a metal at small wave-vectors $k \ll k_F$ where k_F is the wave-vector at the Fermi surface, within the random phase approximation (for the single particle excitations), yields

$$\varepsilon(\omega, \mathbf{k}) = 1 - \frac{\omega_p^2}{\omega^2} \frac{1}{1 - 3/5(\mathbf{k} \cdot \mathbf{v}_F/\omega)^2}, \tag{8.21}$$

where $\mathbf{v}_F = \hbar k_F/m^*$ is the Fermi velocity (m^* being the effective electron mass). This expression yields a quadratic wavenumber dependence to a first approximation. The random phase approximation holds until there is sufficient screening and is valid even at lengthscales of a few atomic sizes ($\sim$2 nm). But given the spatial dispersion, we would not aim at resolving atoms or small molecules with visible or IR light. Some effects of spatial dispersion on the image resolution are discussed in Section 8.6.

Much before the limits caused by the graininess of the structure of matter or the metamaterials set in, other mechanisms come into play to limit the ability of this lens to focus the waves with large transverse wave-vectors, and the image resolution. Foremost is the effect of absorption in the meta-material comprising the lens. As we have seen in earlier chapters, dissipation always accompanies the large dispersion in metamaterials (due to the Kramers-Kronig relationship) which gives rise to negative refractive index. Thus, there would always be some amount of dissipation in the lens medium, although in principle, it can be made arbitrarily small by using lesser and lesser dissipative materials in the construction of the metamaterials. The levels of absorption are maximal in the regions where there are large electro-magnetic fields present (let us remind ourselves that absorption is proportional to $[\text{Im}(\varepsilon)|E|^2 + \text{Im}(\mu)|H|^2]$). The exponential growth of the evanescent waves results in extremely large fields at the far-side interface. Thus, very large absorption occurs for the larger wave-vector components, essentially implying that the source emitting a limited amount of power cannot sustain the amplifi-cation of these waves with a large transverse wave-vectors.[¶] By regulating the energy flow, absorption sets its own large wave-vector cutoff: the larger the absorption rate, the smaller the corresponding cutoff wave-vector. Thanks to this limit, the divergence in the region $d < z < 3d/2$ is avoided (theoretically, an arbitrarily small amount of dissipation suffices to lift the divergence).

The effects of absorption can be accounted for by including the imaginary part of ε_- and μ_- in the models. Consider the transmission coefficient for the P-polarized light across a symmetric slab of a negative refractive index medium (obtained by setting $\varepsilon_3 = \varepsilon_1, \mu_3 = \mu_1$, and $k_{z3} = k_{z1}$ in Eq. (8.20)):

$$T_p(k_x) = \frac{4\left(\frac{k_{z1}}{\varepsilon_+}\right)\left(\frac{k_{z2}}{\varepsilon_-}\right)}{\left(\frac{k_{z1}}{\varepsilon_+} + \frac{k_{z2}}{\varepsilon_-}\right)^2 \exp(-ik_{z2}d) - \left(\frac{k_{z1}}{\varepsilon_+} - \frac{k_{z2}}{\varepsilon_-}\right)^2 \exp(ik_{z2}d)}. \tag{8.22}$$

For evanescent waves, note that the exponential in the second term decays in amplitude while the exponential in the first term increases with k_x. Under the perfect lens conditions, the first term in the denominator vanishes for the evanescent waves and there is no exponential increase of the amplitude. However, if there is a mismatch of the material parameters ($\varepsilon_- = -1 + i\delta\varepsilon$ and $\varepsilon_+ = +1$, say), the first term is no longer zero, and beyond some value of k_x, it begins to dominate the behavior of the denominator due to the growing exponential term. When this happens, the transmission coefficient begins to decay exponentially. Thus, the wave-vector for which the two coefficients are approximately equal sets a cutoff for k_x beyond which the evanescent waves are no longer effectively amplified across the slab. In the quasi-static limit ($k_x \gg \omega/c$), we find that the two terms in the denominator are approximately equal when

$$k_x d = -\ln|\delta\varepsilon/2|, \tag{8.23}$$

which can be adopted as being the largest wave-vector for which there is an effective amplification.

[¶]Note that the Poynting vector is zero for purely evanescent waves in a non-dissipative medium. The dissipation causes a phase shift that now drives an energy flow.

Upon defining the subwavelength resolution as the ratio of the optical wavelength to the linear size ($\Delta_{\min}$) of the smallest resolved feature, we obtain

$$\text{res} = \frac{\lambda}{\Delta_{\min}} = \frac{-\ln|\delta\varepsilon/2|}{2\pi}\frac{\lambda}{d}. \tag{8.24}$$

Thus, the resolution depends logarithmically on the deviations of the material parameters and inversely on the width of the slab, implying that the perfect lens effect is indeed very sensitive to the material imperfections.

For the silver slab lens with a thickness of one tenth the wavelength of radiation, this expression yields a minimum resolved feature of about $\lambda/2.5$. In the case of the asymmetric lens, we note that the resolution becomes

$$\text{res} = \lambda_0/\lambda_{min} = -\frac{\ln|\varepsilon_2''/2\varepsilon_3|\lambda_0}{4\pi d} \tag{8.25}$$

(assuming $\varepsilon_3 \gg \varepsilon_2''$ and $\varepsilon_3 \gg \varepsilon_1$) in the favorable case of low reflection when $\varepsilon_2 \simeq -\varepsilon_3$. In the limit of large ε_3, the resolution is actually enhanced (Ramakrishna et al. 2002). Note that the enhancement in the image resolution obtained by increasing ε_3 results from a smaller percentage mismatch in the material parameters arising from a given value of the imaginary part of ε or μ. It does not arise from the fact that waves with larger transverse wave-vectors propagate in the high index medium as in an oil-immersion lens.

Due to dispersion in the material parameters, in practice, one can also end up with deviations in the real parts of ε and μ due to a mismatch in the choice of the operating frequency. In general, the perfect lens effect turns out to be rather sensitive to any deviations from the lens conditions $\varepsilon_- = -\varepsilon_+$ and $\mu_- = -\mu_+$ (Shamonina et al. 2001, Ramakrishna et al. 2002, Shen and Platzman 2002, Fang and Zhang 2003, Smith et al. 2003, Merlin 2004). The deviation considered above, $\delta\varepsilon$, could be in the real part of ε_- and the same expression given by Eq. (8.24) would be obtained for the image resolution. The fields at the image plane for a point source in the limit of small $\delta\varepsilon$ has been analytically obtained in Merlin (2004) and it has been shown that the transverse width of the image is effectively given by the same number predicted above. A similar result holds for the limit on image resolution caused by the deviation in $\delta\mu$ from the ideal conditions for S-polarized light. A detailed analysis of this condition for deviations in the real parts of the material parameters is given in Section 8.3.1.

Actually the image formed by using P(S)-polarized radiation is more severely affected by deviations in ε_- (μ_-) than deviations in μ_- (ε_-) from their ideal theoretical values. This is a true reflection of the electric nature of P-polarized light compared to the magnetic nature of S-polarized light. However, a deviation in μ_- from the ideal conditions would also set an ultimate limit on the image resolution possible for P-polarized light even if $\varepsilon_- = -\varepsilon_+$ in an ideal manner and with no dissipation. Even in the limit of large k_x, the image resolution would be limited via retardation effects – the surface plasmons on both sides of the slab just cannot keep pace with each other. For example, let us examine the transmission coefficient for P-polarized light through a slab with $\varepsilon_- = -\varepsilon_+$, but with $\mu_- = \mu_+ = \mu$, which is a rather large deviation in the magnetic permeability from the ideal conditions. Let us determine the value of k_x for which the two terms in the denominator become approximately equal in Eq. (8.22). As we expect this to happen at reasonably large values of the wave-vectors, let us approximate the normal components of the wave-vector in the two media by

$$k_{z\pm} = i\sqrt{k_x^2 - \mu\varepsilon_\pm \frac{\omega^2}{c^2}} \simeq i(k_x - \mu\varepsilon_\pm \frac{\omega^2}{2k_x c^2}), \qquad (8.26)$$

i.e. we only retain terms to first order in k_x^{-1}. In this case, we obtain the corresponding condition on the wave-vector in the form of a transcendental equation

$$k_x d \simeq -\ln\left(|\varepsilon_\pm \mu| \frac{k_0^2}{2k_x^2}\right), \qquad (8.27)$$

where $k_0 = \omega/c$ as usual. For a lossless medium with $\varepsilon = \pm 1$, $\mu = 1$, and slab thickness of $k_0 d = 0.35$, we can estimate the cutoff wave-vector to be $k_x \sim 18.7$, which agrees very well with the transmission function plotted in Fig. 8.6 (curve (1) in the left panel). Thus we find that the effects of even such a large mismatch in the μ is rather limited compared to the effects of a mismatch in ε when imaging with P-polarized light. However, for very thin slabs the limit obtained here due to the retardation effects eventually makes the amplification ineffective for larger wave-vectors and sets a fundamental limit on the process.

The limit on image resolution caused by deviations in the real parts of the material parameters essentially arises from the finite detuning from the resonance frequency for off-resonantly exciting the slab plasmon polaritons. For the perfect lens the detuning becomes exponentially small for large wave vectors in comparison. Thus, the slab plasmon polaritons for large transverse wave-vectors ($k_x \rightarrow \infty$) cannot be excited sufficiently strongly in this case and the exponential enhancement of the electromagnetic fields required for the large wave-vectors cannot occur. For any small amount of deviation in the real parts of the parameters, slab plasmon polaritons are, instead, resonantly excited at certain wave-vectors as evidenced in Fig. 8.6. These resonant excitations of the slab would imply a very large transmission at the corresponding wave vector (for evanescent waves $T > 1$ is allowed). The direct excitation of these slab modes for imaging applications is undesirable, as the fields due to these resonances are disproportionately represented in the image. Yet, the existence of these resonances is essential, since the recovery of the evanescent modes can be seen as the result of driving the surface plasmons off-resonance.

The absorption mainly causes an exponentially decaying transmission for large wave-vectors. Absorption actually damps these transmission resonances and makes possible the very lensing action with some subwavelength resolution

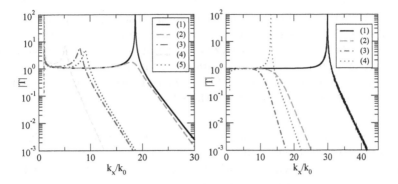

Figure 8.6 Absolute value of the transmitted field at the image plane $|T\exp(ik_z^{(1)}d)|$, for the P-polarization as a function of k_x, showing the transmission resonances caused by the resonant excitations of the slab plasmon polaritons. On the left panel, $\mu = +1$ and (1) $\varepsilon = -1 + \mathrm{i}0.0$, $k_0 d = 0.35$; (2) $\varepsilon = -1 + \mathrm{i}0.01$, $k_0 d = 0.35$; (3) $\varepsilon = -0.9 + \mathrm{i}0.01$, $k_0 d = 0.35$; (4) $\varepsilon = -0.9 + \mathrm{i}0.01$, $k_0 d = 0.5$, and (5) $\varepsilon = -1.1 + \mathrm{i}0.01$, $k_0 d = 0.35$. On the right panel, (1) $\mathrm{Re}\,\varepsilon = -1$ and (1) $\mathrm{Im}(\varepsilon) = 0$, $\mu = -1.1$, $k_0 d = 0.35$; (2) $\mathrm{Im}(\varepsilon) = 0.01$, $\mu = -1.1$, $k_0 d = 0.35$; (3) $\mathrm{Im}(\varepsilon) = 0.01$, $\mu = -1.1$, $k_0 d = 0.5$; (4) $\mathrm{Im}(\varepsilon) = 0$, $\mu = -2$, $k_0 d = 0.35$. The absorption broadens and softens the resonances. Deviation in ε affects the P-polarization more than the deviations in μ. (Reproduced with permission from Ramakrishna (2005). © 2005, Institute of Physics Publishing, U.K.)

capabilities in the case of the silver lens with $\mu = +1$. Although the limitation they impose on the resolution is similar, the detailed effects of absorption and the deviations in the real part of the μ or ε which introduce retardation effects are quite different. The perfect lens solution in the absence of losses is very special, and can lead to paradoxical interpretations. The Poynting vector at the location of a point focus is strictly ill-defined. However, the presence of a small and finite absorption makes the focus a small blob instead and the Poynting vector also becomes well defined everywhere.

8.3.1 Analysis of the lens transfer function for mismatched material parameters

It has already been discussed above that the mismatch in the constitutive parameters between the lens material (ε_-, μ_-) and its surrounding medium (ε_+, μ_+) is a key factor in deteriorating the performance. The inherently dispersive and dissipative nature of left-handed media implies that both ε_- and μ_- cannot be purely real. An additional reason, less fundamental but important in practice, is that exactly achieving a value of (-1) for both the permittivity and the permeability simultaneously is quite difficult in practice.

Consequently, the previous section has examined the importance of the impact of small deviations from the perfect case $(\varepsilon_-, \mu_-) = (-1, -1)$ on the resolution of the lens. We have already analyzed the addition of losses so that we shall focus here on a more analytical treatment of the mismatch in the real parts. Note that in order to keep the effects separate, we suppose here that the imaginary parts are zero, and study the effect of a mismatch in the real part only.

The configuration is the symmetric one shown in Fig. 8.3: a slab of left-handed medium, characterized by constitutive parameters (ε_-, μ_-), is surrounded by free-space, whose constitutive parameters are denoted by (ε_+, μ_+). Since the problem is effectively two-dimensional, we shall use a two-dimensional source in the form of a line source expressed by

$$\mathbf{J}(\mathbf{r}) = \hat{y} I \delta(x) \delta(z). \tag{8.28}$$

For a S-polarized incidence, the field in the three regions (regions 1 and 3 being to the left and to the right of the lens, region 2 being the lens itself) can be expressed in integral form as

$$E_{y1} = \int_{-\infty}^{+\infty} dk_x \left(e^{ik_{z1}z} + R e^{-ik_{z1}z} \right) E_\ell \, e^{ik_x x}, \tag{8.29a}$$

$$E_{y2} = \int_{-\infty}^{+\infty} dk_x \left(A \, e^{ik_{z2}z} + B e^{-ik_{z2}z} \right) E_\ell \, e^{ik_x x}, \tag{8.29b}$$

$$E_{y3} = \int_{-\infty}^{+\infty} dk_x \, T e^{ik_{z1}z} E_\ell \, e^{ik_x x}. \tag{8.29c}$$

In the previous equations, k_x is the phase-matched transverse wave-vector component and is the same in all three media, k_{z1} and k_{z2} are the longitudinal components of the wave-vector in regions 1 and 2, respectively (the longitudinal component in the third region is equal to the one in the first region). In addition, the reflection and transmission coefficients (R and T) as well as the coefficients of the left-propagating and right-propagating waves inside the slab are obtained in the usual fashion by matching the boundary conditions, and are given by (see Appendix C)

$$R = e^{ik_{z1}d/2} \, \frac{r_{21} + r_{32} e^{2ik_{z2}d}}{1 + r_{21} r_{32} e^{2ik_{z2}d}}, \tag{8.30a}$$

$$T = \frac{4 \, e^{i(k_{z2} - k_{z1})d}}{(1 + p_{21})(1 + p_{32})(1 + r_{21} r_{32} e^{2ik_{z2}d})}, \tag{8.30b}$$

$$A = \frac{2 \, e^{-i(k_{z2} - k_{z1})d/2}}{(1 + p_{21})(1 + r_{21} r_{32} e^{2ik_{z2}d})}, \tag{8.30c}$$

$$B = \frac{2 r_{32} \, e^{-i(k_{z2} - k_{z1})d/2} \, e^{2ik_{z2}3d/2}}{(1 + p_{21})(1 + r_{21} r_{32} e^{2ik_{z2}d})}, \tag{8.30d}$$

where

$$p_{21} = \frac{\mu_+ k_{z2}}{\mu_- k_{z1}}, \qquad\qquad p_{32} = 1/p_{21}, \qquad\qquad (8.31a)$$

$$r_{21} = \frac{\mu_- k_{z1} - \mu_+ k_{z2}}{\mu_- k_{z1} + \mu_+ k_{z2}}, \qquad\qquad r_{32} = -r_{21}. \qquad\qquad (8.31b)$$

Finally, the term E_ℓ represents the spectral kernel of a line source and is given by

$$E_\ell = -\frac{\omega \mu_0 \mu_+ I}{4\pi k_z}. \qquad\qquad (8.32)$$

Consequently, it can be seen for example that in the case of no reflection, $R = 0$ and the expression of E_{1y} reduces to the integral representation of a Hankel function, which is what we expect from an infinitesimal source in two dimensions (the electric field is in this case directly analogous to the Green's function in two dimensions).

¿From the previous expressions, it can be seen that all field components contain a similar term in their denominator, $1 - r_{21}^2 \, e^{2ik_{z2}d}$. When this term is zero, for specific values of k_{z2}, the electric fields in all three regions diverge, i.e. we have a mathematical pole. The location of this pole usually needs to be determined numerically but the analysis can be carried out analytically in the following situation: let us introduce a mismatch $\delta > 0$ (δ is thus real) in the constitutive parameters of the slab while keeping the wave-number constant so that (Lu et al. 2005a)

$$\mu_- = -(1 + \delta), \qquad \varepsilon = -\frac{1}{1 + \delta}, \qquad\qquad (8.33)$$

and take $(\varepsilon_+, \mu_+) = (1, 1)$. Since all the wave-numbers have the same magnitude, k_x is identical in all three regions and $k_z = -k_{2z}$ (although the solution is independent of the choice of the sign of k_{z2}). Under these circumstances, the reflection coefficient R_{12} simply becomes $R_{12} = \delta/(2+\delta)$ and the electric field in the third region (where we expect the image of the line source to be formed) is written as

$$E_{3y} = -\frac{2\omega \mu_- \mu_0 I}{\pi} \frac{1 + \delta}{(2 + \delta)^2} \int_0^\infty \frac{1}{k_{z1}} \frac{e^{ik_{z1}z}}{e^{2ik_{z1}d} - \left(\frac{\delta}{2+\delta}\right)^2} \cos(k_x x) \, \mathrm{d}k_x, \qquad (8.34)$$

where we have folded the integration from $[-\infty, +\infty]$ to $[0, +\infty]$ by symmetry. The poles of the integrand are obtained by letting the denominator go to zero, which trivially yields

$$k_{z1} = \frac{m\pi}{d} + \frac{i}{d} \ln\left(\frac{2 + \delta}{\delta}\right), \qquad\qquad (8.35)$$

where m is an arbitrary integer and where we find again a logarithmic dependency as in Eq. (8.24). In order to understand the influence of this pole on the value of the electric field, we need to transform the previous integral from the k_x plane to the k_z plane. This is immediately done using the dispersion relation which reveals that values of $k_x \in [0, \omega/c]$ map onto values of $k_z \in [\omega/c, 0]$, whereas $k_x > \omega/c$ map onto the imaginary axis $k_z \in [\mathrm{i}0, +\mathrm{i}\infty]$. We therefore see that the pole corresponding to $m = 0$ in Eq. (8.35) falls exactly onto the integration path, which therefore needs to be deformed in the complex k_z plane. In addition, this pole is representative of the surface plasmon excited at the boundary between the free-space and the medium characterized by (ε_-, μ_-).

Upon transforming the integration from the k_x plane to the k_z plane, the path needs to be deformed in order to avoid the pole. Nonetheless, the kernel of the integrand is increasing up to the value of the pole, making the numerical evaluation of the integral potentially challenging. In particular, when the parameters of the slab approach those required by the perfect lens condition $(\varepsilon_-, \mu_-) \to (-1, -1)$, δ decreases, the pole moves away toward infinity, and the integral becomes increasingly difficult to evaluate, being divergent in the limit of $\delta = 0$. Note that the numerical integration of Eq. (8.34) is very much akin to the numerical integration of Sommerfeld integrals and similar techniques can be used such as the weighted average algorithm (Michalski 1998), an adaptive Simpson's rule or a Romberg's rule on a modified integration path (Chew 1990).

It is important to repeat the previous analysis in the case when ε_- and μ_- are positive, in order to verify that a similar mode is not sustained by this configuration. Upon doing so, we find that the poles are given by the condition

$$k_z = \frac{m\pi}{d} - \frac{\mathrm{i}}{d} \ln\left(\frac{2+\delta}{\delta}\right), \tag{8.36}$$

which clearly shows that the imaginary part of k_z is negative in this case and therefore that the pole does not influence the spectral integration of the electric field which extends over the positive imaginary axis only. In other words, no surface plasmon is supported at the interface between the two positive index media, as expected.

As we have mentioned previously, the pole is directly related to the surface plasmon excited at the interface between the two media (for a symmetric configuration like here, a surface plasmon is excited at each interface). Since the pole has a longitudinal wave-vector (obtained for $m = 0$)

$$k_z^{(o)} = \frac{\mathrm{i}}{d} \ln\left(\frac{2+\delta}{\delta}\right) \tag{8.37}$$

the wavelength of the surface mode along the $\hat{y}$ direction ($\lambda_y^{(o)} = 2\pi/k_y^{(o)}$) can be directly computed and yields

$$\lambda_y^{(o)} = \frac{2\pi}{\sqrt{(\omega/c)^2 + \frac{1}{d^2} \ln\left(\frac{2+\delta}{\delta}\right)}}. \tag{8.38}$$

It is seen that as the configuration approaches the perfect lens, $\delta \to 0$ and $\lambda_y^o \to 0$, indicating that the lens is able to discern finer and finer details. When $\lambda_y^o = 0$, the lens is perfect.

Interestingly, this argument is very much akin to the principle of uncertainty of quantum mechanics. If we write it as $\Delta k_y \Delta y \leq 2\pi$, then it is clear that in order to have $\Delta y \to 0$, we must make sure that $\Delta k_y \to \infty$, which corresponds to a pole that moves to infinity. As a direct consequence, we see that a perfect resolution is achieved only when *all* the evanescent waves are included in the spectrum (in addition to the propagating waves of course), which can only be achieved by a perfect lens with $(\varepsilon_-, \mu_-) = (-1, -1)$ and of infinite lateral extent.

8.3.2 Focussing properties of a finite slab of NRM

The previous discussion on the perfect lens has highlighted a few unphysical assumptions necessary for the achievement of perfect resolution. The most obvious one, discussed at length, is the requirement of a lossless medium perfectly matched to its surrounding, i.e. of constitutive parameters $(-\varepsilon_0, -\mu_0)$ in the case of a free-space background. The degradation of the image resolution as function of deviations from this ideal situation have been studied and have led to the more physical concept of *super-lenses*. In this section, we examine the second most stringent assumption of the perfect lens, namely, its infinite extent along the transverse direction. Being impractical, this theoretical idealization naturally breaks down when physical systems are realized, and it appears important to provide a methodology for the study of the resolution degradation as function of the finite transverse extent of the lens.

Results of numerical calculations based on the Finite Element Method for the imaging of two line sources through a finite slab of about 0.8λ width and $\varepsilon = -1 + i10^{-4}$, and $\mu = -1$ are shown in Fig. 8.7. At the image plane, the images of the two line sources placed about 0.2λ apart are clearly resolved although the slab itself is of subwavelength size. The field profile along the imaging direction clearly reflects the excitation of an anti-symmetric plasmon, although this can differ depending on the overall size of the slab and the relative transverse placement of the sources. Overall, while some information is lost due to the loss of some propagating modes, much of the near-field information is still preserved. Effectively the slab of finite transverse extent appears to also have the effect of an aperture whereby propagating waves incident at large angles from a localized source are lost. But the transport of evanescent components does not appear to be substantially affected and they appear to be able to contribute to image formation effectively. Similar calculations reveal that unless the transverse size becomes comparable to the size of the objects imaged, it does not significantly affect the image. It is also seen that large surface plasmon excitations form standing waves along the length of the slab. The finite levels of dissipation quickly damp out the surface plasmon excitations with large wave-vectors while allowing the excitation of surface

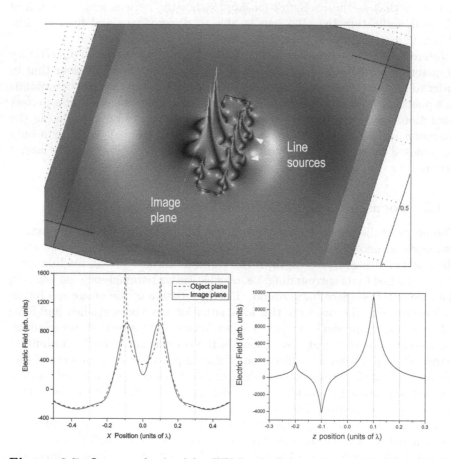

Figure 8.7 Images obtained by FEM calculations for a slab lens of finite transverse extent and $\varepsilon = -1 + i10^{-4}$, $\mu = -1$. The images of the two line sources placed at a subwavelength distance apart (0.2λ) are clearly resolved in the image plane although the slab itself is subwavelength in size (0.8λ). The object plane is at $z = -0.2$, the image plane is at $z = 0.2$, while the lens occupies the region $-0.1 < z < 0.1$ and $-0.4 < x < 0.4$. Top: Gray-scale plot of the electric field. The outer lines show the boundaries for the FEM calculations where Perfectly Matched Layers (PML) are imposed to have no reflections. Bottom left: The fields at the object and image planes. Bottom right: The field pattern along the imaging direction shows the excitation of an antisymmetric slab plasmon. The $\hat{z}$ axis runs through one of the sources.

plasmons of smaller wave-vectors. The image resolution in these calculations is also constrained by the levels of dissipation. The fact that the slab lens can be finite and actually even smaller than the wavelength is very significant for imaging and lithographic applications.

While the above numerical example shows that subwavelength resolution is possible with a slab of finite transverse width, it does not offer much insight into the physics of the perfect imaging process and its degradation. In addition, since some numerical techniques are prone to errors when applied to the case of the perfect lens (see Section 8.4), they should be taken cautiously, merely providing a ground of comparison and cross validation with more analytical approaches.

We have shown in Section 8.3.1 that the resolution of a left-handed medium lens is directly related to a pole in the complex k_z plane, which sets a cutoff point: the evanescent waves below it are amplified and their information is used in the image reconstruction whereas the evanescent waves above it are practically lost. This defines the *spectrum* of the lens (a broader spectrum implies a better the resolution), which is a good metric to quantify the quality of the lens. In the extreme case of a perfect lens, the spectrum is infinitely broad and all the evanescent waves are reconstructed. The purpose of this section is therefore to examine the influence of the finite transverse extent of the lens on its spectrum.

The configuration of the problem is similar to Fig. 8.3 except that the length in the $\hat{x}$ direction is finite and denoted by L. Although analytical solutions to the Maxwell equations exist in the case of some canonical geometries such as sphere, cylinders, spheroids, etc., it is not the case for a finite length slab so that we have to resort to some well-justified approximations.

We approach the problem from the Huygens' equivalence principle, by which the fields at the second boundary, denoted by $z = d_2$ for the sake of generality, can be replaced by radiating currents, which in turn can be used to obtain the field in the transmitting region at $z > d_2$, where we expect an image point to be formed. The problem is therefore reduced to the deduction of the equivalent currents at the second boundary and asserting their similarities and differences between configurations with finite and infinite boundaries.

Let us first consider the situation of an infinite slab. The electric field in the region beyond the second boundary is given by Eq. (8.29c), whereas the magnetic field is directly obtained from the Maxwell equations (we assume $\mu_+ = -\mu_- = 1$):

$$E_y(x, z = d_2) = -\frac{\omega \mu_0}{4\pi} \int_{-\infty}^{\infty} dk_x \frac{T(k_z)}{k_z} e^{ik_z d_2} e^{ik_x x}, \tag{8.39a}$$

$$H_x(x, z = d) = -\frac{1}{4\pi} \int_{-\infty}^{\infty} dk_x T(k_z) e^{ik_z d_2} e^{ik_x x}, \tag{8.39b}$$

where we have supposed that $I = 1$ without loss of generality. The field beyond the slab boundary is obtained from the Huygens' principle written in the Stratton-Chu form as

$$\mathbf{E}(x, z) = \int_S d x' \left\{ i \omega \mu_0 \left[\hat{n} \times \mathbf{H}(\mathbf{r}') \right] g(\mathbf{r}, \mathbf{r}') + \left[\hat{n} \cdot \mathbf{E}(\mathbf{r}') \right] \nabla' g(\mathbf{r}, \mathbf{r}') \right.$$
$$\left. + \left[\hat{n} \times \mathbf{E}(\mathbf{r}') \right] \times \nabla' g(\mathbf{r}, \mathbf{r}') \right\} \tag{8.40}$$

where the surface can be taken over a line only due to the translational invariance of the problem in the $\hat{y}$ direction, $g(\mathbf{r}, \mathbf{r}')$ is the two-dimensional Green's function given by $g(\mathbf{r}, \mathbf{r}') = i/4 H_0^{(1)}(k_0 |\rho - \rho'|)$, and where $\hat{n}$ is the normal to the boundary. Introducing Eqs. (8.39) into Eq. (8.40) yields two terms for the electric field:

$$E_y(x, z) = \frac{\omega \mu_0}{16 \pi} \int_{-\infty}^{+\infty} d x' \int_{-\infty}^{+\infty} d k_x T(k_z) e^{i k_z z} e^{i k_x x} H_0^{(1)}(k \sqrt{(x - x')^2 + z^2})$$
$$- \frac{i \omega \mu_0}{16 \pi} \int_{-\infty}^{+\infty} d x' \int_{-\infty}^{+\infty} d k_x T(k_z) e^{i k_z z} e^{i k_x x} \times$$
$$\frac{\partial}{\partial z'} H_0^{(1)}(k \sqrt{(x - x')^2 + (z - z')^2})|_{z'=d_2}. \tag{8.41}$$

Upon using the integral representation of the Hankel function, it can be shown that the terms are equal, indicating that the contribution from the electric current sheet is identical to the contribution from the magnetic current sheet. The electric field in the spectral domain is therefore given by

$$E_y(k_z) = -\frac{\omega \mu_0}{4 \pi} \frac{e^{2 i k_z (d_2 - d_1)}}{k_z} T(k_z), \tag{8.42}$$

where $T(k_z)$ is the transmission coefficient of the slab. ¿From these equations, we see that the spectrum of the image is composed of the spectrum of the line source multiplied by the transmission coefficient (Chen et al. 2006c).

This approach, developed for an infinite slab, can be approximately extended to the slab of finite transverse width. As a matter of fact, we assume that the electric field can be written as in Eq. (8.39a) but weighted by a spatial window function (we reiterate that this is only an approximation and it is not the exact solution):

$$E_y(x, z = d_2) \approx -\frac{\omega \mu_0}{4 \pi} f(z) \int_{-\infty}^{+\infty} d k_x \frac{T(k_z)}{k_z} e^{i k_z d_2} e^{i k_x x}, \tag{8.43}$$

where the functions f, still to be determined, is the window function. Following the same procedure as previously, it can be shown that the spectrum of the image is given by

$$E(k_x) \approx -\frac{\omega \mu_0}{4 \pi} \frac{e^{2 i k_z (d_2 - d_1)}}{k_z} [T(k_z) \otimes F(k_z)], \tag{8.44}$$

where the $\otimes$ symbol represents the convolution operator and F is the Fourier transform of f. Consequently, the spectrum of the image due to a finite-size

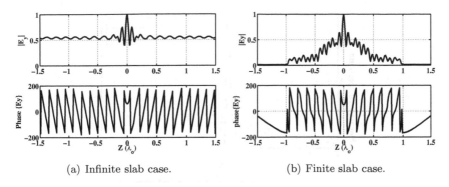

(a) Infinite slab case.　　　　(b) Finite slab case.

Figure 8.8 Comparison between the electric fields at the image plane obtained from an infinite and a finite slab. The top plots represent the amplitudes whereas the bottom plots represent the phases. (Reproduced with permission from Chen et al. (2006c). ©2006 by The American Physical Society.)

width is approximately given by the spectrum of the source itself multiplied by the convolution of the transmission coefficient and the Fourier transform of the window function f. In this way, the finite size of the slab explicitly appears in the image spectrum and is thus directly related to the resolution of the lens.

Let us work out an example in order to illustrate how the methodology just presented can be applied to the study of the resolution of a finite width slab of thickness $d = 0.2\lambda$ and of transverse dimension $L = 2\lambda$, where λ is the free-space wavelength. The above dimensions are chosen to typically represent the ratio of most of the experiments realized to date.

Fig. 8.8 represents the fields at the second boundary of both the finite and the infinite lens, where the first ones have been obtained using the FDTD and the second ones have been obtained using Eq. (8.39a), i.e. from the Huygens principle and the infinite current sheet. Note that the constitutive parameters of the finite lens studied by the FDTD have been chosen with a slight deviation from the ideal situation of $(\varepsilon_-, \mu_-) = (-1, -1)$ in order to account for the discretization error in the FDTD (a discussion on this issue is offered in Section 8.4).

Comparing the fields, it is seen that the phases are very similar within the region $-L/2 < x < L/2$. This result is very significant since it allows us to immediately conclude that the resolutions of the finite and the infinite lens are comparable in the case studied. As a matter of fact, the phase similarity indicates that the fields at the second boundary are dominated by the surface plasmon related to the pole in the complex k_z plane, in both the finite and infinite lens cases, in turn indicating that the spectra of the two lenses present a similar cutoff and therefore that the lenses present a similar

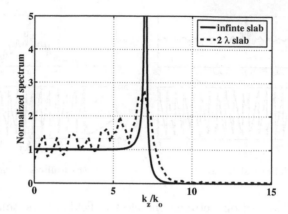

Figure 8.9 Comparison between the spectra of the electric field at the image plane as produced by an infinite slab and a slab of length 2λ. In both cases the thickness of the slab is 0.2λ and the spectrum is normalized to the line-source spectrum. (Reproduced from Chen et al. (2006c). © 2006 The American Physical Society.)

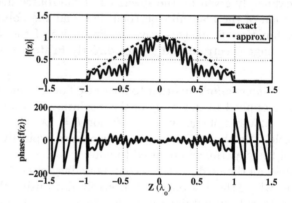

Figure 8.10 Amplitude (top) and phase (bottom) of the window function f. The exact value has been obtained using the FDTD whereas the approximate function is chosen to have a truncated Gaussian shape. (Reproduced with permission from Chen et al. (2006c). © by The American Physical Society.)

resolution (see Section 8.3.1). It should be mentioned that a corroborating conclusion was also reported in Lagarkov and Kissel (2004), although using different arguments. This conclusion is also confirmed in Fig. 8.9, which compares the two spectra and clearly shows the similar cutoffs. This similarity suggests that the surface plasmon is only weakly perturbed by the finite extent of the lens, at least for the aspect ratio of the example.

Although it is out of the scope of this chapter to present an exhaustive study of the impact of lens size on the resolution, we still want to propose a fast methodology for doing so. Indeed, both Fig. 8.8 and Fig. 8.9 present important discrepancies which may become important as the lateral dimension of the lens is reduced. In Fig. 8.8 for example, it is seen that the amplitude of the fields tapers off much faster in the finite lens case, whereas in Fig. 8.9, spectral oscillations are seen that are not reproduced by the infinite lens approximation. Both effects are in fact related to the aperture function f which still needs to be determined and imposed. Fig. 8.10 shows the ratio of the fields of Fig. 8.8 and reveals that f could be a real function with a Gaussian tapered amplitude $\exp(-x^2/g^2)$ truncated at the aperture of the lens. The shape of the function f is shown as the dashed line in Fig. 8.10 and resolves both issues mentioned above: the field amplitude now naturally tapers off at the edge of the lens and the spectrum presents the oscillatory behavior obtained from the FDTD (not shown but easily verified). Consequently, the simple method that consists of tapering the current distribution obtained from an infinite boundary is seen to yield very similar characteristics to those obtained from a full-wave numerical study of a real finite lens, and can therefore be used for a more systematic study of various lens widths.

8.4 Issues with numerical simulations and time evolution

Soon after the demonstration of the perfect lens effect (Pendry 2000) there were several attempts to model this effect numerically. Some calculations obtained focusing (Paul et al. 2001, Kolinko and Smith 2003), while some others (Ziolkowski and Heyman 2001) did not show any steady focusing at all, while yet others obtained focusing, but not subwavelength image resolution (Karkkainen and Maslovski 2002, Loschialpo et al. 2003). In view of the controversy surrounding the effect at that time, it was felt that some degree of numerical verification based on exact simulations of the Maxwell equations was necessary. On afterthought though, it is realized that one was attempting to verify an exact analytical solution with numerical simulations that are necessarily approximate due to the finite differencing.

It was first pointed out in Cummer (2003) that the discrete fields on a lattice do not obey the same dispersion relations as the continuous fields do and that many of the discrepancies noted are an artifact due to this fact. To understand the sensitivity of the solutions to numerical errors and excitation of spurious resonances, consider the dispersion of the finite difference equations used for a spectral method (for a specified ω) on a simple cubic lattice in media with spatially constant material parameters (Pendry 1994):

$$\omega^2(k) = \frac{c^2}{\varepsilon\mu a^2} 4\sin^2(\frac{1}{2}ka) \simeq \frac{c^2}{\varepsilon\mu}k^2 \left[1 - \frac{1}{12}(ka)^2\right], \qquad (8.45)$$

where a is the numerical grid size. This evidently tends to the dispersion-Maxwell equations for the continuum Maxwell equations only when $ka \ll 1$, or particularly, for very small a when we consider subwavelength wave-vectors. We can rewrite the above dispersion as

$$k^2 = \varepsilon\mu\frac{\omega^2}{c^2}\left[1 + \frac{1}{12}(ka)^2\right], \qquad (8.46)$$

where the second term can be effectively regarded as arising from a deviation, $\delta\mu$ (or a deviation $\delta\varepsilon$ in ε) from the ideal case of $\varepsilon = \mu = \pm 1$. For S-polarized light, this deviation,

$$\delta\mu = \frac{\mu k^2 a^2}{12}, \qquad (8.47)$$

in the real part of μ gives rise to a resonant excitation on a slab of NRM of thickness d with a parallel wave-vector (k_x) given by Eq. (8.24),

$$k_x d = -\ln\left(\frac{\delta\mu}{2}\right) = -\ln\left(\frac{k_x^2 a^2}{24}\right), \qquad (8.48)$$

in the limit of large parallel wave-vector $k_x \gg k_0$. In the limit $k_x a \ll 1$, the parallel wave-vector for this spurious resonance induced by the discretization is $k_x \sim$ (a large number)$/d$. In the ideal continuum limit of $a \to 0$, we have $k_x \to \infty$ as expected from the continuum equations. Thus, at any level of discretization, one always excites a spurious resonance of the system for some finite k_x in a numerical calculation. The value of the parallel wave-vector $k_x^{(s)}$ for the spurious resonance gives the range of k_x significantly smaller than $k_x^{(s)}$ for which the calculations would have converged.

Thus, any discrete method has a wavenumber dependence and an inbuilt higher momentum cutoff depending on the level of discretization that automatically precludes the perfect focus with infinite image resolution. In the case of the perfect lens, as it has been pointed out, the focusing is very sensitive to meeting the prescribed conditions on $\varepsilon = -1$ and $\mu = -1$. The different dispersion for the discrete fields is equivalent to having slightly different material parameters, which thus lead to quite imperfect focusing as some investigators reported. Particularly for large wave-vectors, the different dispersion is acutely felt and it was found to be only remedied by resorting to a much larger degree of discretization (Pendry and Ramakrishna 2003). The level of subwavelength focusing that can be obtained by a numerical calculation is a function of the finiteness of the differencing scheme even under the perfect conditions of zero dissipation.

For the FDTD technique, the dispersion for the finite difference equations on a cubic lattice would be given similarly by

$$\left(\frac{1}{c\delta t}\right)^2 \sin^2\left(\frac{\omega\,\delta t}{2}\right) = \left(\frac{1}{\delta x}\right)^2 \sin^2\left(\frac{k_x\,\delta x}{2}\right) + \left(\frac{1}{\delta y}\right)^2 \sin^2\left(\frac{k_y\,\delta y}{2}\right)$$
$$+ \left(\frac{1}{\delta z}\right)^2 \sin^2\left(\frac{k_z\,\delta z}{2}\right), \tag{8.49}$$

where δt is the time step involved while δx, δy, and δz are the sizes of the numerical grid used to represent the fields. Similar problems to the ones outlined for the spectral calculations would result in FDTD calculations as well due to the difference in the dispersion for the discrete equations. Cummer (2003) also pointed out that the resonant surface plasmon modes at the interfaces for $n = -1$ are always excited by any causal incident wave or pulse with a finite bandwidth even when the perfect lens conditions are satisfied at the carrier frequency. In a FDTD calculation, these resonances with large electromagnetic fields at the two interfaces would ring indefinitely and dissipation is always introduced into the material parameters in any such calculation to reach a steady state in a reasonable amount of time. This probably explains the results of some research reports, such as in Ziolkowski and Heyman (2001), where it was found that the FDTD simulations never reached a steady behavior for non-lossy NRM. The loss also damps out the amplification for large wave-vectors and the image resolution has been shown to be consistent with the estimate in the previous section. These conclusions have also been confirmed by FDTD calculations (Li et al. 2003b, Rao and Ong 2003) where different boundary conditions were used to simulate evanescent waves.

For example, the discretized Drude model over an FDTD mesh yields a relative permittivity expressed as

$$\varepsilon_r = 1 - \frac{\omega_p^2}{4\sin^2(\omega\Delta t/2)/(\Delta t)^2}, \tag{8.50}$$

which clearly converges to the continuous form $\varepsilon_r = 1 - \omega_p^2/\omega^2$ when $\Delta t \to 0$. However, at the theoretical value of $\omega = \omega_p/\sqrt{2}$ where the continuous permittivity is equal to (-1), the discrete value of Eq. (8.50) presents a slight variation from (-1) as soon as $\Delta t \neq 0$. For example, a value of about -1.0003 is obtained for the typical grid size of $\lambda/100$ [where the grid size and time step are related by the Courant condition (Taflove and Hagness 2005)]. This small perturbation has almost no effect on the propagating waves, but it has a critical impact on the amplification of the evanescent waves, which require an exact matching between the lens and its surrounding medium. Consequently, even in the hypothetical situation of a lossless lens, the resolution obtained from a method such as FDTD is inherently limited because of the method itself. It is important to realize, however, that this is merely a numerical artifact and not a physical limitation of the configuration.

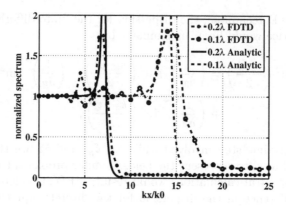

Figure 8.11 Comparison between the spectra obtained with FDTD and using an analytical calculation for two slab thicknesses. The FDTD simulation was run in both cases with a step size of $\lambda/100$. (Reproduced with permission from Chen et al. (2005b). © 2005 Optical Society of America.)

An illustration of this phenomenon is provided in Fig. 8.11, where the spectra of the electric fields computed both analytically and numerically at the image plane are shown for two slab thicknesses. The numerical results have been obtained using a modified FDTD implementation as suggested in Chen et al. (2005b), whereas the analytical results have been obtained as described previously but on slightly mismatched constitutive parameters in order to match the discrete values inherently obtained by the FDTD (the values of both the permittivity and the permeability are set to -1.000297). The excellent agreement between the two spectra, as well as the large variations of the spectra between the two slab thicknesses (and thus the different resolutions obtained), prove that the analytical results based on the mismatched constitutive parameters indeed match those obtained numerically, although the later ones have been originally designed to reproduce the perfect lens configuration (which is obviously not reproduced since the spectra are cut off and the resolution is not infinite).

Consequently, it is important to keep in mind the inherent numerical limitations of discrete methods such as the FDTD in simulating a highly idealized and theoretical configuration as the perfect lens. Usually, the deviation between the two solutions (discrete and continuous) can be minimized and neglected, but the ideal situation of the perfect lens is so stringent that numerical deviations may yield inaccurate results nonetheless. Losing this from sight presents the danger of attributing to physics what in fact comes from mere numerical artifacts.

8.4.1   Temporal evolution of the focus

As briefly noted before, the temporal evolution of the perfect focus is interesting as the focus refines in time, and under the ideal conditions, refines to a point focus at infinite time which is what we obtain with the time-harmonic solutions. A calculation based on Laplace transform methods (Smith et al. 2003) indicated that for a source that is sharply switched on at $t = 0$, the lens forms a steady image only at times that are of the order of the absorption time $(t \sim 1/\text{Im}(\varepsilon\mu)\omega)$.

It was pointed out in Gomez-Santos (2003) that the resonant excitation of the plasmons on the two surfaces is identical to the problem of two coupled, identical oscillators when one of the oscillators is forced by an external force. The equations of motion for such a system of oscillators are

$$\ddot{x}_l + \gamma\dot{x}_l\omega_0^2 x_l + k_c x_r = f(t), \tag{8.51}$$

$$\ddot{x}_r + \gamma\dot{x}_r\omega_0^2 x_r + k_c x_l = 0, \tag{8.52}$$

where x_l and x_r are the amplitudes for the left and right oscillators and k_c is the coupling between them. First note that forcing the system at ω_0 with no damping makes only x_r non-zero in the steady state. The left oscillator does not move at all, and all the excitation energy is passed on the right oscillator. This corresponds to the zero reflection in the case of the perfect lens and all the electromagnetic energy seems to be concentrated at the interface far from the source. In the steady state, the solutions for a time-harmonic excitation $f_0 \exp(-i\omega t)$ are

$$x_l = \frac{(\omega_0^2 - \omega^2 + i\gamma\omega)f_0}{(\omega_0^2 - \omega^2 + i\gamma\omega)^2 - k_c^4}, \qquad x_r = \frac{-k_c^2 f_0}{(\omega_0^2 - \omega^2 + i\gamma\omega)^2 - k_c^4}, \tag{8.53}$$

and it can be seen for the case of non-zero dissipation at resonance, the left oscillator is also excited. For large dissipation, the right oscillator is no longer excited, which corresponds to the non-effective amplification in the case of the NRM slab. From this model we can further consider the temporal evolution of the imaging process by considering the rate at which energy is transferred between the two surfaces. Note that the slab plasmon polariton modes are decoupled from the single plasmon frequency. However, a source with a sharp onset has a frequency finite bandwidth and would excite them. The transmitted field across the lossless slab then has an amplitude (Gomez-Santos 2003)

$$E_t(t) = \Theta(t)A(t)\exp(-i\omega_0 t)\exp(k_x d), \tag{8.54}$$

where $\Theta(t)$ is the Heaviside step function. In the quasi-static limit, $A(t) \simeq 1/2(\Delta\omega_{kx}t)^2$ where $\Delta\omega_{kx}$ is the frequency spacing between the two slab plasmon polariton modes at a wave-vector k_x. As pointed out before, the spacing between the slab plasmon modes monotonically decreases and is exponentially small at k_x. Thus the large wave-vector components are very small at short times and a cutoff time can be defined by $\Delta\omega_{kx}t_{kx} \sim 1$ for every wave-vector

k_x, before it does not appreciably contribute to the image formation. Thus the image initially is ill-defined and increasingly improves with the passage of time until it reaches the optimal resolution that is possible given the levels of absorption in the system. Also note that without absorption, the slab plasmon polariton modes would ring indefinitely, never allowing the formation of a steady image. Thus, absorption is essential to the very process of the image formation with the perfect lens.

8.5   Negative stream of energy in the perfect lens geometry

Let us return to the perfect lens configuration and examine more precisely the energy distribution at the second interface. Upon taking a point source or a line source in two dimensions, it has been reported that a peculiar phenomenon occurs by which negative time average energy stream can be witnessed at the second boundary (Chen et al. 2004b). By "negative energy stream," it is meant here that the energy propagates backward toward the source: for a point source at $z = 0$ and a slab between $z = d_1$ and $z = d_2 > d_1$, it is found that the z component of the Poynting vector S_z can take negative values beyond the second boundary. In general, this situation is perfectly acceptable when scatterers are present: in the first region for example, for $0 < z < d_1$, the first boundary of the slab reflects the wave so that some energy is returned to the source, after propagating away from it. Beyond the second boundary, however, for $z > 2d$, no such scatterer is present since the medium is semi-infinite, so that the phenomenon of negative energy flow is at first unexpected.

It is important to realize that when a single plane wave impinges onto the slab, no negative energy stream is witnessed, i.e. all the energy propagates away from the source.[‖] This suggests that the latter effect is due to the interaction between multiple waves, as originating from the line source.[**] Let us therefore consider these interactions. Three situations need to be examined, all present under a line source excitation: the interaction between two propagating waves, between two evanescent waves, and between a propagating and an evanescent wave (Chen et al. 2006c).

[‖]There has been some debate about a source-sink-source solution to the perfect lens configuration, whereby the image inside the slab acts as a sink and the image outside the slab acts as a second source. Such a solution seems physically unsatisfactory and we therefore do not pursue it here. Note, however, that the concept of negative energy stream would not be clearly defined in this case.

[**]We remind the reader that the spectrum of a line source can be expressed as a continuous superposition of weighted plane waves.

For the first situation, we consider the two normalized electric fields expressed by

$$E_y^\ell = e^{ik_x^{(\ell)}y + ik_z^{(\ell)}z},\tag{8.55}$$

where $\ell = \{1, 2\}$ identifies the two plane waves, and where $k^{(\ell)^2} = k_x^{(\ell)^2} + k_z^{(\ell)^2} = \omega^2 \varepsilon_+ \mu_+/c^2$, for which both $k_x^{(\ell)}$ and $k_z^{(\ell)}$ are real. A straightforward calculation of the time average Poynting power yields the $\hat{z}$ component as

$$\langle S_z \rangle = \frac{1}{2\omega\mu_+}(k_z^{(1)} + k_z^{(2)})\left[1 + \cos(\Delta k_x x + \Delta k_z z)\right],\tag{8.56}$$

where $\Delta k_x = k_x^{(1)} - k_x^{(2)}$ and $\Delta k_z = k_z^{(1)} - k_z^{(2)}$. In the third medium, the z components of the wave-vector are all positive, as well as the permeability, so that $\langle S_z \rangle$ is always positive and hence the interaction between propagating waves always yields a positive longitudinal power. This explains for example why, when only propagating waves are taken in the spectrum of the line source, no negative energy stream is witnessed, as seen by some authors.

In the second situation, we need to compute the interaction between two evanescent waves expressed as

$$E_y^\ell = e^{ik_x^{(\ell)}x} e^{-\alpha_z^{(\ell)}z},\tag{8.57}$$

where $\alpha_z^{(\ell)} = \sqrt{k^2 - k_x^{(\ell)^2}}$. In such case, the time average z component of the Poynting vector becomes

$$\langle S_z \rangle = \frac{1}{2\omega\mu_+}(\alpha_z^{(1)} - \alpha_z^{(2)})\sin(\Delta k_x x)e^{-(\alpha_z^{(1)} + \alpha_z^{(2)})z}.\tag{8.58}$$

As can be seen, this time average power flow has a sine form along the $\hat{x}$ direction, yielding positive and negative values and apparently offering a possible explanation for the negative energy stream. The case of a line source, however, is more complex than the simple interaction of two evanescent waves and therefore requires a more careful examination. As a matter of fact, the spectrum of a line source is composed of evanescent waves that need to be all included by carrying out the integration shown in Eq. (8.39a). Upon doing so, we see that we are integrating an odd function of $(k_x^{(1)} - k_x^{(2)})$ over the entire spectrum with symmetric amplitude distribution, effectively yielding no real power. Consequently, for a line source, the interaction between the evanescent waves does not offer the explanation on the negative energy stream observed at the second interface.

Finally, the third situation we shall examine concerns the interaction between evanescent and propagating waves. The power along the $\hat{x}$ direction is found in this case to be

$$\langle S_z \rangle = \frac{1}{2\omega\mu_+}\Big\{k_z^{(1)} + k_z^{(1)}e^{-\alpha_z^{(2)}z}\cos[\Delta k_x x - k_z^{(1)}z]$$
$$\big\{+\alpha_z^{(2)}e^{-\alpha_z^{(2)}z}\sin[\Delta k_x x + k_z^{(1)}z]\big\},\tag{8.59}$$

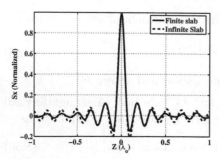

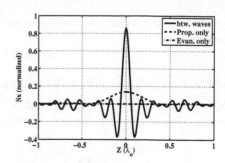

(a) Poynting power along the direction of propagation at the image location in the case of an infinite slab and a finite slab.

(b) Separate contributions from the interactions between propagating waves only, evanescent waves only, as well as propagating and evanescent waves together in the case of the infinite lens.

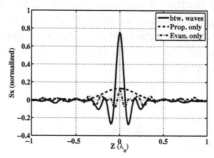

(c) Same as case (b) but for a finite lens (results obtained from FDTD simulations).

Figure 8.12 Poynting power in various configurations and from various wave interactions. Results are taken from Chen et al. (2006c). ©2006 by The American Physical Society.)

which is no longer an odd function of $(k_x^{(1)} - k_x^{(2)})$ as in the previous case. Moreover, $\langle S_z \rangle$ can take negative values if the $e^{-\alpha_z^{(2)}z}$ term is compensated by the amplitude of the evanescent wave. Such amplification, or growth, is precisely one of the key properties of left-handed media, whereas it is not achieved with conventional media. Consequently, a slab of left-handed medium amplifies the evanescent waves to a degree enough that it perturbs the near-field energy distribution at the second boundary of the perfect lens configuration and the perturbation is so important that it is responsible for the negative energy stream. The amplification of the evanescent waves being so crucial, it is also clear why this effect had not been observed in configuration involving standard media.

As a consequence, we see that the negative energy stream observed at the second interface of the perfect lens has a simple explanation based on simple electromagnetic properties of wave interaction, and is not due to numerical artifacts or cavity effects. A vivid illustration of the interactions between propagating and evanescent waves is provided in Fig. 8.12, where we include the case of a lens of finite transverse extent for the sake of comparison. Fig. 8.12(a) shows the total $\langle S_z \rangle$ component at the image plane and illustrates that $\langle S_z \rangle$ can take negative values despite the absence of scatterers beyond the image plane. Fig. 8.12(b) shows the separate contributions of the three interactions studied above. It it seen that the interaction between propagating waves yields a positive $\langle S_z \rangle$ only, that the interaction between evanescent waves yields no real power flow as expected, and that most of the power flow comes from the interaction between propagating and evanescent waves, which is also seen to be the only mechanism responsible for a negative $\langle S_z \rangle$. These conclusions can be generalized to the case of a finite lens case as well as shown in Fig. 8.12(c) (see Section 8.3.2), except for the evanescent waves that start to diffract from the edge of the lens and can yield a real power, including negative values as well (Chen et al. 2006c).

8.6 Effects of spatial dispersion

We now briefly approach the problem of spatial dispersion in the material parameters and its effects on the imaging process of the flat lens. As mentioned earlier in this chapter, once the graininess of the material (or the metamaterial) begins to get probed by the electromagnetic fields, the response is no longer isotropic and, more importantly, is also no longer local. That is to say, the electric field applied elsewhere begins to make a contribution to the polarization in the medium at a given point. The charge distributions are not properly screened out and this can give rise to spatial dispersion. The inadequate screening causes the material parameters to also become functions of the wave-vector k to a first approximation. Here we assume that only the absolute distance between the given point and the point where the field is applied is important and that there is no preferred point in the medium. This can be rarely violated in some metamaterials where there may be very special points of symmetry. The material parameters are now functions of both the frequency and the wavenumber: $\varepsilon(\omega, k)$ and $\mu(\omega, k)$, when the wave-vector becomes a significant fraction of $2\pi/l_s$ where l_s is a typical screening length below which the presence of individual constituents (whether atoms and molecules or metamaterial units) becomes felt by the radiation. It must be said here that non-local effects in metamaterials have not yet been well studied.

Let us now consider the effects of spatial dispersion on the silver lens. It has been shown for example (Ruppin 2005) that spatial dispersion can actu-

320 *Veselago's lens is a perfect lens*

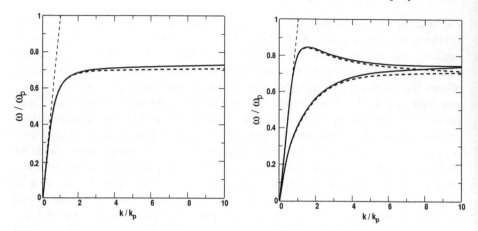

Figure 8.13 Left: Dispersion of surface plasmon modes at the interface of a semi-infinite plasma with vacuum. Right: Dispersion of the slab plasmon polaritons on a slab of thickness $k_p d = 0.5$. In both cases, the solid lines show the dispersion for the non-local Lindhardt form of ε while the dotted curves show the dispersion for a local plasma. (Taken with permission from Ruppin (2005). © 2005 Institute of Physics Publishing, U.K.)

ally improve the image resolution at intermediate lengthscales while setting an ultimate limit on the resolution at smaller lengthscales. First consider Eq. (8.21) which gives the wave number dependence of the dielectric permittivity of a metal. As a direct consequence, the bulk plasmon in the system become dispersive with a dispersion given by ($\varepsilon = 0$ being the condition)

$$\omega^2(k) = \omega_P^2 + \frac{3}{5}k^2 v_F^2. \tag{8.60}$$

The dispersive nature of the bulk plasmons now renders it possible to excite the standing bulk plasmons in thin films of metals using radiation of the appropriate frequency. The reflection and transmission of light from the surface of such a non-local medium has been calculated (Melnyk and Harrison 1970) by including the excitation of a third longitudinal component, which showed that there were small differences in the reflectivity from that of a local medium. The dispersion of surface plasmons on the interface of a non-local metal with vacuum as well as the slab plasmon polaritons in a thin film of a non-local metal has been calculated in Ruppin (2005) and the results are presented in Fig. 8.13: it can be seen that while there are no substantial differences from that of a local medium at small wave numbers, the salient effect of the spatial dispersion is that the SPP modes disperse linearly at large wave-vectors. They are not degenerate at the surface plasmon frequency for large wave numbers as in a local medium. Since the perfect lens focusing action requires an exponentially small detuning from the conditions of resonance for SPP modes at larger wave-vectors (see Section 8.2.1), detuning the dispersion at large wave-vectors affects the image resolution at subwavelength lengthscales. At any frequency of operation, the detuning of the SPP modes begins to increase beyond some wave-vector, setting a large wave-vector cutoff beyond which the amplification of evanescent waves ceases to be effective. By choosing an appropriate frequency near $\omega_p/\sqrt{2}$, however, it might be possible to enhance the image resolution from that of the local medium due to a better overlap with the SPP modes at intermediate wave-vectors. Fig. 8.14 shows the image resolution offered by a local medium and a non-local medium for a slab of thickness of about 50 nm where it is clear that the non-local metal slab offers far better image resolution. Thus non-locality might actually be utilized to improve image resolution and may explain the experimental image resolution of almost $\lambda/6$ obtained by Fang et al. (2005) whereas one would predict only about $\lambda/3$ for a local medium.

Among other non-local effects, one is that the assumption of a sudden change in the charge density at the interfaces of the lens breaks down. One can incorporate a non-abrupt change in the electron density for the plasma by assuming an extra surface charge layer at the boundary with different charge density than the bulk one. Bratkovsky et al. (2005) have attempted to incorporate such boundary layers with different ε and μ at the boundaries of metamaterials and shown that its presence usually lowers the image resolution. There are no universal non-local models for the material parameters of metamaterials, such as the local Lorentz model, with which one could study non-local effects in the imaging process. But to a first approximation, non-local contributions have a quadratic dependence on the wave number in general, and the validity of the above study can be extended to imaging using non-local negative index media as well. The signs of the non-local contributions can, of course, be different for different metamaterial designs.

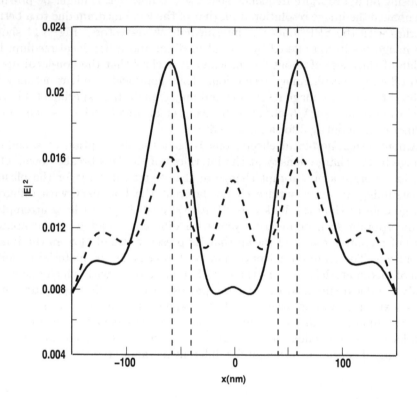

Figure 8.14 Images of two slits placed at subwavelength distance of about 100 nm obtained by a thin film of silver with a thickness of 50 nm at an operating frequency corresponding to 3.68eV when silver approximately has a dielectric constant of about $-1+i0.4$. The solid curves are for the calculations that incorporate non-local effects while the dotted curves assume a local model for ε. (Taken with permission from Ruppin (2005). © 2005 Institute of Physics Publishing, U.K.)

9

Designing super-lenses

The previous chapter focused on the essential ideas of the *Perfect Lens*. The perfect lens is an ideal imaging device that requires ideal materials with $\varepsilon = \mu = -1$ for image resolution without limit. It was shown that very small deviations of the material parameters from these ideal conditions could lead to the excitation of resonances that cause deterioration of the performance of the lens. In addition, finite amounts of dissipation and other imperfections that occur in actual materials have been shown to also limit the resolution of this device. Substantial subwavelength resolution is, however, possible in spite of dissipation, and the lenses exhibiting some degree of subwavelength image resolution capabilities have been termed *Super-Lenses*.

It turns out that the Veselago lens is only one of a whole class of perfect lenses or super-lenses that are possible. Negative refractive media (NRM) with ($\varepsilon < 0$, $\mu < 0$) are the optical analogue of anti-matter, in the sense that the effects on radiation (amplitude and phase change) upon passage through negative refractive index media nullifies the effects (amplitude and phase change) of passing through an equal thickness of positive media with the same magnitudes of ε and μ (Pendry and Ramakrishna 2003). This concept of complementarity has lead to the generalization of the Veselago lens to slab pairs of complementary optical media (Pendry and Ramakrishna 2003) as well as to other geometries via a geometric transformation technique (Ward and Pendry 1996). One can also achieve magnification of the near-field images in the cylindrical (Pendry and Ramakrishna 2002, Pendry 2003) and spherical (Ramakrishna and Pendry 2004) geometries. In addition, super-lens behavior is also exhibited in structures involving sharp corners and wedges of negative refractive index media.

Another interesting class of imaging devices are based on anisotropic media where some components of the material tensors such as $\bar{\bar{\varepsilon}}$ and $\bar{\bar{\mu}}$ are negative while others are positive – the so-called *indefinite media* (see Section 5.3.1). Waves that are evanescent in ordinary media would become propagating inside such media due to the hyperbolic nature of their dispersion. Thus one can project out the evanescent waves from a source across large thicknesses of these media and even outcouple the near-field information to the far-field. Such a super-lens has been called a *Hyperlens* (Jacob et al. 2006), after the hyperbolic nature of the dispersion that makes its functioning possible. The hyperlens, which outcouples to the far field, has enormous potential applications in near-field imaging and has already been demonstrated experimentally (Liu et al.

2007).

In this chapter, we discuss a variety of super-lenses with different geometries, and with an intention to reduce the deteriorating effects of dissipation and imperfections in the materials used in the construction of such super-lenses. We note that any deviations from the ideal conditions for perfect lensing leads to a drastic reduction in the possible subwavelength resolution of the super-lens. In fact, it is absorption that is the main culprit responsible for reducing the resolution. In the case of the silver lens, one merely manages to obtain some subwavelength image information. Minimizing the effects of absorption is therefore crucial in order to make these imaging devices work. Restructuring the lens to reduce these effects is possible to some extent and we first discuss some such strategies. Eventually it does, however, appear that the composite structures would need to incorporate media with active gain (as in a laser) in order to counter dissipation (Tretyakov 2001, Ramakrishna and Pendry 2003, Pendry and Smith 2004).

9.1   Overcoming the limitations of real materials

Real materials with negative real parts of ε and μ pose several challenges for the implementation of super-lenses. Of course, it is imperative to ensure that the materials practically have excellent chemical composition and smooth surfaces, so that the metamaterials have as few imperfections as possible, and also meet the perfect lens conditions for the real parts of the material parameters as accurately as possible at the operating frequency. In particular, the main restriction for obtaining high levels of image resolution with most metamaterials or metals is the presence of large levels of dissipation. Given the fact that materials with negative material parameters usually depend on a resonance for their property, there is necessarily at least some level of absorption near the resonance.* Hence if the level of absorption in the metamaterial cannot itself be reduced, the principal task at hand is to see how to reduce the effects of absorption that degrade the image quality. Potentially a different geometry or frequency could enhance the image resolution. For example, in Section 8.2.3, we could increase the resolution of the slab lens just by choosing the asymmetric configuration with a large dielectric constant on the far side of the slab. It has also been shown that in the asymmetric lens, one could obtain a better image resolution at a slightly lower frequency than when $\varepsilon_2 = -\varepsilon_3$ because then the frequency $\omega = \omega_p/\sqrt{1 + \varepsilon_3}$ comes into

---

*The Kramers-Kronig (see Eqs. (1.9)) relations which relate the real and imaginary parts of the material parameters impose that the imaginary part would be large at frequencies near the resonance where the real part disperses violently and can be negative.

better overlap with the resonant excitations of the surface plasmon dispersion curve at larger transverse wave-vectors (Ramakrishna et al. 2002).

9.1.1 Layering the lens

One possibility to reduce the effects of absorption is to decrease the slab thickness since the evanescent fields amplify in magnitude to much smaller levels in thinner slabs. As absorption is maximal in regions of large electromagnetic fields, the total absorption immediately reduces as a consequence of smaller electromagnetic fields. Larger wave-vectors can then contribute to the image thereby enhancing the image resolution. Note that it is the ratio λ/d (wavelength to slab thickness) that dominates the resolution, the logarithm term being a relatively weakly varying function (see Eq. (8.24)). For example, if $\lambda/d = 3$, we find that $\delta\varepsilon$ must be no greater than about $\sim 10^{-11}$ to achieve a resolution factor of 10. Comparatively, $\delta\varepsilon$ can be as large as ~ 0.002 for $\lambda/d = 10$ while still achieving the same resolution. In this case, however, the actual distance over which the image is transferred becomes smaller. Hence just making the slab thinner is not a satisfactory solution in most cases.

With ideal lossless negative index media satisfying the perfect lens conditions, we could just take the original thick slab, divide it into thin layers, and redistribute the layers alternatively between the source and the image planes without affecting the result of perfect imaging thanks to the perfect impedance matching (Shamonina et al. 2001, Ramakrishna et al. 2003). The evanescent fields alternatively grow within the NRM layers and decay in the positive layers as shown in Fig. 9.1. This rearrangement, however, makes an enormous difference to the dissipation in the system. As the fields do not amplify to the same extent as they would have to in the single slab, the associated dissipation is significantly decreased. However, we note that the (object plane – lens-edge) and the (lens-edge – image plane) distances have reduced considerably. In the limit of very small layer thicknesses, these distances become extremely small and the layered system merely transfers the image of the source from one edge to the other in the manner of an optical fiber bundle. Note, however, that the image transfer here involves both the near-field and the radiative modes.

While the idea of layering the lens is straightforward for negative index media, it is not absolutely clear if this configuration works in the case of the silver lens (see Section 8.2.2) where there is a finite impedance mismatch at the boundaries. Admittedly in the limit of small lengthscales (quasi-static limit), the near-field solutions for the silver lens should appear to be similar to the negative index medium. In Fig. 9.2, the transfer function for the intensity from the object plane to the image plane for P-polarized light has been plotted as a function of the parallel component of the wave-vector. Since only the relative material parameters are relevant for the single frequency problem, we assume without loss of generality that the positive medium has $\varepsilon_+ = 1$ and $\mu_+ = 1$. In the case of a hypothetical negative dielectric medium with no

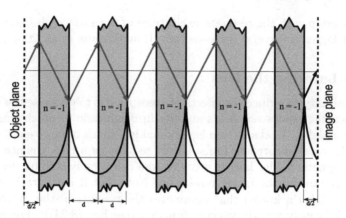

Figure 9.1 Chopping the slab of NRM into thin slices and distributing them around does not affect the image. The field distributions for a system of five such pieces is depicted schematically. The evanescent fields amplify along the direction normal to the layers ($\hat{z}$) within the negative index slabs and decay exponentially along $\hat{z}$ in the normal media. The rays corresponding to the propagating waves that undergo negative refraction at each interface and come to a focus at the image plane.

losses ($\varepsilon_- = -1$, $\mu = +1$), we see that the transmittance is nearly unity for a significant interval of the subwavelength wave-vectors and is dominated by the presence of a series of resonances at subwavelength wave-vectors. These are due to the hybridized resonances of the layered stack that arise from the coupling of the surface plasmons on the individual interfaces. In the limit of a periodic layered medium, these would give rise to continuous bands for the surface plasmons in the system. Because the individual layer thickness is very small, these coupled excitations have large wave-vectors for $\omega = \omega_p/\sqrt{2}$ as they are detuned only slightly from the surface plasmon frequency of the semi-infinite negative medium. The cutoff wave-vector is large for the imaging process and only depends on the thickness of the individual layers. The cutoff wave-vector is given by the value of k_x for the intersection between the $\omega = \omega_p/\sqrt{2}$ line and the most detuned asymmetric stack plasmon branch. This turns out to be slightly larger than the limit given in Section 8.3 for a given thickness of the individual negative dielectric layers with $\mu = 1$. Thus, the idea of layering up the silver lens gets us closer to the electrostatic limit where μ becomes irrelevant for the P-polarized light. In the presence of dissipation, however, these modes are damped and the resonances broaden out. Absorption causes deterioration of the image resolution and then the layered stack yields an image resolution far worse than the limit set by retardation for the individual thin layers (see Eq. (8.27)). Nonetheless, the transfer function shows that much higher image resolution can be obtained by using such a

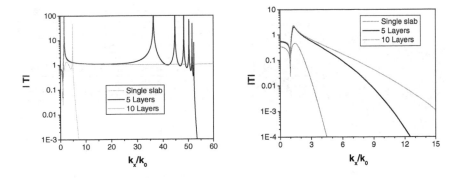

Figure 9.2 Modulus of the transmitted fields for P-polarized light at the image plane for a stack of alternating equally thick layers of a negative dielectric medium (silver) and a positive dielectric medium of equal thickness. The total distance from the object plane to the image plane is 100 nm. The other distances are (i) for a single slab of silver, layer thickness = 50 nm; (ii) for 5 layers of silver in-between, the layer thickness is 10 nm; and (iii) for 10 layers of silver in-between, the layer thickness is 5 nm. Left: The fields at the image plane for hypothetical lossless negative medium with $\varepsilon = -1$. Right: The fields at the image plane for silver layers with $\varepsilon = -1 + i0.4$. The positive media are assumed to have $\varepsilon = +1$ in both cases. The wavelength assumed is 358 nm about where $\mathrm{Re}(\varepsilon) = -1$ for silver.

layered stack than the single slab lens in spite of dissipation in the medium. Experiments have indicated that while the image resolution is better for a double layer than for a single layer lens, the image fidelity suffers due to enhanced surface roughness in the experimental implementation of the bi-layer lens (Melville and Blaikie 2006). Clearly, surface roughness and evenness of the layers become important technological issues that need to be resolved for the multilayer lens.

9.1.2   A layered stack to direct radiation

When the thickness δ of the individual layers becomes very small compared to the wavelength, it is fruitful to think of the layered stack as a slab of effective medium (Fig. 9.3). The effective response of a layered medium has been considered in Section 2.5 and has been shown to be anisotropic. For the response of a layered stack of a medium with equal layer thickness of the negative permittivity medium and the positive medium, one obtains

$$\varepsilon_x = \frac{(\varepsilon_+ + \varepsilon_-)}{2}, \tag{9.1}$$

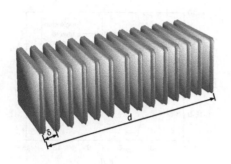
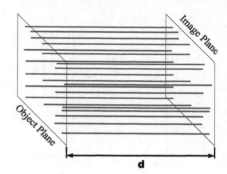

Figure 9.3 Left: Layered medium consisting of alternately negative and positive dielectric layers of equal thickness $\delta/2$. The layered stack behaves as a slab of an effective medium in the limit of very small layer thickness. Right: The picture schematically depicts how the anisotropic medium is equivalent to the electrostatic case of the two ends of the slab being connected point by point with perfectly conducting wires going through an insulating medium.

if the electric field is applied parallel to the layers and the tangential component of the **E** field is continuous across the layer boundaries. Similarly, considering the response to an electric field applied normal to the layers, we obtain

$$\varepsilon_z = \frac{2}{(\varepsilon_+^{-1} + \varepsilon_-^{-1})}, \tag{9.2}$$

where the continuity of the normal component of the **D** field is implied. Under the perfect lens conditions, we have the effective medium dielectric permittivity components

$$\varepsilon_x \to 0, \qquad \varepsilon_z \to \infty, \tag{9.3}$$

and the layered stack acts as a slab of an unusual anisotropic medium. Similar relations can be derived for the magnetic permeability using the continuity of $\mathbf{H}_\parallel$ and $\mathbf{B}_\perp$.

The wave propagation for the P-polarized radiation in such an uniaxially anisotropic medium has the dispersion similar to the S-polarization given by Eq. (5.6),

$$\frac{k_x^2}{\varepsilon_z} + \frac{k_z^2}{\varepsilon_x} = \mu_y \frac{\omega^2}{c^2}, \tag{9.4}$$

where we assume $k_y = 0$ without loss of generality. In the limit of the perfect lens conditions, $\varepsilon_x \to 0$ and $\varepsilon_z \to \infty$, we have $k_z = 0$ as the only solution to the dispersion equation. Thus, every wave passes through this anisotropic material slab without any change in amplitude or phase (Ramakrishna et al. 2003). In the static limit and no dissipation, the situation corresponds to infinitely thin and perfectly conducting wires embedded in an insulating medium with $\varepsilon = 0$, and running down the slab connecting the two end-faces of the slab of the effective medium point to point as depicted schematically in Fig. 9.3. Thus, the potential on one face is just directly transferred onto the other face and the layered stack acts as an optical fiber bundle, although it also transmits the near-field information (unlike a conventional optical fiber bundle). When there is dissipation, the wires acquire a finite thickness and consequently give a finite image resolution.

Having shown that the layered stack is equivalent to a slab of an unusual anisotropic effective medium, we now investigate the robustness against dissipation of the image resolution obtained by this slab. Since the dissipation mainly comes in via the negative dielectric medium, we consider the case $\varepsilon_- = -\varepsilon_+ + i\delta\epsilon$ for the negative dielectric layers. Hence we have the different components of the effective permittivity as $\varepsilon_x = i\,\delta\epsilon/2$ and $\varepsilon_z = -2i\varepsilon_+(-\varepsilon_+ + i\,\delta\epsilon)/(\delta\epsilon)$. The transmission coefficient for P-polarized light across the anisotropic slab of thickness d is given by (see Appendix B for a more general bianisotropic case)

$$T_p = \frac{1}{\cos(k_{z2}d) - \frac{i}{2}(K + \frac{1}{K})\sin(k_{z2}d)},\tag{9.5}$$

where $K = k_{z2}/(\varepsilon_x k_{z1})$, k_{z2} is the wave-vector in the slab, and k_{z1} is the wave-vector in the medium (with ε_+) outside the slab. Substituting for the wave-vectors in the quasi-static limit of $\omega \to 0$ or $k_x \to \infty$, we obtain $k_{z1} \to ik_x$, $k_{z2} \to i(\delta\epsilon)/(2\varepsilon_+)\,k_x$, and $K = 1/(i\varepsilon_+)$. Hence the transmission coefficient in this limit becomes (Ramakrishna et al. 2003)

$$T_p = \left[\cosh(\frac{\delta\epsilon}{2\varepsilon_+}d) + \frac{i}{2}\left(\varepsilon_+ - \frac{1}{\varepsilon_+}\right)\sinh(\frac{\delta\epsilon}{2\varepsilon_+}d)\right]^{-1}.\tag{9.6}$$

For the case when $\varepsilon_+ = 1$ as for vacuum, we have $T_p = \mathrm{sech}(\delta\epsilon\,k_x d/2)$. It is easy to see that the transmittance is of the order of unity for $\delta\epsilon\,k_x d \sim 1$, and hence that the resolution only depends inversely on the dissipation in the medium

$$\mathrm{res} = \frac{\lambda}{2\pi\,d\,\delta\epsilon}.\tag{9.7}$$

This should be compared to the slowly varying logarithmic dependence on the dissipation in the case of the single negative refractive index slab Eq. (8.25). The scope for improvement in the image resolution by making the metamaterials less dissipative is much higher in this case of the anisotropic layered medium. One should note that Wiltshire et al. (2003b) have experimentally obtained a similar imaging for S-polarized radio frequency waves at about 21 MHz using Swiss roll cylindrical metamaterials with the cylinders aligned along the imaging axis, which give a large value of μ_z near the resonance frequency and $\mu_x = 1$. It was found that the resolution is limited to 10 mm, approximately equal to the diameter of the individual Swiss rolls.

Actually the wave dispersion in such anisotropic media can be utilized much more creatively. *Indefinite* media with different signs for the components of the material parameter tensors were discussed in Section 5.3.1 and it was shown that negative refraction is possible at the interface of such media. It was pointed out in (Smith and Schurig 2003) that the hyperbolic nature of the dispersion in certain indefinite media could imply that evanescent waves in vacuum could become propagating inside these media. As an example, let us rewrite the dispersion for the z-component of the wave-vector for P-polarized waves given by Eq. (9.4) as

$$k_z^2 = \varepsilon_x \mu_y \frac{\omega^2}{c^2} - \frac{\varepsilon_x}{\varepsilon_z} k_x^2, \qquad (9.8)$$

where k_x is determined by the incident radiation. It is immediately clear that if $\varepsilon_x \mu_y > 0$ and $\varepsilon_x/\varepsilon_z < 0$, there are no evanescent waves in this medium (x-hyperbolic dispersion of Section 5.3.1). It was shown how to obtain an indefinite anisotropic medium by using alternating layers of a metal and dielectric with suitable volume fractions in Section 2.5. A look at Fig. 2.5 reveals that while $\mu_y = 1$, $\varepsilon_x < 0$, and $\varepsilon_z > 0$ at smaller frequencies in such media. In such a metamaterial, there is a cutoff wave-vector below which the waves in the effective medium are evanescent and above which waves are propagating (z-hyperbolic dispersion of Section 5.3.1). Incident evanescent waves with large parallel component of the wave-vector would couple to propagating modes in this medium. In the quasi-static limit, one can write

$$k_z = \pm \sqrt{\frac{-\varepsilon_x}{\varepsilon_z}} k_x. \qquad (9.9)$$

The waves with large k_x have almost a specific direction of propagation in the anisotropic medium along the angle $\theta = \pm \tan^{-1}(\sqrt{\varepsilon_z/\varepsilon_x})$ to the $\hat{z}$-axis, which is the directional property of the anisotropic medium.

Theoretical studies on such slabs have been reported (Wood et al. 2006) where it has been shown that the location of the image moves outward in the transverse direction due to this preferred direction of propagation for large k_x, and can also get periodically repeated due to multiple reflections in the slab. Very good subwavelength image resolution of about $\lambda/10$ was shown to be obtainable with thin layers of silver (about few nanometers thick) and in spite of the large levels of dissipation in silver ($\text{Im}(\varepsilon) \sim 0.4$). Resonant modes corresponding to the slab resonances of the anisotropic slab were excited by incident evanescent waves and measured in experiments (Wiltshire et al. 2003b) with Swiss-roll slabs of $\mu_z < 0$, $\mu_x = 1$. The resonances essentially occurred at values of $k_x d/\sqrt{\mu_z} = m\pi$ where m is an integer and d is the thickness of the slab. Of course, there would be a cutoff limit on the largest k_x that can be transported in this manner. This would either again be limited by spatial lengthscales of the underlying metamaterial units. It has pointed out in Wood et al. (2006) that for a periodic layered medium with period δ, a bandgap would essentially develop once the k_z in the medium became large enough, i.e., of the order $2\pi/\delta$ which sets a cutoff for the parallel wave-vector as there would not be any propagating modes for larger wave-vectors at the given frequency. Indefinite media represent an interesting opportunity for subwavelength imaging as they are not as susceptible to dissipation since the electromagnetic fields have been distributed around more uniformly in space in comparison to the perfect lens.

9.1.3 Use of amplifying media to reduce dissipation

Although the image resolution of the layered lens is less susceptible to dissipation, the latter still plays a fundamental role in defining the performance of all the imaging devices. Since most metamaterials have appreciable levels of dissipation, we need to consider whether there is some way by which the energy lost by absorption can be reintroduced into the system. One intuitive idea would be to use optical amplification for this purpose (Ramakrishna and Pendry 2003).

Let us note that the dissipation in materials with negative material parameters actually represents a mismatch in the perfect lens conditions. One way to view it would be to regard the perfect lens conditions as conditions on the complex material parameters:

$$\varepsilon_+ = \varepsilon'_+ - i\varepsilon''_+, \qquad\qquad \varepsilon_- = -\varepsilon'_+ + i\varepsilon''_+, \qquad\qquad (9.10a)$$
$$\mu_+ = \mu'_+ - i\mu''_+, \qquad\qquad \mu_- = -\mu'_+ + i\mu''_+, \qquad\qquad (9.10b)$$

where the quantities $\mu''_+ \geq 0$ and $\varepsilon''_+ \geq 0$. We note that $\mathrm{Im}(\varepsilon_+) < 0$ and $\mathrm{Im}(\mu_+) < 0$ correspond to amplifying media. Consequently, the use of amplifying media to counter absorption using stimulated emission is implied in the perfect lens conditions itself.

In principle, surface plasmon excitations can be sustained across such an interface of a positive and amplifying medium with a negative and absorbing medium and have an interesting aspect to them: there is a continuous flow of electromagnetic energy from the amplifying medium into the absorbing medium that serves as a sink for the energy. Reduction in the damping for the surface plasmons propagating on the interface of silver and a laser pumped organic dye (Rhodamine 101 or Cresyl violet) solution has been reported (Seidel et al. 2005). The actual reason that amplification can compensate for the loss of information via absorption is because both absorption and stimulated emission are coherent processes. In the limit of large intensities (photon numbers), operation by the (quantum) annihilation operator does not alter the phase of a coherent photon state. There is, of course, some generation of noise via the spontaneous emission, which is an expression of the non-commutativity of the creation and annihilation operators. But as long as the amplification does not lead to self-sustaining oscillations of some modes in the lens structure, the amplification in one region can compensate for the absorption in another region of space. If self-sustaining oscillations occur, however, one is only amplifying the spontaneously generated photons, which swamp out all the image information. The amplification levels are also likely to get saturated in the presence of intense field enhancements that are expected in the presence of surface plasmon resonances. The largest field enhancements occurs for the largest transverse wave-vectors. If we make the layers very thin, the local field enhancements are not as intense and it is likely that the gain will not get completely bleached.

More realistically, we can have optical amplification primarily for the electric dipole transitions. The incorporation of gain through the magnetic permeability would necessarily be through electronic amplifiers in the metamaterials for the microwave and radio frequencies. Let us consider here the case of alternating thin layers of silver with positive amplifying dielectric media layers of the same thickness in-between. Since only thin layers are being considered, we shall concentrate on the P-polarized light for which the negative dielectric permittivity is sufficient to make a super-lens in the quasi-static limit. Note that $\mu = 1$ in all such cases and the requirements for resonant surface plasmon excitations are met only in the quasi-static limit. Hence the incorporation of positive media with gain would be mainly for the situations where the quasi-static approximation works well, for example, in the layered lens rather the single slab lens. An implementation could be to use a semiconductor laser material such as GaN or AlGaAs for the positive medium and silver for the negative medium. Using blue/ultraviolet (UV) light to pump the AlGaAs,[†] one can make the AlGaAs optically amplifying in the red region of the spectrum, where one can satisfy the perfect lens condition for the real parts of the dielectric permittivity. Alternatively one could also use other high gain processes such as Raman gain for this purpose.

As a proof of principle, the transmission function for the layered silver lens, with and without amplification to exactly compensate for the losses, and the images of two closely spaced sources as resolved by the respective lenses is shown in Fig. 9.4. For comparison, we also show the case of the original single slab of silver as the lens (solid line and $\delta/2 = 40$ nm) and a layered, but gain-less system. The two peaks in the image for the single slab can hardly be resolved, while they are clearly resolved in the case of the layered system with no gain. The improvement in the image resolution for the layered system with gain over the corresponding gain-less systems is obvious with the sharp edges of the slits becoming visible. The transmittance of the layered lens with gain does not decay rapidly with the increase in the number of layers, even when the total stack thickness is about a few wavelengths (Ramakrishna and Pendry 2003). Although the transmission function is not constant with the wave-vector due to the excitation of the layer plasmon resonances, it should be emphasized that the high spatial frequency components are efficiently transferred across. Knowledge of the transfer function of the lens would therefore enable one to computationally recover a clean image from the observed image.

[†]The layers of silver will be transparent to UV radiation.

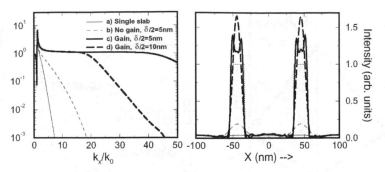

Figure 9.4 Transmission function (left) and electromagnetic field intensity at the image plane (right) obtained (a) with a single slab of silver of 40 nm thickness; (b) when the slab is split into 8 thin layers of $\delta/2 = 5$ nm thicknesses; (c) layered silver-dielectric stack with optical gain and $\delta/2 = 5$ nm and (d) layered silver-dielectric stack with optical gain and $\delta/2 = 10$ nm. $\varepsilon_\pm = \pm 1 \mp i0.4$ in (c) and (d). (Reproduced with permission from Ramakrishna (2005). © 2005, Institute of Physics Publishing, U.K.)

The levels of gain to exactly compensate for the absorption in the metal are extremely large, and perhaps can just be achieved in some very high gain physical systems. But the incorporation of even some lower levels of gain in the positive parts of the lens will certainly lead to some enhancement of the image resolution. With the single slab super-lens, the enhancement in image resolution due to incomplete compensation of absorption can be marginal as the resolution depends logarithmically on the deviations from the perfect lens conditions. Nevertheless, for the layered lens with very thin layers and the lenses based on indefinite media where the image resolution depends inversely on the dissipation, even a partial compensation of the absorption by gain can lead to large improvements in the image resolution.

9.2   Generalized perfect lens theorem

The perfect slab lens as well as the layered lenses that we have previously described have a translational invariance along the transverse directions. It is this translational invariance that ensures that the image is identical in size to the source. In the case of the finite transverse lens, we broke the translational invariance and found that cavity-like effects can become important. We therefore ask ourselves whether the translation invariance in the transverse directions is essential to the functioning of the lens. It turns out that the translational invariance can be broken, the material parameters can have

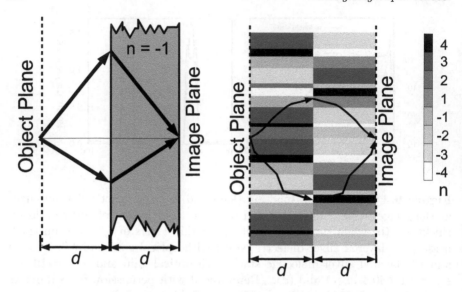

Figure 9.5 A flat slab of $n = -1$ of thickness d along with a slab of vacuum of thickness d is a special case of optical complementary media. A pair of complementary media nullifies the effects of each other for the passage of light. The gray scales for figure on the right suggests a possible scale for the refractive indices. The paths for the propagating modes in the media are not straight lines in general.

an arbitrary variation along the transverse direction, and yet we can have perfect focusing if certain conditions are satisfied.

Let us consider the region $0 < z < d$ to be filled by a medium of one kind and the region $d < z < 2d$ to be filled with a second medium. It has been shown in Pendry and Ramakrishna (2003) that the electromagnetic fields on the $z = 0$ object plane are reproduced at the image plane at $z = 2d$, if the perfect lens conditions are generalized to

$$\varepsilon_1 = +\varepsilon(x,y), \qquad \mu_1 = +\mu(x,y), \qquad \forall \quad 0 < z < d, \qquad (9.11\text{a})$$
$$\varepsilon_2 = -\varepsilon(x,y), \qquad \mu_2 = -\mu(x,y), \qquad \forall \quad d < z < 2d, \qquad (9.11\text{b})$$

where $\varepsilon(x,y)$ and $\mu(x,y)$ are some arbitrary functions of x and y and can take positive or negative values (See Fig. 9.5 [right] for a schematic picture). We term such media *Complementary Media*. The Veselago lens shown on the left is only a special case of complementary media with no transverse variation. Upon transport through a pair of complementary media of equal thickness, radiation undergoes no change in phase or amplitude. Thus, to the world on the right side of the complementary media pair, all sources on the left side of the lens just appear transposed by a distance $2d$ forward. The presence of the complementary media pair is not felt at all and corresponds to zero optical thickness. This gives an alternative and fruitful physical picture of negative

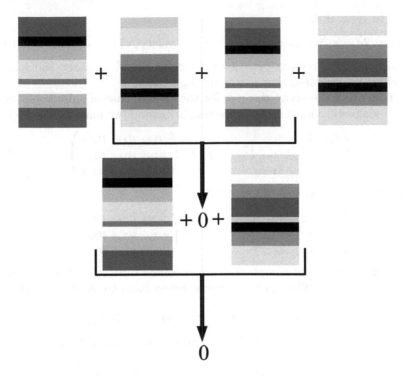

Figure 9.6 A pair of complementary media nullifies the effects of each other for the passage of light. The complementary media sum to an optical null pairwise if the system has a mirror antisymmetry about the central plane, and the entire system sums to a null. The same gray scale as in Fig. 9.5 is used to indicate the refractive indices of different regions.

refractive index media: negative refractive index media are the equivalent of "optical anti-matter" which nullify the effects of normal media on radiation. Note that we have not defined the media outside the region $0 < z < 2d$ as that is immaterial to the imaging process. Also note that the path of light in these media is not along straight lines as in homogeneous media and is schematically depicted in the figure. Effectively, this theorem states that the variation in the transverse directions is immaterial as long as the same sense of variation is present in the complementary slab.

In fact, if the media on either side of the $z = d$ plane also had material parameters that depend on z, the focusing would still be obtainable if the $z = d$ plane were a plane of anti-symmetry along the $\hat{z}$ direction. In other words, if the system is anti-symmetric about the $z = d$ plane, the complementary media can be eliminated to a null in a pairwise fashion as shown in Fig. 9.6, where the meaning of this theorem is expressed pictorially for piecewise continuous media. In general, if

$$\varepsilon_2(x, y, 2d - z) = -\varepsilon_1(x, y, z), \qquad \mu_2(x, y, 2d - z) = -\mu_1(x, y, z), \quad (9.12)$$

then also the two regions behave as optically complementary and have no total effect on radiation that passes through them.

We now prove the above theorem for the general case of anisotropic dielectric and magnetic complementary media. In general, we have for the region-1,

$$\bar{\bar{\varepsilon}}_1 = \begin{pmatrix} \varepsilon_{1xx} & \varepsilon_{1xy} & \varepsilon_{1xz} \\ \varepsilon_{1yx} & \varepsilon_{1yy} & \varepsilon_{1yz} \\ \varepsilon_{1zx} & \varepsilon_{1zy} & \varepsilon_{1zz} \end{pmatrix}, \qquad \bar{\bar{\mu}}_1 = \begin{pmatrix} \mu_{1xx} & \mu_{1xy} & \mu_{1xz} \\ \mu_{1yx} & \mu_{1yy} & \mu_{1yz} \\ \mu_{1zx} & \mu_{1zy} & \mu_{1zz} \end{pmatrix}. \quad (9.13)$$

Similarly, for the region-2 $(d < z < 2d)$,

$$\bar{\bar{\varepsilon}}_2 = \begin{pmatrix} \varepsilon_{2xx} & \varepsilon_{2xy} & \varepsilon_{2xz} \\ \varepsilon_{2yx} & \varepsilon_{2yy} & \varepsilon_{2yz} \\ \varepsilon_{2zx} & \varepsilon_{2zy} & \varepsilon_{2zz} \end{pmatrix}, \qquad \bar{\bar{\mu}}_2 = \begin{pmatrix} \mu_{2xx} & \mu_{2xy} & \mu_{2xz} \\ \mu_{2yx} & \mu_{2yy} & \mu_{2yz} \\ \mu_{2zx} & \mu_{2zy} & \mu_{2zz} \end{pmatrix}. \quad (9.14)$$

Note that the tensorial components are specified functions of the transverse coordinates (x, y). The electromagnetic fields have to satisfy the Maxwell equations:

$$\nabla \times \mathbf{E} = i\omega\mu_0\bar{\bar{\mu}}\,\mathbf{H}, \qquad \nabla \times \mathbf{H} = -i\omega\varepsilon_0\bar{\bar{\varepsilon}}\,\mathbf{E}. \quad (9.15)$$

Decomposing the fields into the Fourier components in the two slabs,

$$\mathbf{E}_1(x, y, z) = \exp(ik_{1z}z) \sum_{k_x,k_y} \exp[i(k_x x + k_y y)] \begin{pmatrix} E_{1x}(k_x, k_y) \\ E_{1y}(k_x, k_y) \\ E_{1z}(k_x, k_y) \end{pmatrix}, \quad (9.16a)$$

$$\mathbf{E}_2(x, y, z) = \exp(ik_{2z}z) \sum_{k_x,k_y} \exp[i(k_x x + k_y y)] \begin{pmatrix} E_{2x}(k_x, k_y) \\ E_{2y}(k_x, k_y) \\ E_{2z}(k_x, k_y) \end{pmatrix}, \quad (9.16b)$$

where the Bloch conditions can be assumed. Substituting these fields into the Maxwell equations and separating out the components, we have

$$k_y E_{1z}(k_x, k_y) - k_{1z} E_{1y}(k_x, k_y) = -\omega\mu_0 \sum_{k'_x,k'_y} [\mu_{1xx}(k_x, k_y; k'_x, k'_y) H_{1x}(k'_x, k'_y)$$

$$+ \mu_{1xy}(k_x, k_y; k'_x, k'_y) H_{1y}(k'_x, k'_y) + \mu_{1xz}(k_x, k_y; k'_x, k'_y) H_{1z}(k'_x, k'_y)], \quad (9.17a)$$

$$k_{1z} E_{1x}(k_x, k_y) - k_x E_{1z}(k_x, k_y) = -\omega\mu_0 \sum_{k'_x,k'_y} [\, \mu_{1yx}(k_x, k_y; k'_x, k'_y) H_{1x}(k'_x, k'_y)$$

$$+\mu_{1yy}(k_x, k_y; k'_x, k'_y) H_{1y}(k'_x, k'_y) + \mu_{1yz}(k_x, k_y; k'_x, k'_y) H_{1z}(k'_x, k'_y)], \quad (9.17b)$$

$$k_x E_{1y}(k_x, k_y) - k_y E_{1x}(k_x, k_y) = -\omega\mu_0 \sum_{k'_x,k'_y} [\mu_{1zx}(k_x, k_y; k'_x, k'_y) H_{1x}(k'_x, k'_y)$$

$$+ \mu_{1zy}(k_x, k_y; k'_x, k'_y) H_{1y}(k'_x, k'_y) + \mu_{1zz}(k_x, k_y; k'_x, k'_y) H_{1z}(k'_x, k'_y)], \quad (9.17c)$$

and

$$k_y H_{1z}(k_x, k_y) - k_{1z} H_{1y}(k_x, k_y) = \omega \varepsilon_0 \sum_{k_x', k_y'} [\varepsilon_{1xx}(k_x, k_y; k_x', k_y') E_{1x}(k_x', k_y')$$

$$+ \varepsilon_{1xy}(k_x, k_y; k_x', k_y') E_{1y}(k_x', k_y') + \varepsilon_{1xz}(k_x, k_y; k_x', k_y') E_{1z}(k_x', k_y')],$$

$$(9.18a)$$

$$k_{1z} H_{1x}(k_x, k_y) - k_x H_{1z}(k_x, k_y) = \omega \varepsilon_0 \sum_{k_x', k_y'} [\varepsilon_{1yx}(k_x, k_y; k_x', k_y') E_{1x}(k_x', k_y')$$

$$+ \varepsilon_{1yy}(k_x, k_y; k_x', k_y') E_{1y}(k_x', k_y') + \varepsilon_{1yz}(k_x, k_y; k_x', k_y') E_{1z}(k_x', k_y')],$$

$$(9.18b)$$

$$k_x H_{1y}(k_x, k_y) - k_y H_{1x}(k_x, k_y) = \omega \varepsilon_0 \sum_{k_x', k_y'} [\varepsilon_{1zx}(k_x, k_y; k_x', k_y') E_{1x}(k_x', k_y')$$

$$+ \varepsilon_{1zy}(k_x, k_y; k_x', k_y') E_{1y}(k_x', k_y') + \varepsilon_{1zz}(k_x, k_y; k_x', k_y') E_{1z}(k_x', k_y')].$$

$$(9.18c)$$

Now consider the substitutions for the field quantities:

$$E_{2x}(k_x, k_y) = E_{1x}(k_x, k_y), \tag{9.19a}$$
$$E_{2y}(k_x, k_y) = E_{1y}(k_x, k_y), \tag{9.19b}$$
$$E_{2z}(k_x, k_y) = -E_{1z}(k_x, k_y), \tag{9.19c}$$
$$H_{2x}(k_x, k_y) = H_{1x}(k_x, k_y), \tag{9.19d}$$
$$H_{2y}(k_x, k_y) = H_{1y}(k_x, k_y), \tag{9.19e}$$
$$H_{2z}(k_x, k_y) = -H_{1z}(k_x, k_y), \tag{9.19f}$$
$$k_{2z} = -k_{1z}, \tag{9.19g}$$

and for the material parameters, the substitutions

$$\bar{\bar{\varepsilon}}_2(k_x, k_y; k_x', k_y') =$$
$$\begin{pmatrix} -\varepsilon_{1xx}(k_x, k_y; k_x', k_y') & -\varepsilon_{1xy}(k_x, k_y; k_x', k_y') & +\varepsilon_{1xz}(k_x, k_y; k_x', k_y') \\ -\varepsilon_{1yx}(k_x, k_y; k_x', k_y') & -\varepsilon_{1yy}(k_x, k_y; k_x', k_y') & +\varepsilon_{1yz}(k_x, k_y; k_x', k_y') \\ +\varepsilon_{1zx}(k_x, k_y; k_x', k_y') & +\varepsilon_{1zy}(k_x, k_y; k_x', k_y') & -\varepsilon_{1zz}(k_x, k_y; k_x', k_y') \end{pmatrix},$$

$$(9.20a)$$

$$\bar{\bar{\mu}}_2(k_x, k_y; k_x', k_y') =$$
$$\begin{pmatrix} -\mu_{1xx}(k_x, k_y; k_x', k_y') & -\mu_{1xy}(k_x, k_y; k_x', k_y') & +\mu_{1xz}(k_x, k_y; k_x', k_y') \\ -\mu_{1yx}(k_x, k_y; k_x', k_y') & -\mu_{1yy}(k_x, k_y; k_x', k_y') & +\mu_{1yz}(k_x, k_y; k_x', k_y') \\ +\mu_{1zx}(k_x, k_y; k_x', k_y') & +\mu_{1zy}(k_x, k_y; k_x', k_y') & -\mu_{1zz}(k_x, k_y; k_x', k_y') \end{pmatrix}.$$

$$(9.20b)$$

These fields resolve the Maxwell equations in region-2 for the new material parameter tensors and the boundary conditions on the continuity of the electromagnetic fields across the interface of the slabs at $z = d$ are also satisfied.

Hence we have the field

$$E(x, y, z = 2d) = E(x, y, z = d) \exp(-ik_z d) = E(x, y, z = 0), \qquad (9.21)$$

i.e., the fields at $z = 0$ are reproduced at $z = 2d$. The fields are exactly repeated in region-2 but in the opposite order along $\hat{z}$. For an arbitrary source in general, we have a sum over the wave-vector k_z. Since the theorem holds individually for each k_z, it also holds for the sum. This completes the proof of this theorem that generalizes the perfect lens effect to spatially varying complementary media. Thus the sufficient condition for the optical complementarity is that the material parameters for the second complementarity slab should be related to the first by

$$\bar{\bar{\varepsilon}}_2 = \begin{pmatrix} -\varepsilon_{1xx} & -\varepsilon_{1xy} & +\varepsilon_{1xz} \\ -\varepsilon_{1yx} & -\varepsilon_{1yy} & +\varepsilon_{1yz} \\ +\varepsilon_{1zx} & +\varepsilon_{1zy} & -\varepsilon_{1zz} \end{pmatrix}, \quad \bar{\bar{\mu}}_2 = \begin{pmatrix} -\mu_{1xx} & -\mu_{1xy} & +\mu_{1xz} \\ -\mu_{1yx} & -\mu_{1yy} & +\mu_{1yz} \\ +\mu_{1zx} & +\mu_{1zy} & -\mu_{1zz} \end{pmatrix}, \quad (9.22)$$

where all the tensorial components are arbitrary well-behaved functions of the transverse coordinates. Numerical simulations have confirmed the focusing for complementary media with transverse spatial variation up to the accuracy of the calculations (Pendry and Ramakrishna 2003).

9.2.1 Proof based on the symmetries of the Maxwell equations

A simple and elegant proof for the generalized perfect lens theorem with anisotropic media can be deduced from the symmetry of the Maxwell equations. Consider the Maxwell equations

$$\nabla \times \mathbf{E} = i\omega \mathbf{B}, \qquad (9.23a)$$

$$\nabla \times (\mu^{-1} \mathbf{B}) = -i\omega \varepsilon \mathbf{E}, \qquad (9.23b)$$

where by ε and μ we imply the actual permittivity and actual magnetic permeability and not the relative ones (for use in this proof alone). Some transformations that leave these equations invariant are:

S1: $\mathbf{E} \to \mathcal{A}\mathbf{E}, \mathbf{B} \to \mathcal{A}\mathbf{B}, \mu^{-1} \to \mathcal{A}\mu^{-1}\mathcal{A}^{-1}, \varepsilon \to \mathcal{A}\varepsilon\mathcal{A}^{-1}$, where $\mathcal{A}$ is invertible and an element of $GL_3(\mathbf{R})$ (the group of 3×3 linear operators). ... *Generalized conformal invariance.*

S2: $\mathbf{E} \to -\varepsilon^{-1}\mathbf{B}, \mathbf{B} \to \mu\mathbf{E}$, or $\mathbf{E} \to \mu^{-1}\mathbf{B}, \mathbf{B} \to -\varepsilon\mathbf{E}$ (*iff* $\mu = \varepsilon$) ... *Generalized duality.*

S3: $\mu \to \alpha\mu, \varepsilon \to \varepsilon/\alpha$ where α is a non-zero scalar.

S4: $\mathbf{r} \to -\mathbf{r}$, where $\mathbf{r} = [x, y, z]$, $\mathbf{E} \to -\mathbf{E}$ (**B** is a pseudo vector) ... *Inversion.*

S5:   Any additional space-time symmetries such as translational invariance etc.

The combination of any of these symmetries is again a symmetry of the system of equations. Then we can assert that *if the fields in a particular region of space can be mapped onto another region of space through the symmetry transformations S1 – S5, while preserving the respective boundary conditions, then the transformed fields solve the field equations whenever the original fields do.*

The boundary conditions here include, of course, the conditions of continuity of the tangential components of $\mathbf{E}$ and the normal component of $\mathbf{B}$ across any charge free and current free boundaries.

Let us apply these transformations to our slab in the region $-d < z < 0$ with dielectric permittivity and magnetic permeability tensors

$$\bar{\bar{\varepsilon}}_1 = \begin{pmatrix} \varepsilon_{xx} & \varepsilon_{xy} & \varepsilon_{xz} \\ \varepsilon_{yx} & \varepsilon_{yy} & \varepsilon_{yz} \\ \varepsilon_{zx} & \varepsilon_{zy} & \varepsilon_{zz} \end{pmatrix}, \qquad \bar{\bar{\mu}}_1 = \begin{pmatrix} \mu_{xx} & \mu_{xy} & \mu_{xz} \\ \mu_{yx} & \mu_{yy} & \mu_{yz} \\ \mu_{zx} & \mu_{zy} & \mu_{zz} \end{pmatrix}. \tag{9.24}$$

For propagation along the z direction and origin at the interface, let us use the symmetry operations of a reflection $S4$ ($z \to -z$) and $S3$ ($\alpha = -1$), followed by $S1$ ($\mathcal{A}$) with

$$\mathcal{A} = \begin{pmatrix} -1 & 0 & 0 \\ 0 & -1 & 0 \\ 0 & 0 & 1 \end{pmatrix}. \tag{9.25}$$

We call this sequence of operations a mirror operation. This choice of $\mathcal{A}$ preserves the continuity of $\mathbf{E}_\parallel$ and $\mathbf{B}_\perp$ across the $z = 0$ interface. Then the resulting complementary medium on the right for $0 < z < d$ is obtained as

$$\bar{\bar{\varepsilon}}_2 = \begin{pmatrix} -\varepsilon_{xx} & -\varepsilon_{xy} & +\varepsilon_{xz} \\ -\varepsilon_{yx} & -\varepsilon_{yy} & +\varepsilon_{yz} \\ +\varepsilon_{zx} & +\varepsilon_{zy} & -\varepsilon_{zz} \end{pmatrix}, \qquad \bar{\bar{\mu}}_2 = \begin{pmatrix} -\mu_{xx} & -\mu_{xy} & +\mu_{xz} \\ -\mu_{yx} & -\mu_{yy} & +\mu_{yz} \\ +\mu_{zx} & +\mu_{zy} & -\mu_{zz} \end{pmatrix}, \tag{9.26}$$

which is the result we obtained before.

9.2.2 Contradictions between the ray picture and the full wave solutions

The generalized perfect lens theorem is a very powerful statement about the behavior of electromagnetic fields in complementary media. It is actually counter-intuitive in many ways and, interestingly, contradicts the results of a ray analysis in many cases. For example, consider the rays that are incident on a pair of slabs of complementary media shown in Fig. 9.7. The first one on the left consists of alternating rectangular regions of $n = \pm 1$. The rays can be seen to either transmit through or retro-reflect depending on the point and on the angle at which they are incident. The generalized lens theorem, however, predicts only one thing: the complementary slab pair has a transmission coefficient of unity for all plane waves regardless of their direction of incidence.

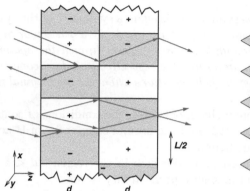 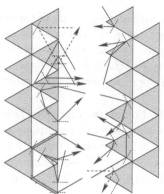

Figure 9.7 Left: Pair of complementary media slabs of thickness d each. The refractive index takes the values $n = +1$ and $n = -1$ periodically along the transverse axis (effectively part of a rectangular checkerboard) with a period of $L/2$. Rays can be seen to be either transmitted or retro-reflected. Right: Extreme example of a complementary media slab pair where each slab consists of alternating equilateral triangular regions with $n = +1$ and $n = -1$, which can be seen to reflect all rays incident upon them. Invariance along the axis normal to the plane of the figure is assumed in both cases.

The second example on the right of Fig. 9.7 shows a complementary medium slab pair composed of alternating equilateral triangular regions with $n = \pm 1$ and the invariance assumed in the direction normal to the plane is more extreme. The ray analysis reveals that every ray incident on the slab should be reflected, regardless of the direction and the point (x) on the slab at which it is incident. We do neglect here the points of singularities such as the corners for the ray analysis. The generalized lens theorem, however, predicts a perfect transmission for every wave. Thus, there is a total contradiction between the predictions of the ray picture and that of the generalized lens theorem that is based on the full solutions to the Maxwell equations. Another case consisting of a spherical cavity in an otherwise homogeneous slab of $n = \pm 1$ and its complementary slab has also been pointed out (Pendry 2004b).

The contradiction between the generalized lens theorem and the ray analyses can be worrisome at first sight. However, one has to bear in mind that the ray picture stems from a high frequency approximation of the Maxwell equations, based on the Eikonal equation (Born and Wolf 1999), while the generalized lens theorem is an exact statement. In fact, there are other instances in Optics where the ray analysis is known to fail. For example, in order to explain the case of complete reflection of light incident at an angle on a stratified medium with slowly and monotonically decreasing refractive index, one has to invoke either the total internal reflection (which involves evanescent waves for which we have no corresponding ray) or the curvature of the wave-front associated with the wave, which again is not considered in the ray approximation.

The properties of complementary media strongly depend on the resonant surface plasmon states on the interfaces and on their interactions, whereas these localized resonances are not accounted for in the ray picture. The strange transmittive power of the piecewise continuous complementary media presented here can be traced to the propagation of surface plasmons on their interfaces and the scattering of these interface states from the corners. These are another instance of extraordinary transmission that is possible with meta-materials and which holds for all incident waves – propagating and evanescent. The mechanism of plasmonic guidance involved here via the interfaces between positive and negative index media differs substantially from the extraordinary transmission through subwavelength holes in thick metallic films that have become so popular in recent times (Ebbesen et al. 1998, Barnes et al. 2003).

Given the simplicity of the ray picture compared to full wave calculations, one is likely to be tempted to use the ray analysis as the first line of attack. In many cases the ray picture does yield some useful information about the modes of the system. However, we have seen that the ray analysis of situations involving localized resonances can lead to incorrect conclusions. Hence, any prediction based on the ray analysis in the case of complementary media has to be carefully checked by a full wave analysis.

9.3 The perfect lens in other geometries

Beyond a point, the planar geometry of the slab lens and the layered lens can be quite constraining. For example, the size of the image is bound to be the same as the object, and the system cannot focus a beam of light into a tight spot. Lenses in Optics are used for a multitude of effects and most importantly to magnify or demagnify the image of a source. In the slab lens the unit magnification comes about primarily due to the translational invariance along the transverse directions due to which the parallel wave-vector components (k_x, k_y) are conserved. That is why traditional Optical lenses have curved surfaces – the translational invariance is broken, and the distribution of parallel wave-vectors can be changed. Then the image can have a different transverse size than the source.

Obtaining super-lenses with curved surfaces is not straightforward. The performance and subwavelength image resolution of a super-lens is directly related to the near-degeneracy of the surface plasmons on the interfaces of the super-lens. In general, curved surfaces have a very different surface plasmon spectrum than the flat surfaces discussed so far. The only known exception

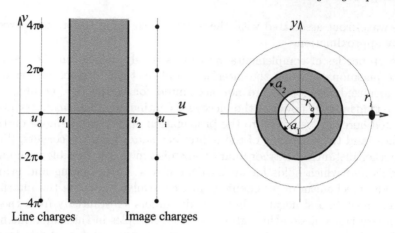

Line charges   Image charges

Figure 9.8 Left: Imaging by a slab lens of a set of periodically placed charges at $u = u_o$ and $v = 2m\pi$ in the z' plane. Right: Upon a conformal mapping $z' = \ln z$, the parallel lines go into concentric circles and the periodically placed charges map into one charge. The image of the charge is formed outside the cylindrical annulus of negative dielectric permittivity indicated by a gray shadow.

is approximate: a cylindrical surface with negative, homogeneous dielectric permittivity has a nearly degenerate plasmon dispersion in the quasi-static limit, i.e., when the cylinder is very thin compared to the wavelength. It has been shown (Pendry and Ramakrishna 2002) that a cylindrical shell of negative dielectric permittivity ($\varepsilon = -1$) can act as a super-lens in the quasi-static limit for TM polarization whereby it transfers inside (outside) the image of a line charge placed outside (inside) it. The proof of this is most easily given by utilizing a conformal map

$$z' = \ln z, \tag{9.27}$$

where $z = x + iy$ and $z' = u + iv$ for the solutions of the Laplace equation (extreme near-field limit) for the slab geometry. Here in the z'-plane, u is the optical axis of the slab lens and v is the transverse direction. The transformation maps lines parallel to the v axis into concentric circles in the z plane, hence mapping a set of periodic line charges along the line $u = u_0$ into a single charge of the same magnitude. The positions of the source, image, and the radii of the cylindrical shell are given by

$$u_o = \ln r_o, \quad u_1 = \ln a_1, \quad u_2 = \ln a_2, \quad u_i = \ln r_i, \tag{9.28}$$

as shown in Fig. 9.8. This lens can also be shown to have a spatial magnification of the near-field image by a factor of $(a_2/a_1)^2$.

While the Laplace equation is invariant under a conformal mapping and the solutions so obtained are valid in the extreme near-field approximation, the Maxwell equations and the Helmholtz equations are notinvariant under a conformal map. However, we can utilize the known solutions in one geometry such as the slab geometry and map them to another geometry in a similar manner using a coordinate transformation. In the remainder of this chapter, we discuss the possibility of obtaining focusing with cylindrical lenses (Pendry 2003), spherical lenses (Pendry and Ramakrishna 2003, Ramakrishna and Pendry 2004), two-dimensional sharp corners (Pendry and Ramakrishna 2003), checkerboards and three-dimensional corners (Guenneau et al. 2005b).

9.3.1 A transformation technique

We should note that a geometric mapping via a coordinate transformation would distort space, inducing a change in the material tensors ε and μ in the transformed geometry.[‡] Thus, the material parameters in the transformed geometry would be spatially varying (heterogeneous) and anisotropic in general.

Consider a general transformation of the coordinates from the Cartesian coordinates:

$$q_1 = q_1(x, y, z), \quad q_2 = q_2(x, y, z), \quad q_3 = q_3(x, y, z). \tag{9.29}$$

A cube in the new coordinate system would appear highly distorted in the Cartesian coordinate system. In the new coordinate system, Maxwell equations have been shown to take the form (Ward and Pendry 1996)

$$\nabla_{\mathbf{q}} \times \tilde{\mathbf{E}} = i\omega\mu_0\tilde{\mu}\tilde{\mathbf{H}}, \tag{9.30a}$$

$$\nabla_{\mathbf{q}} \times \tilde{\mathbf{H}} = -i\omega\varepsilon_0\tilde{\varepsilon}\tilde{\mathbf{E}}, \tag{9.30b}$$

where $\tilde{\varepsilon}$ and $\tilde{\mu}$ are, in general, some (frequency-dependent) tensors, and $\tilde{\mathbf{E}}$ and $\tilde{\mathbf{H}}$ are renormalized fields. The Maxwell equations hence preserve their form in terms of these new renormalized quantities. If (q_1, q_2, q_3) are assumed to be orthogonal, then the renormalized $\tilde{\varepsilon}$ and $\tilde{\mu}$, and fields in the new coordinate system are related to the actual material parameters and fields by

$$\tilde{\varepsilon}_i = \varepsilon_i \frac{Q_1 Q_2 Q_3}{Q_i^2}, \quad \tilde{\mu}_i = \mu_i \frac{Q_1 Q_2 Q_3}{Q_i^2}, \tag{9.31a}$$

$$\tilde{E}_i = Q_i E_i, \quad \tilde{H}_i = Q_i H_i, \tag{9.31b}$$

where

$$Q_i^2 = \left(\frac{\partial x}{\partial q_i}\right)^2 + \left(\frac{\partial y}{\partial q_i}\right)^2 + \left(\frac{\partial z}{\partial q_i}\right)^2. \tag{9.32}$$

[‡]The same principle can be used to cloak regions of spaces from external electromagnetic radiation and achieve electromagnetic invisibility, as discussed in Chapter 10.

The essential idea is to map new geometries into geometries where we know the solutions, for example, the Veselago slab lens. We can then use the known solutions of the fields to recover the fields in the new geometries as well as to define the actual inhomogeneous and anisotropic material parameter tensors in the new geometries. The essential condition for this procedure to work is, of course, that the transformation should be invertible. Since we have already assumed that we only transform to orthogonal coordinate systems, the boundary conditions on the fields are satisfied in both geometries.

In the remainder of this section, we use this coordinate transformation technique to generate super-lenses in new geometrical configurations.

9.3.2 Perfect lenses in curved geometries: cylindrical and spherical lenses

Cylindrical lenses:

To generate cylindrical lenses, let us make a transformation from the Cartesian to cylindrical coordinates:

$$x = r_0 e^{\ell/\ell_0} \cos\phi, \qquad y = r_0 e^{\ell/\ell_0} \sin\phi, \qquad z = Z, \qquad (9.33)$$

where $r = r_0 e^{\ell/\ell_0}$ and r_0, ℓ_0 are some scaling parameters. We obtain

$$Q_\ell = \frac{r_0}{\ell_0} e^{\ell/\ell_0}, \qquad (9.34a)$$

$$Q_\phi = r_0 e^{\ell/\ell_0}, \qquad (9.34b)$$

$$Q_Z = 1 \qquad (9.34c)$$

for the transformation. Hence the renormalized material parameters come out to be

$$\tilde{\varepsilon}_\ell = \ell_0 \varepsilon_\ell, \qquad \tilde{\varepsilon}_\phi = \ell_0^{-1} \varepsilon_\phi, \qquad \tilde{\varepsilon}_Z = \frac{r_0^2}{\ell_0^2} \exp\left(\frac{2\ell}{\ell_0}\right) \varepsilon_z, \qquad (9.35a)$$

$$\tilde{\mu}_\ell = \ell_0 \mu_\ell, \qquad \tilde{\mu}_\phi = \ell_0^{-1} \mu_\phi, \qquad \tilde{\mu}_Z = \frac{r_0^2}{\ell_0^2} \exp\left(\frac{2\ell}{\ell_0}\right) \mu_z. \qquad (9.35b)$$

Choosing the scale factor $\ell_0 = 1$ and explicitly defining the material parameters in the cylindrical geometry as

$$\left.\begin{array}{l} \varepsilon_r = \mu_r = +1, \\ \varepsilon_\phi = \mu_\phi = +1, \\ \varepsilon_z = \mu_z = +1/r^2, \end{array}\right\} \qquad \forall \ r < a_1, \qquad (9.36a)$$

$$\left.\begin{array}{l} \varepsilon_r = \mu_r = -1, \\ \varepsilon_\phi = \mu_\phi = -1, \\ \varepsilon_z = \mu_z = -1/r^2, \end{array}\right\} \qquad \forall \ a_1 < r < a_2, \qquad (9.36b)$$

$$\left.\begin{array}{l} \varepsilon_r = \mu_r = +1, \\ \varepsilon_\phi = \mu_\phi = +1, \\ \varepsilon_z = \mu_z = +1/r^2, \end{array}\right\} \qquad \forall \ a_2 < r, \qquad (9.36c)$$

we obtain that

$$\left.\begin{array}{l} \tilde{\varepsilon}_\ell = \tilde{\mu}_\ell = +1, \\ \tilde{\varepsilon}_\phi = \tilde{\mu}_\phi = +1, \\ \tilde{\varepsilon}_z = \tilde{\mu}_z = +1, \end{array}\right\} \qquad \forall \ \ell < \ell_0 \ln(a_1/r_0), \qquad (9.37a)$$

$$\left.\begin{array}{l} \tilde{\varepsilon}_\ell = \tilde{\mu}_\ell = -1, \\ \tilde{\varepsilon}_\phi = \tilde{\mu}_\phi = -1, \\ \tilde{\varepsilon}_z = \tilde{\mu}_z = -1, \end{array}\right\} \qquad \forall \ \ell_0 \ln(a_1/r_0) < \ell < \ell_0 \ln(a_2/r_0), \qquad (9.37b)$$

$$\left.\begin{array}{l} \tilde{\varepsilon}_\ell = \tilde{\mu}_\ell = +1, \\ \tilde{\varepsilon}_\phi = \tilde{\mu}_\phi = +1, \\ \tilde{\varepsilon}_z = \tilde{\mu}_Z = +1, \end{array}\right\} \qquad \forall \ \ell_0 \ln(a_2/r_0) < \ell \qquad (9.37c)$$

in the ℓ, ϕ, Z coordinates. The renormalized material parameters satisfy the same conditions as the slab perfect lens in Cartesian coordinates. Hence we have actually specified a perfect lens in the cylindrical geometry through our choice of the material parameters. Note that our choice of the spatial dependence of the material parameters was done so as to obtain homogeneous renormalized $\tilde{\varepsilon}$ and $\tilde{\mu}$. Note that the z-components of the material parameter tensors are inhomogeneous and have $1/r^2$ dependence.

The new system with the specified parameters must accordingly act as a cylindrical super-lens, and transfer images in and out of the cylindrical annulus. Let us suppose that the line source is located at (r_o, ϕ_o) inside the cylindrical annulus (see Fig. 9.8). The location of the image can be obtained from the condition

$$(\ell_1 - \ell_o) - (\ell_i - \ell_2) = (\ell_2 - \ell_1) \qquad \Rightarrow \qquad r_i = r_o(a_2/a_1)^2, \phi_i = \phi_o. \qquad (9.38)$$

All lengthscales on the image are magnified (angular magnification $\Delta\phi$) by a factor of

$$\mathcal{M} = \left(\frac{a_2}{a_1}\right)^2, \qquad (9.39)$$

which is the magnification of this system. Note that the magnification or de-magnification of the image holds for the near-field image as well. For example, if the fast-varying features are not amenable to interrogation by a scanning near-field microscope due to resolution limitations, one could instead try to image these fast-varying features in the magnified image. Another possibility is if $r_i = a_1$ and $r_o = a_2$, then the magnification of the image could allow coupling to the propagating modes in the region $r > a_2$ and hence allow us to project out the near-field. Another important point is that the demagnified images inside the cylindrical annulus would be more intense: this has poten-tial consequences when we want to concentrate weak incident fields. Note that these cylindrical lens are also short-sighted in the same manner as the

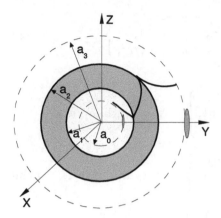

Figure 9.9 A spherical shell with negative $\varepsilon_-(r) \sim -1/r$ and $\mu_-(r) \sim -1/r$ images a source located inside the shell into the external region. The media outside have positive refractive index, but $\varepsilon_+(r) \sim 1/r$ and $\mu_+(r) \sim 1/r$. The amplification inside the spherical shell of the otherwise decaying field is schematically shown. (Reproduced with permission from Ramakrishna (2005). © 2005, Institute of Physics Publishing, U.K.)

slab lens is: they can only focus sources from inside to the outside only when $a_1^2/a_2 < r < a_1$, and the other way around from outside to the inner world when the source is located in $a_2 < r < a_2^2/a_1$. Furthermore, note that for the optical axis along the ℓ direction, the generalized lens theorem predicts that the variations in the transverse ϕ, Z directions are irrelevant. Hence the medium parameters, in general, could also be arbitrary functions of ϕ and Z.

Spherical lenses:

In an analogous manner to the cylindrical lens, we can show that a spherical shell of negative index media can also act as a super-lens. Consider a shell of negative refractive index material embedded in a positive index medium as shown in Fig. 9.9. Transforming to the spherical geometry from the Cartesian geometry:

$$x = r_0 e^{\ell/\ell_0} \sin\theta \, \cos\phi, \qquad y = r_0 e^{\ell/\ell_0} \sin\theta \, \sin\phi, \qquad z = r_0 e^{\ell/\ell_0} \cos\theta,$$
$$(9.40)$$

where ℓ is oriented along the radial direction, $r = r_0 e^{\ell/\ell_0}$, and r_0, ℓ_0 are some scaling parameters as before. We obtain

$$Q_\ell = \frac{r_0}{\ell_0} e^{\ell/\ell_0}, \tag{9.41a}$$

$$Q_\theta = r_0 e^{\ell/\ell_0}, \tag{9.41b}$$

$$Q_\phi = r_0 e^{\ell/\ell_0} \sin\theta \tag{9.41c}$$

for the transformation. Hence the renormalized material parameters are

$$\tilde{\varepsilon}_\ell = r_0 \ell_0 e^{\ell/\ell_0} \sin\theta \, \varepsilon_\ell, \qquad \tilde{\varepsilon}_\theta = \frac{r_0}{\ell_0} e^{\ell/\ell_0} \sin\theta \, \varepsilon_\theta, \qquad \tilde{\varepsilon}_\phi = \frac{r_0 e^{\ell/\ell_0}}{\ell_0 \sin\theta} \, \varepsilon_\phi,$$

$$\tag{9.42a}$$

$$\tilde{\mu}_\ell = r_0 \ell_0 e^{\ell/\ell_0} \sin\theta \, \mu_\ell, \qquad \tilde{\mu}_\theta = \frac{r_0}{\ell_0} e^{\ell/\ell_0} \sin\theta \, \mu_\theta, \qquad \tilde{\mu}_\phi = \frac{r_0 e^{\ell/\ell_0}}{\ell_0 \sin\theta} \, \mu_\phi.$$

$$\tag{9.42b}$$

Note that the imaging direction that we seek is along the radial (ℓ) direction. Choosing the scale factor $\ell_0 = 1$, we explicitly define the material parameters in the spherical geometry as

$$\varepsilon_\ell = \varepsilon_\theta = \varepsilon_\phi = \varepsilon e^{-\ell/\ell_0} = \varepsilon \frac{r_0}{r} \tag{9.43a}$$

$$\mu_\ell = \mu_\theta = \mu_\phi = \mu e^{-\ell/\ell_0} = \mu \frac{r_0}{r}. \tag{9.43b}$$

Then we obtain renormalized material parameters that have no ℓ dependence:

$$\tilde{\varepsilon}_\ell = +\varepsilon r_0 \sin\theta, \qquad \tilde{\varepsilon}_\theta = +\varepsilon r_0 \sin\theta, \qquad \tilde{\varepsilon}_\phi = \frac{+\varepsilon r_0}{\sin\theta}, \tag{9.44a}$$

$$\tilde{\mu}_\ell = +\mu r_0 \sin\theta, \qquad \tilde{\mu}_\theta = +\mu r_0 \sin\theta, \qquad \tilde{\mu}_\phi = \frac{+\mu r_0}{\sin\theta}. \tag{9.44b}$$

If the quantities ε and μ take the values $+1$ in the the regions $\ell < \ell_0 \ln(a_1/r_0)$ and $\ell_0 \ln(a_2/r_0) < \ell$, and the value -1 for $\ell_0 \ln(a_1/r_0) < \ell < \ell_0 \ln(a_2/r_0)$, then the renormalized material parameters have the required complementary behavior with no variation along the radial (imaging) direction and only a variation along the transverse (θ) direction. Hence we conclude that this system, which has isotropic but inhomogeneous ($\sim 1/r$) material parameters, acts as a spherical super-lens. Note here that it is the generalized lens theorem for complementary media with transverse variation that allows us to make this conclusion. Once again it is straightforward to show that the image of a source located at (r_o, θ_o, ϕ_o) appears at (r_i, θ_i, ϕ_i) where

$$r_i = r_o(a_2/a_1)^2, \qquad \theta_i = \theta_o, \qquad \phi_i = \phi_o, \tag{9.45}$$

and that the system has a magnification given by $\mathcal{M} = \left(\frac{a_2}{a_1}\right)^2$.

Actually this result can also be obtained by a more conventional but tedious calculation. In order to convince the reader of the considerable simplicity and power of the transformation technique, we present this more conventional calculation (Ramakrishna and Pendry 2004) here which confirms our earlier result for the spherical shell. We consider a spherical system whose ε and μ have a variation of $1/r$ along the radial direction. Under the circumstances of spherical symmetry, it is sufficient to specify the quantities $(\mathbf{r} \cdot \mathbf{E})$ and $(\mathbf{r} \cdot \mathbf{H})$, which constitutes a full solution to the problem.

Consider the TM polarized modes $\mathbf{r} \cdot \mathbf{H} = 0$, in which case only the electric fields have a radial component E_r. Operating on the Maxwell equation by ∇,

$$\nabla \times \nabla \times \mathbf{E} = i\omega\mu_0 \nabla \times [\mu(\mathbf{r})\mathbf{H}],$$

$$= \frac{\omega^2}{c^2}\mu(\mathbf{r})\varepsilon(\mathbf{r})\mathbf{E} + i\omega\frac{\nabla\mu(\mathbf{r})}{\mu(\mathbf{r})} \times \nabla \times \mathbf{E}, \tag{9.46}$$

and we have

$$\nabla \cdot \mathbf{D} = \nabla \cdot [\varepsilon(\mathbf{r})\mathbf{E}] = \nabla\varepsilon(\mathbf{r}) \cdot \mathbf{E} + \varepsilon(\mathbf{r})\nabla \cdot \mathbf{E} = 0. \tag{9.47}$$

If we assume $\varepsilon(\mathbf{r}) = \varepsilon(r)$ and $\mu(\mathbf{r}) = \mu(r)$, we have

$$\nabla \cdot \mathbf{E} = -\frac{\varepsilon'(r)}{r\varepsilon(r)}\mathbf{r} \cdot \mathbf{E} = -\frac{\varepsilon'(r)}{r\varepsilon(r)}(rE_r), \tag{9.48}$$

where the prime indicates the derivative with respect to the argument shown in the parenthesis. We note the following two useful identities for subsequent use:

$$\nabla \times \nabla \times \mathbf{E} = \nabla(\nabla \cdot \mathbf{E}) - \nabla^2\mathbf{E}, \tag{9.49}$$

and

$$\nabla^2(\mathbf{r} \cdot \mathbf{E}) = \mathbf{r} \cdot \nabla^2\mathbf{E} + 2\nabla \cdot \mathbf{E}. \tag{9.50}$$

Noting that ε only depends on r, we obtain that

$$\mathbf{r} \cdot \nabla(\nabla \cdot \mathbf{E}) = -\frac{\partial}{\partial r}\left(\frac{\varepsilon'(r)}{\varepsilon(r)}(rE_r)\right) + \left(\frac{\varepsilon'(r)}{\varepsilon(r)}E_r\right). \tag{9.51}$$

Using the above four equations, we can obtain an equation for the electric field as

$$\nabla^2(rE_r) + \frac{\partial}{\partial r}\left[\frac{\varepsilon'(r)}{\varepsilon(r)}(rE_r)\right] + \frac{\varepsilon'(r)}{r\varepsilon(r)}(rE_r) + \varepsilon(r)\mu(r)\frac{\omega^2}{c^2}(rE_r) = 0. \tag{9.52}$$

This equation is separable in (r, θ, ϕ) and the Spherical Harmonic functions are a solution to the angular part. Hence, the solution is $(rE_r) = U(r)Y_{lm}(\theta, \phi)$ where the radial part $U(r)$ satisfies

$$\frac{1}{r^2}\frac{\partial}{\partial r}\left(r^2\frac{\partial U}{\partial r}\right) - \frac{l(l+1)}{r^2}U + \frac{\partial}{\partial r}\left[\frac{\varepsilon'(r)}{\varepsilon(r)}U\right] + \frac{\varepsilon'(r)}{r\varepsilon(r)}U + \varepsilon(r)\mu(r)\frac{\omega^2}{c^2}U = 0. \tag{9.53}$$

If we choose $\varepsilon(r) = \alpha r^p$ and $\mu(r) = \beta r^q$, we can use a trial solution $U(r) \sim r^n$ and obtain

$$[n(n+1) - l(l+1) + p(n-1) + p]r^{n-2} + \alpha\beta\omega^2/c^2 r^{p+q+n} = 0, \qquad (9.54)$$

implying $p + q = -2$ and

$$n_\pm = 1/2 \left[-(p+1) \pm \sqrt{(p+1)^2 + 4l(l+1) - 4\alpha\beta\omega^2/c^2} \right]. \qquad (9.55)$$

Hence the general solution can be written as

$$E_r(\mathbf{r}) = \sum_{l,m} \left[n_+ A_{lm} r^{n+-1} + n_- B_{lm} r^{n--1} \right] Y_{lm}(\theta, \phi), \qquad (9.56)$$

and a similar solution can be obtained for the TE modes with $\mathbf{r} \cdot \mathbf{E} = 0$.

Assuming an arbitrary source at $r = a_0$, we can now write the electric fields of the TM modes in the different regions for the negative spherical shell of Fig. 2 as

$$\mathbf{E}^{(1)}(\mathbf{r}) = \sum_{l,m} \left[n_+ A_{lm}^{(1)} r^{n+-1} + n_- B_{lm}^{(1)} r^{n--1} \right] Y_{lm}(\theta, \phi), \qquad a_0 < r < a_1,$$

$$(9.57a)$$

$$\mathbf{E}^{(2)}(\mathbf{r}) = \sum_{l,m} \left[n_+ A_{lm}^{(2)} r^{n+-1} + n_- B_{lm}^{(2)} r^{n--1} \right] Y_{lm}(\theta, \phi), \qquad a_1 < r < a_2,$$

$$(9.57b)$$

$$\mathbf{E}^{(3)}(\mathbf{r}) = \sum_{l,m} \left[n_+ A_{lm}^{(3)} r^{n+-1} + n_- B_{lm}^{(3)} r^{n--1} \right] Y_{lm}(\theta, \phi), \qquad a_2 < r < \infty,$$

$$(9.57c)$$

and similarly for the magnetic fields. Note that the $B_{lm}^{(1)}$ correspond to the field components of the source located at $r = a_0$. For causal solutions, $A_{lm}^{(3)} = 0$. Satisfying the boundary conditions for the fields, and under the conditions $p = -1$, $q = -1$, $\varepsilon_+(a_1) = -\varepsilon_-(a_1)$, and $\varepsilon_+(a_2) = -\varepsilon_-(a_2)$, we have

$$A_{lm}^{(1)} = 0, \qquad (9.58a)$$

$$A_{lm}^{(2)} = \left(\frac{1}{a_1^2} \right)^{\sqrt{l(l+1) - \alpha\beta\omega^2/c^2}} B_{lm}^{(1)}, \qquad B_{lm}^{(2)} = 0, \qquad (9.58b)$$

$$B_{lm}^{(3)} = \left(\frac{a_2^2}{a_1^2} \right)^{\sqrt{l(l+1) - \alpha\beta\omega^2/c^2}} B_{lm}^{(1)}. \qquad (9.58c)$$

The lens-like property of the system becomes clear by writing the field outside the spherical shell as

$$E_r^{(3)} = \frac{1}{r} \left[\frac{a_2^2}{a_1^2} r \right]^{\sqrt{l(l+1) - \alpha\beta\omega^2/c^2}} B_{lm}^{(1)} Y_{lm}(\theta, \phi). \qquad (9.59)$$

Hence apart from a scaling factor of $1/r$, the fields on the sphere $r = a_3 = (a_2^2/a_1^2)a_0$ are identical to the fields on the sphere $r = a_0$. We also have a spatial magnification in the image by a factor of a_2^2/a_1^2. Similar to the cylindrical lens, this spherical lens is able to image sources that are also located only within a finite distance of the negative spherical shell. Note that the positive medium outside the shell is also required to have spatially varying material parameters that vary as $1/r$. In the presence of finite dissipation, using the ideas of the asymmetric lens (Ramakrishna et al. 2002), we can terminate the spatially varying media at some finite but large distance away from the spherical shell.

Let us note a couple of points about the above perfect lens solutions in the spherical geometry. First, for $r > a_3$, i.e., points outside the image surface, the fields appear as if the source were located on the spherical image surface ($r = a_3$). However, this is not true for points $a_2 < r < a_3$ within the image surface. Second, note that our imaging direction is along r and ε and μ can be an arbitrary function of θ and ϕ with the only condition of complementarity between the negative and positive regions. We can, however, reach this conclusion only by recourse to the generalized lens theorem and it cannot be straightforwardly obtained from the conventional calculation. Third, given that $\varepsilon_-(a_2) = -\varepsilon_+(a_2)$, we have the perfect lens solutions if and only if $n_+ = -n_-$, which implies that $p = -1$ in Eq. (9.55). Although the solutions given by Eq. (9.56) occur in any medium with $\varepsilon\mu \sim 1/r^2$, the perfect lens solutions work for both polarizations only if $\varepsilon \sim \mu \sim 1/r$. Here we have written down the solutions for the TM modes. The solutions for the TE modes can be similarly obtained. A choice of $\varepsilon \sim 1/r^2$ enables focusing for only the TM modes, and $\mu \sim 1/r^2$ enables focusing of only the TE modes.

Again there is a considerable simplification that is possible in the extreme near-field limit (Ramakrishna and Pendry 2004) in which case the imaging for the TM modes becomes independent of the magnetic permeability and only requires $\varepsilon_- \sim -1/r^2$. Similarly, the TE modes become independent of the ε and the only condition is that $\mu_- \sim -1/r^2$. Further, the constraint that the positive media outside also be spatially varying can be dropped. In this limit it can be shown that the largest multipole that can be resolved is again limited by dissipation in the NRM and is approximately (Ramakrishna and Pendry 2004)

$$l_{\max} \simeq \frac{\ln\{3\varepsilon_1\varepsilon_3/[\varepsilon_i(a_1)\varepsilon_i(a_2)]\}}{2\ln(a_2/a_1)}, \qquad (9.60)$$

where $\varepsilon_1 = -\mathrm{Re}(\varepsilon_-(a_1))$ and $\varepsilon_3 = -\mathrm{Re}(\varepsilon_-(a_2))$ for the perfect lens conditions and $\varepsilon_i(r) \sim 1/r^2$ is the imaginary part of the permittivity for the negative medium shell.

A fundamental issue that comes up for the spherical or cylindrical lenses is: what do we mean by complementarity and equal thickness? For example, how would the medium within the complementary layers appear to an observer external to the complementary spherical regions? Consider the two

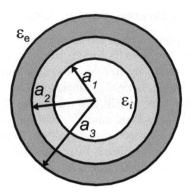

Figure 9.10 A pair of spherical complementary shells cancel each other. The two complementary shells are shown by different shades of gray.

complementary spherical regions shown in Fig. 9.10 by different shades of gray. Let (ε_e, μ_e) and (ε_i, μ_i) be the material parameters in the regions external and internal to the complementary layers, respectively. We first note that the condition of complementarity in the (ℓ, θ, ϕ) space would just imply that $\ell_3 - \ell_2 = \ell_2 - \ell_1$, which simply translates to $a_3 = a_2^2/a_1$. In the context of the generalized lens theorem in the Cartesian geometry, for an observer to the right of the layers, we could just cut out the complementary layers of equal thicknesses and move the media and the sources on the left of the layers by a distance $2d$ toward the right. In the context of the spherical lens, we could carry the operation out for the equivalent slab pair in the Cartesian geometry and then transform back into the spherical geometry. For the region beyond the complementary layers, i.e., $r > a_2$, this only amounts to undoing the original transformation and we obtain

$$\left.\begin{array}{l} \varepsilon_e^{eq} = \varepsilon_e(r), \\ \mu_e^{eq} = \mu_e(r), \end{array}\right\} \quad \forall \quad r > a_3. \tag{9.61}$$

However for the points $r < a_3$, the material that was originally inside the inner sphere $r < a_1$ has to now shift into the region $r < a_3$. Hence the new equivalent parameters for this region $r < a_3$ would be

$$\left.\begin{array}{l} \varepsilon_i^{eq} = \frac{a_1}{a_3}\varepsilon_i(\frac{a_1}{a_3}r), \\ \mu_i^{eq} = \frac{a_1}{a_3}\mu_i(\frac{a_1}{a_3}r), \end{array}\right\} \quad \forall \quad r < a_3. \tag{9.62}$$

We see that not only does the spatial dependence need to be rescaled but also the magnitudes of ε_i and μ_i need to be adjusted to account for this transformation.

9.3.3 Hyperlens: a layered curved lens

We have seen that super-lenses that utilize the excitation of the surface plasmons have a limitation in that the image resolution is deteriorated by the presence of dissipation. One way to overcome this limitation is to simply use the idea of the layered lens in curved geometries. Even if we have resonances, dissipation does not affect the obtainable resolution significantly because the amplitudes of the modes are kept small. Another possibility is to use the indefinite anisotropy of the layered system as was explained in Section 9.1.2. A lens whose function crucially stems from this anisotropy and depends on the hyperbolic nature of the dispersion has been termed a hyperlens (Jacob et al. 2006).

Similar to the expansion in plane waves $\exp(ik_x x)$ in Cartesian coordinates, the fields on a cylindrical surface $r = r_o$ can be expanded in the periodic functions $\exp(\pm im\phi)$ where m is a positive integer. In an isotropic dielectric medium, one can write for the dispersion in terms of the radial and azimuthal components of the wave vectors:

$$k_r^2 + k_\phi^2 = \varepsilon \omega^2/c^2. \tag{9.63}$$

The azimuthal wave-vector $k_\phi = m/r$ and decays in magnitude at large r. For any given m, there is a radius below which the k_r becomes imaginary and the amplitude of the mode is exponentially small for points within this radial distance. Hence if a small scattering object (with fine variation along ϕ) is placed at a small radial distance, the coupling to the large m modes that accurately describes the fast variation with ϕ becomes small and the near-field information does not propagate away. However, this would be different if one had a hyperbolic dispersion in the medium.

Consider a concentrically stratified medium made of alternative layers of silver (a negative dielectric medium) and a positive index medium as shown in Fig. 9.11. The layering creates an anisotropy for both field components, normal and parallel to the interfaces. The medium can be considered to be anisotropic with different ε_r and ε_ϕ in the effective medium limit of small layer thickness. It has been shown before that these two components can have opposite signs. We have seen in Section 2.5 that at low frequencies, $\varepsilon_\phi < 0$ and $\varepsilon_r > 0$, while at high frequencies the converse is true. In either case, the dispersion equation becomes hyperbolic, but has a low wave-vector cutoff in the first low frequency case, while there is no cutoff in the second high frequency case. In the second case, the dispersion equation is

$$\frac{k_r^2}{\varepsilon_\phi} - \frac{m^2}{r^2|\varepsilon_r|} = \mu_z \frac{\omega^2}{c^2}. \tag{9.64}$$

Thus, the large m modes do not become evanescent here for any value of m (this would be valid as far as the effective medium picture works). Hence the higher order m modes of the scattering object can couple to propagating modes in the cylinder and can propagate radially away, thus transferring

Figure 9.11 A cylindrically layered medium with alternate cylindrical layers of a negative dielectric medium (gray) and a positive medium (white) acts as an indefinite effective medium. This can be utilized to project out the near-field information of two small sources embedded within the innermost layer to the space outside the layered cylindrical system. The image so produced has a magnification and may be outcoupled to the far-field propagating modes easily.

the near-field information to much larger radial points. There would be a magnification in the image formed at larger radial distances by the ratio of the tangential wave-vectors at the source and image points, which translates to the number $\mathcal{M} = r_i/r_o$. At much larger radial points, since k_ϕ is now a small quantity, one can actually outcouple these modes to propagating modes and image them using conventional optics. The hyperlens is limited again only by the levels of dissipation in the silver, which damps out the waves as they propagate radially outward. Another critical issue is the quality and the surface roughness of the silver layers, which need to be exceptionally good for implementing the hyperlens. The hyperlens that combines the properties of the layered effective medium and the cylindrical (and in principle, spherical) lens represents an exciting implementation that utilizes the hyperbolic nature of dispersion in indefinite media to full advantage and allows for the far-field imaging of highly subwavelength sized sources.

The idea of the hyperlens has been demonstrated experimentally (Liu et al. 2007) with 16 alternating layers of silver and alumina (35 nm thickness each) deposited on the inside of a semi-cylindrical cavity on a quartz substrate. The system with radiation of about 364 nm wavelength could produce a magnified image outside the layered cylinder of two 35-nm-thick lines placed 150 nm apart inside the cylindrical layers. The spacing between the lines in the magnified image was 350 nm, which is resolvable by conventional optical microscopes. Thus, subwavelength information of $\lambda/2.5$ was propagated out by a distance of more than 3λ, which is a very significantly large distance.

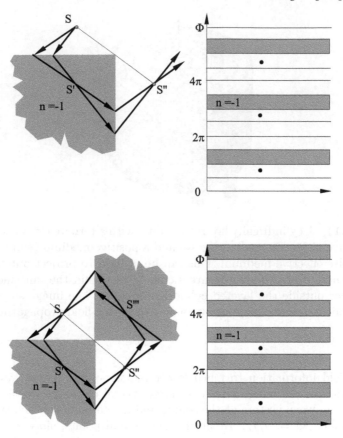

Figure 9.12 Top: (Left) A 90° wedge (corner) of negative refractive index acts as a lens with an image forming at a point in free space across the corner and inside the negative index wedge. This can be mapped into the periodic layered system with periodically placed sources shown on the right. Bottom: (Left) Two 90° wedges of negative refractive index $(n = -1)$ with the corners touching can act as an open resonator. Images form in each quadrant of space and the fourth image gets focused back onto the source. This system can also be mapped into a periodic layered system of negative index slabs with periodically placed sources shown on the right.

9.3.4 Perfect two-dimensional corner lens

In the case of the cylindrical and the spherical lens we chose the radial direction as the imaging direction. An interesting variation can be obtained if one chooses instead the azimuthal ϕ direction for imaging. Sharp corners have very interesting electromagnetic properties. Even the electrostatic properties are rich, with the electrostatic fields showing a divergence in as one approaches the singular point (Jackson 1999). Consider a wedge-shaped piece of a semi-infinite negative refractive index material with $n = -1$ shown in Fig. 9.12 with invariance along the axis normal to the plane of the figure. A ray diagram shows that the properties of negative index wedges are interesting and reveals the possibility of imaging a source across the 90° corner as shown in Fig. 9.12. Negative refraction reverses the direction of the rays and causes a real image to form on the other side of the corner. Compare the similarities of this system to a corner reflector where a source placed inside the corner would generate an apparent image across the corner.

In fact combining two such wedges of negative refractive index medium that just touch at the corners as shown in Fig. 9.12 (bottom panel) makes it even more interesting: classical rays emanating from a source placed in any one of the quadrants are returned back to the source point (Notomi 2002). Thus, the ray picture suggests that the double corner can act as an open resonator. However, we have seen that the ray picture cannot be completely relied upon with complementary media. Hence it is desirable to have a full wave calculation to confirm the imaging properties of such a system.

Using the transformation technique and the generalized lens theorem for complementary media, it has been demonstrated (Pendry and Ramakrishna 2003) that the images formed in the above two cases are perfect in the sense that they involve both the propagating and the near-field modes of the source. Consider the periodic arrangements of slabs shown beside the corners with material parameters:

$$\varepsilon(\phi) = \mu(\phi) = \begin{cases} +1 & \forall \quad -\pi/2 < \phi < \pi, \\ -1 & \forall \quad -\pi < \phi < -\pi/2 \end{cases} \tag{9.65}$$

in the first case (a) of a single corner, and

$$\varepsilon(\phi) = \mu(\phi) = \begin{cases} +1 & \forall \quad -\pi/2 < \phi < 0 \text{ and } \pi/2 < \phi < \pi, \\ -1 & \forall \quad -\pi < \phi < -\pi/2 \text{ and } 0 < \phi < \pi/2, \end{cases} \tag{9.66}$$

in the second case (b) of the double corner lens. We assume invariance in the direction normal to the plane. Note that a source is placed in the second quadrant. It is easily seen that the layered medium periodic stack of negative refractive index can be mapped onto the corners of negative index using the transformation to cylindrical coordinates:

$$x = r_0 e^{\ell/\ell_0} \cos\phi, \qquad y = r_0 e^{\ell/\ell_0} \sin\phi, \qquad z = Z. \tag{9.67}$$

Using the transformation properties, we have

$$\tilde{\varepsilon}_\ell = \tilde{\mu}_\ell = \tilde{\varepsilon}_\phi = \tilde{\mu}_\phi = \begin{cases} +1, & \tilde{\varepsilon}_Z = \tilde{\mu}_Z = r_0^2 e^{2\ell/\ell_0}, \\ -1, & \tilde{\varepsilon}_Z = \tilde{\mu}_Z = -r_0^2 e^{2\ell/\ell_0} \end{cases} \tag{9.68}$$

in the respective positive and negative quadrants. The conditions of complementarity are satisfied and the dependence along the r axis is irrelevant as

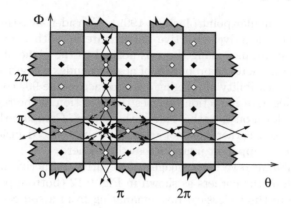

Figure 9.13 A periodic arrangement of cells in checkerboard fashion with adjacent cells having $n = +1$ (white) or $n = -1$ (gray) is shown schematically. If a source is placed in one of the cells, an image forms in every other cell. The properties of the checkerboard are dominated by the modes at the corners. A ray picture only predicts a few of the images as shown.

it is orthogonal (transverse) to the imaging axis along ϕ. Hence we conclude that the images in both the cases of the homogeneous single corner (a) and the homogeneous double corner (b) are perfect and the imaging involves both the propagating and near-field evanescent modes of radiation.

Note that case (b) is unusually singular as the source and the image are imaged onto each other and the fields grow indefinitely in time. It has also been shown that all the surface modes of this system are degenerate at the surface plasmon frequency and the density of modes in the system at this frequency should be infinite (Pendry and Ramakrishna 2003). This mapping into periodic layered media can easily be extended to show that several identical intersecting wedges would also have interesting focusing properties and can image a source onto itself (Guenneau et al. 2005a). It has been shown in He et al. (2005) that there are modes that are essentially trapped at the corners and do not propagate away from them. Thus, the touching corners actually behave like open resonators and can be used to localize light for some time.

9.3.5 Checkerboards and a three-dimensional corner lens

We finally briefly examine an even more singular super-structure of negative refractive index materials obtained by putting together the touching double corners onto a lattice. Rectangular checkerboards of alternating regions of positive and negative refractive index as shown in Fig. 9.13 have been shown to have several interesting electromagnetic properties (Guenneau et al. 2005b). Let us call the axes along the checkerboard directions $\hat{\theta}$ and $\hat{\phi}$, and the direction normal to the plane as ℓ which is a direction of invariance. Consider that the system is periodic with a period of π along both the $\hat{\theta}$ and $\hat{\phi}$ directions.

Let us define a checkerboard of homogeneous regions with $\varepsilon = \mu = +1$ and $\varepsilon = \mu = -1$ in alternating periodic regions of the checkerboard as shown in the figure. A ray analysis shows that the rays from a source at (θ_1, ϕ_1) can either be trapped around corners or can flow out to the infinities in the crystal while undergoing negative refraction at every interface. The ray analysis obtains a series of images that are formed periodically along the two lines $\theta = \theta_1$ and $\phi = \phi_1$. This decomposition, however, would not be complete for waves for which the interactions with the singular corner points would also be very important. Using the generalized lens theorem, it can be deduced that an image should be formed in each cell of the checkerboard lattice. To show this, let us consider the imaging along, say, the $\hat{\theta}$ direction and use the generalized lens theorem, whereby the condition of complementarity is satisfied for the layers along the imaging $\hat{\theta}$ direction with varying refractive index n in the $\hat{\phi}$ direction transverse to the imaging direction. For a source at $\theta = \theta_1$, $\phi = \phi_1$ in the first positive cell, we have a set of images along the $\phi = \phi_1$ line at $\theta = \pm m\pi \pm \theta_1$ and $-\theta_1$ where m is a positive integer. Similarly, applying the generalized lens theorem along the $\hat{\phi}$ direction, we can show that the entire set of image points would be reproduced along the $\hat{\phi}$ direction. Hence we have an image point in every cell of the checkerboard structure corresponding to the source placed in any one cell. The contradictions that can arise between the ray analysis and a full wave prediction have already been pointed out in Section 9.2.2 and we have one more such case here where the ray analysis reveals only part of the real field distribution.

The presence of a very large number of corners with a high density of modes renders the system very singular. Each corner can support an infinite density of degenerate surface plasmons (Guenneau et al. 2005a). In principle, all the fields of the source would get coupled into the eigenmodes of the corners and the exponentially large fields at the singular points would completely swamp out the inhomogeneous solutions due to the source. Only absorption prevents the actual divergence of the fields in an infinite lattice. On the other hand, a finite checkerboard would imply that all modes (however highly evanescent) in the system are leaky modes and in time escape out of the checkerboard. Numerical studies of checkerboard structures have indeed confirmed this behavior both with respect to the total size of the checkerboard superstructure and with respect to the levels of absorption in the system (Ramakrishna et al. 2007b). Dissipation or finite size regularizes any such unphysical divergence, in which case it becomes meaningful to talk about the properties of checkerboards.

The properties of an extended two-dimensional checkerboard, however, can be utilized to make an important deduction, i.e., the eight touching three-dimensional corners (octants in three dimensions) of negative refractive index can focus light in a similar manner as two-dimensional intersecting corners

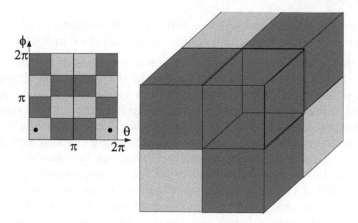

Figure 9.14 The unit cell of the checkerboard on the left with the double periodic set of sources can be mapped into the set of eight alternating three-dimensional corners with alternating positive and negative refractive index ($n = \pm 1$). The corners extend out to the infinities and together act as a three-dimensional open resonator for light.

(quadrants). To show this, consider a checkerboard with the properties

$$\varepsilon_{\pm\theta}(\theta, \phi) = \pm\varepsilon \, \sin\theta, \quad \varepsilon_{\pm\phi}(\theta, \phi) = \pm\frac{\varepsilon}{\sin\theta}, \tag{9.69a}$$

$$\mu_{\pm\theta}(\theta, \phi) = \pm\mu \, \sin\theta, \quad \mu_{\pm\phi}(\theta, \phi) = \pm\frac{\mu}{\sin\theta}, \tag{9.69b}$$

where the positive signs and negative signs occur in a checkerboard alternating fashion in space. It is clear that such a system satisfies the conditions of complementarity and mirror anti-symmetry along the main imaging axis ($\hat{\theta}$ or $\hat{\phi}$ in this case), and therefore it should also have imaging properties similar to the homogeneous checkerboards. Also note that the variation of ε and μ along the $\hat{\ell}$ direction is immaterial to the behavior as ℓ is transverse to the imaging axes along $\hat{\theta}$ and $\hat{\phi}$.

Consider making a transformation to spherical polar coordinates from Cartesian coordinates by Eq. (9.40). We obtain the transformation parameters

$$Q_\ell = \frac{r_0}{\ell_0} e^{\ell/\ell_0}, \quad Q_\theta = r_0 e^{\ell/\ell_0}, \quad Q_\phi = r_0 e^{\ell/\ell_0} \, \sin\theta.$$

The renormalized dielectric permittivity $\tilde{\varepsilon}$ and $\tilde{\mu}$ are given by the Eqs. (9.42a) and (9.42b), respectively, and are reproduced below for convenience:

$$\tilde{\varepsilon}_\ell = r_0\ell_0 e^{\ell/\ell_0} \sin\theta \, \varepsilon_\ell, \quad \tilde{\varepsilon}_\theta = \frac{r_0}{\ell_0} e^{\ell/\ell_0} \sin\theta \, \varepsilon_\theta, \quad \tilde{\varepsilon}_\phi = \frac{r_0 e^{\ell/\ell_0}}{\ell_0 \sin\theta} \varepsilon_\phi,$$

$$\tilde{\mu}_\ell = r_0\ell_0 e^{\ell/\ell_0} \sin\theta \, \mu_\ell, \quad \tilde{\mu}_\theta = \frac{r_0}{\ell_0} e^{\ell/\ell_0} \sin\theta \, \mu_\theta, \quad \tilde{\mu}_\phi = \frac{r_0 e^{\ell/\ell_0}}{\ell_0 \sin\theta} \mu_\phi.$$

The variation along the radial $\hat{\ell}$ (transverse) direction is irrelevant and can be ignored. Consider in addition eight three-dimensional corners of homogeneous materials as shown in Fig. 9.14 with alternate wedges having positive or negative refractive index ($n = \pm 1$). Hence we have

$$\varepsilon_\ell = \mu_\ell = \varepsilon_\theta = \mu_\theta = \varepsilon_\phi = \mu_\phi = \pm 1 \qquad (9.70)$$

in the corresponding regions. Choosing the scale factor $\ell_0 = 1$, Eq. (9.42a) and Eq. (9.42b) become

$$\tilde{\varepsilon}_\ell = \tilde{\mu}_\ell = \tilde{\varepsilon}_\theta = \tilde{\mu}_\theta = \pm r_0 e^\ell \sin\theta, \quad \tilde{\varepsilon}_\phi = \tilde{\mu}_\phi = \pm \frac{r_0 e^\ell}{\sin\theta}, \qquad (9.71)$$

i.e., the system appears just like the two-dimensional checkerboard shown in Fig. 9.14 with spatially varying material parameters considered in the previous paragraph, but with a doubly periodic set of sources with a period 2π along ϕ and θ and a symmetry plane along $\theta = \pi$. Hence, the cubic corner with the homogeneous materials with alternating signs for the refractive index also forms an imaging device with an image point inside every cubic corner. Thus, we have generalized the result of the two-dimensional corner to a three-dimensional corner. An image of the point source in one octant is generated inside all the other seven octants and an image mapped onto the original source point as well. The images are, in a sense, perfect in that both propagating and evanescent waves are involved in the imaging. Thus, a truly three-dimensional open cavity is realized in this configuration. In fact, by considering checkerboards with variations along the $\hat{\theta}$ and $\hat{\phi}$ directions but respecting the condition of a mirror anti-symmetry about the interfaces, one can generate entire classes of interesting configurations of three-dimensional lenses.

We conclude our discussion of checkerboards with an observation on some very interesting properties of triangular (two-dimensional) checkerboards where the cells are equilateral triangles. While a corner of intersection of three finite triangles has some circulating modes about the corner, a ray diagram shows that no ray can escape from a double layer of a triangular checkerboard as shown in Fig. 9.15. This does not, however, hold strictly for a wave, which always has a finite probability of escaping from the finite system. It has been shown that the residence time of waves in the checkerboard can be highly enhanced by layering up the triangles in a concentric manner (Ramakrishna et al. 2007b). The more the number of shells, the better the confinement of radiation. In contrast, this does not hold for a rectangular checkerboard where the ray picture shows that a subset of rays that escapes out of a finite system always exists. Full wave calculations have, indeed, revealed that the residence times in rectangular checkerboards increase comparatively slowly with increase in the number of concentric shells (Ramakrishna et al. 2007b). This confinement holds much potential for enhancing the sensitivity of detectors, or for enhancing non-linearities.

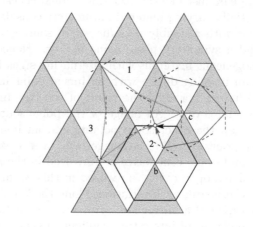

Figure 9.15 Finite checkerboard consisting of alternating regions of positive and negative refractive index media in equilaterial triangular regions. Invariance is assumed normal to the plane of the figure. All rays emitted by a source located in one of the interior triangles are trapped in closed trajectories around one of the three corners.

10

Brief report on electromagnetic invisibility

10.1 Concept of electromagnetic invisibility

Invisibility has been a long-lasting dream of scientists, writers, artists, movie makers, and many more. The notion of invisibility carries a magical aura that sparkles our dreams and make us believe that everything else becomes possible. Certainly achieving invisibility would open the door to a world of applications limited only by our imaginations. The possibility of invisibility, however, has belonged to the realm of mythological tales and science fiction rather than to our real world, until the 21st century. While the invisibility in the tales of yore stemmed purely from "magic" with no attempts to relating the phenomena to actual physics, science fiction has often tried to explore such plausible links. The "invisible man" in Wells (2002) for example used a chemical drink that turned the refractive index of his body identical to that of air thus making him transparent. However, the scientific analogy remained relatively shallow, neglecting the fact that the person would be completely blind if perfectly index matched – the very process of sight crucially depending on absorption of light in the retina, as pointed out in Perelman (1913).

In the scientific quest, two main approaches to achieve invisibility have been taken, active and passive. The active approach is more straightforward and consists of taking real-time pictures of a scene and displaying them on a screen that is actually hiding or blocking the scene itself. For example, the screen can be the coat worn by a person, onto which is displayed in real time the scene behind the person, effectively giving the illusion of transparency to an observer from the front. This virtual reality approach has been reported in (Japanese scientist invents 'invisibility coat'), for example. It has also been reported that the armed forces across the world have toyed with some form or other of active camouflage using virtual reality. However, the actual field use of such virtual reality techniques is not known as yet.

The passive approach, on the contrary, relies on the intrinsic properties of the object or subject to not scatter the incident wave in an usual way and to send confusing information to the receiver. In this regard, two sub-types of invisibility can be distinguished: one where the object is merely not identifiable and the other one where the object does not affect the radiation field (except at very nearby regions of space) so that it cannot be sensed by an observer some distance away and can be considered to be actually invisible. In electromagnetic or optical language, rendering an object not identifiable means to significantly lower its scattering cross-section, with the ideal situation of lowering it down to zero, at which point the object does not scatter anymore the incident fields. Such reduction can for example be achieved using good impedance matching and very absorbing materials: the material (ideally) totally absorbs the incident power and nothing is reflected back to the receiver, which therefore cannot identify the object in front of it.* This is the approach that has been traditionally undertaken for military applications in order to render objects unidentifiable on radar screens. Note that merely having a highly absorbing material does not suffice, as such media are usually poorly impedance matched to vacuum in which case the scattered energy dominates over the absorbed energy. A simple manner of impedance matching for a highly absorbing medium would be to have large imaginary parts (compared to the real parts) for both the permittivity and permeability of the medium so that $(\mu/\varepsilon)^{1/2} \sim [\mathrm{Im}(\mu)/\mathrm{Im}(\varepsilon)] \sim 1$. Ideally one could use anisotropic media to realize impedance matching via perfectly matched layer boundary conditions (Berenger 1994) so that all incident radiation is guided into the medium where it is absorbed.

In the second case of actual invisibility, the object can either allow the light to be seen through (which is, for example, the purpose of the virtual reality process mentioned above or in the sense of the invisible man) or smoothly guide the light around itself like a fluid flowing in streamlines around an obstacle. The second option would be possible by placing around the object another cloaking material that guides the light around the object through the cloak which would have to be a "see through" material. These approaches result in not affecting the radiation at any point some distance away from the object, which is the ultimate goal of invisibility. Note that in the first case in which the object would allow the light to go through itself, the processes of invisibility would crucially depend on the properties of the object. In the second option interestingly, the design of the invisibility cloak becomes independent of the properties of the object itself.

There is, however, a fundamental difference in the two passive invisibilities highlighted above. In the situation of a zero scattering cross-section, the extinction coefficient is still non-zero, i.e. the object is not visible but would still block the rays if it were located between a source and a receiver. The receiver therefore would not know what type of object is in front of it, but it would still know that some perturbing element is present whose location and other properties can be estimated, and perhaps deduced from the shadow. Hence, even

---

*We refer here to active observation, most commonly, where a wave is first sent onto an object, reflects from it, and reaches a receiver. This is, for example, the manner of our vision: light constantly bounces off the objects surrounding us, reaching our eye. In the case of luminous objects, the receiver passively observes radiation emitted directly by the object.

the sole knowledge of the existence of the object is important information. On the other hand, if the object is seen through or excluded, the existence itself of the object disappears since the environment is indistinguishable whether there is or there is no object. Therefore, it is not a situation where the observer receives no information, but instead it is a situation where the observer receives wrong information which leads him to an erroneous conclusion (such as "the space is empty of scatterers") and does not spark any suspicion.

Remarkably, metamaterials have been very instrumental at enabling invisibility and cloaking from electromagnetic waves, primarily due to their ability to make anisotropic metamaterials with the required material parameters.

All the previous chapters have been devoted to the presentation of some aspects of metamaterials: their genesis, their physical implementation, as well as their major properties as they have been investigated and reported in the open literature for about a decade. Many of these properties had already been mentioned in what is considered the first paper directly related to this topic Veselago (1968), although without so much in-depth analysis. These properties revolved around the concepts of negative permittivity ε and permeability μ, negative refraction, and their consequences on otherwise well-known electromagnetic phenomena: refraction, Doppler shift, Čerenkov radiation, imaging, etc.

Over the last decade, the major step forward from Veselago (1968) has been the experimental reality of these media in the form of metamaterials. After the fundamental papers providing ideas on how to design metamaterials to achieve $\varepsilon < 0$ and $\mu < 0$ (Pendry et al. 1998; 1999), various communities of researchers made intense efforts to elaborate, optimize, and further develop upon these first concepts. As a result, it is now possible to tailor the material properties to a level unachieved before, accessing all quadrants of the $(\text{Re}\{\varepsilon\}, \text{Re}\{\mu\})$ plane shown in Fig. 1.10. In addition to only controlling the sign of the constitutive parameters, scientists have learned how to induce and tailor anisotropic properties and even bianisotropic properties to some extent. From this perspective, the unprecedented control of material properties offered by metamaterials is at its infancy. The latter have been discovered by trying to achieve negative constitutive parameters, but have opened a door much wider than the original intended purpose. The quest for invisibility can be viewed as a second domain of application of metamaterials, which may yield as unexpected and revolutionary ideas as those prompted by the quest for negative constitutive parameters,

The concept of invisibility using metamaterials has appeared in the literature very recently, and is currently mainly based on two independent approaches, already highlighted in the paragraphs above. The first, proposed in Alù and Engheta (2005), takes advantage of the negative polarizability of left-handed media in order to compensate for the polarizability of regular dielectrics and to lower their scattering cross-section. This, however, requires a detailed knowledge of the electromagnetic scattering properties of the object in order to design the compensating cloak for it. A more detailed discussion on this technique is provided in Section 10.3.

The second approach proposed in Pendry et al. (2006) to achieve invisibility relies on the proper control of the constitutive parameters within a certain region of space such as to smoothly deviate and guide the incident light around a core region, and to return the rays to their original propagation path upon leaving that same region. In this approach, the cloaked object placed in the core region never interacts with the radiation and the design of the cloak becomes independent of the cloaked object. This approach, where invisibility arises from exclusion from the electromagnetic fields, is discussed subsequently. An analogous approach but based mainly on ray analysis was independently developed by Leonhardt (2006a;b).

10.2   Excluding electromagnetic fields

10.2.1   Principle

The concept of electromagnetic invisibility is fairly simple to grasp and is represented in Fig. 10.1: a region of space is invisible to an observer if it does not affect the light going through it or in its vicinity differently than the medium in which it is embedded. In fact, it has been claimed that true invisibility is impossible on the basis of the uniqueness of the inverse scattering problem: a full knowledge of the phase and amplitudes of the scattered fields in all directions enables a unique specification of the scatterer's spatial distribution of the refractive index (Nachman 1988, Wolf and Habashy 1993). However, this neglects that both magnetic permeability and the dielectric permittivity can be anisotropic, independently specified, and that scattering systems with different local resonant fields can have similar far-field scattering properties.

In common words, this concept of invisibility is described as "see through." It is important to realize that the observer has no knowledge of what happens inside the invisible region of space, insofar as it does not introduce detectable perturbations of the surrounding rays. In Fig. 10.1 for example, the connection between points A_1 and A_2 can be arbitrary, as long as the phase, the amplitude, and the direction of the waves are leaving the region as they would have been if there had been no object. The challenge to achieve invisibility is therefore to use the phenomenon of refraction in order to design a region of space with the right properties so that the rays that enter it seemingly leave it without any perturbation.

It is often the case that natural phenomena provide ideas to physicists and engineers to find solutions to new problems. The present situation is an additional example of such case, and an analogy with a totally unrelated field

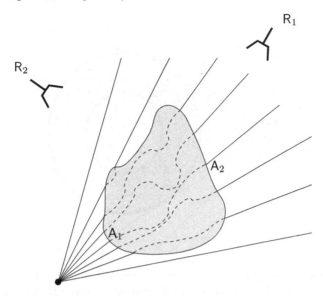

Figure 10.1 Illustration of the concept of invisibility: the gray region of space is invisible to both receivers R_1 and R_2 if it does not perturb the rays outside it. The receivers, however, do not need to know the ray distribution inside the region and the path between A_1 and A_2 can be arbitrary as far as the phase, amplitude, and directions are preserved from the two points.

can be provided, *viz.*, the theory of General Relativity.

General Relativity is a cosmological theory of space and time established by Einstein (Einstein 1916) that explains the trajectory of planets and other celestial bodies as a deformation of the space-time continuum due to mass and energy. Hence, unlike the Newtonian theory of gravitation which involves a force that perturbs the motion of planets from their straight-line trajectories in the Euclidian space, Einstein's theory shows that all trajectories are in fact geodesics but in a Riemann curved space-time. Consequently, the trajectories of planets as we observe them is merely the projection of such geodesic from the real four-dimensional (4D) space (three dimensions of space and one of time) onto our more common three-dimensional (3D) space. In other words, in our 3D Euclidian space, the geodesics, which can be viewed as straight lines in the 4D space, appear curved.

There is a simple leap from these ideas to the concept of invisibility. Suppose that what is achieved by mass and energy in the general theory of relativity can be achieved by altering the material properties of a medium; then we should be able to bend the rays along predefined geodesics inside the medium and return them to straight paths (which are also geodesics in a different space or a different medium in our case) at will. This concept is illustrated in Fig. 10.2: a circular region of space is created with such constitutive parameters to

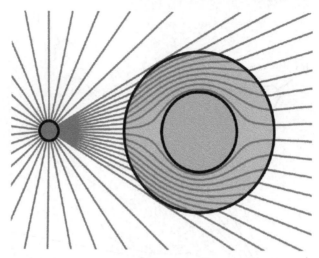

Figure 10.2 Bending the rays along geodesics: the path of rays is modified by the proper material parameters and returned to straight lines upon leaving the medium, thus achieving invisibility. A proper design of geodesics can also cloak an object in the center of the structure. (Reproduced from Pendry et al. (2006). Reprinted with permission from AAAS.)

bend the rays as shown. Again, the rays are bent in this way because we observe them from an Euclidian space, whereas they correspond to geodesics inside the medium. The rays therefore travel along straight paths in free-space, "staight" paths inside the medium (although they appear curved to us observing from free-space), and are returned to regular straight paths upon leaving the medium and re-entering free-space. Comparing the field outside the region with the illustration of Fig. 10.1, we see a striking similarity and the direct connection to the concept of invisibility. This is the approach based on ray analysis developed in Leonhardt (2006a;b) for inhomogeneous but isotropic refractive media.

Notice in addition a striking feature of the configuration in Fig. 10.2: the central region of the medium is totally shielded from any radiation. An object of arbitrary shape can be located in this region and would not perturb the radiation in the surrounding regions, which would then propagate as if the object were not there. The object is therefore cloaked with an invisibility shield and cannot be detected.[†] By reciprocity, any fields emitted by the object inside the shielded volume would be confined to that volume only, without any possibility of propagating out of the cloaked volume. This design requires anisotropic materials and independent ε and μ for its function.

[†] As we shall see subsequently, such cloaking only works perfectly at a single frequency.

10.2.2 Design procedure

The constitutive parameters that achieve a ray diagram of the type shown in Fig. 10.2 are very specific and need to be properly computed. They result from transforming the fields in one geometry (typically the regular 3D free-space) onto another (a sphere in Fig. 10.2) or vice versa, and therefore are obtained by a properly chosen change of coordinates (or mapping). This is a very similar approach to the one already taken in Section 9.3 to realize a perfect lens with geometries other than an infinite slab. The main transformation relation was shown to be Eq. (9.31b), which yields the constitutive parameters $\bar{\bar{\varepsilon}}(\mathbf{r})$ and $\bar{\bar{\mu}}(\mathbf{r})$ in the new coordinate system. As already suggested by the notation and explained in Section 9.3.1, the constitutive parameters become anisotropic and spatially varying. The Maxwell equations (9.30) with the generalized constitutive parameters in place are, however, form invariant so that all the necessary electromagnetic results in homogeneous, isotropic media can be directly carried over.

The design principle is therefore to transform the constitutive parameters such that the electromagnetic fields are excluded from a specified region of space, and packed into another region in another geometry. An example was proposed in Pendry et al. (2006) for excluding a spherical cavity of electromagnetic fields and compressing these fields into a spherical shell of a medium enclosing the cavity. Consider the transformation

$$r' = a_1 + r(a_2 - a_1)/a_2, \qquad \theta' = \theta, \qquad \phi' = \phi \qquad (10.1)$$

which takes every point $r < a_2$ into the region $a_1 < r < a_2$. Hence the field inside the spherical cavity $r < a_1$ and inside the spherical shell $a_1 < r < a_2$ are both compressed into the spherical shell. Application of the transformation relations Eq. (9.31b) indicates that

$$\varepsilon'_{r'} = \mu'_{r'} = \frac{a_2}{a_2 - a_1}\left(\frac{r' - a_1}{r'}\right)^2, \qquad (10.2a)$$

$$\varepsilon_{\theta'} = \mu_{\theta'} = \frac{a_2}{a_2 - a_1}, \qquad (10.2b)$$

$$\varepsilon_{\phi'} = \mu_{\phi'} = \frac{a_2}{a_2 - a_1} \qquad (10.2c)$$

gives one set of parameters for $a_1 < r < a_2$ that can map the fields in the interior of the sphere into the spherical shell. Outside the spherical shell the material parameters take the values of unity. Note that at the outer edge of the spherical shell the material parameters of the shell satisfy

$$\varepsilon_{\theta'} = \varepsilon_{\phi'} = \varepsilon_{r'}^{-1}, \qquad (10.3a)$$

$$\mu_{\theta'} = \mu_{\phi'} = \mu_{r'}^{-1}, \qquad (10.3b)$$

which are precisely the conditions for a perfectly matched layer to vacuum (Berenger 1994). Thus, the spherical shell in this example is also perfectly impedance matched with free-space and hence no reflection or scattering arises at its interface. In this sense, invisibility is achieved in a similar manner to the impedance-matched absorber. However, in a major difference with this system, the spherical shell also contains within itself any fields that emanate from a source and smoothly guides the radiation away from the central core region. Furthermore, by construction this spherical shell does not affect the fields outside the shell. Thus, an observer cannot obtain any information about the concealed object or about the shell itself. Also note that the choices of the parameters for cloaking is not unique but that the above choices of parameters are the only ones that give perfect impedance matching as well.

Consequently, the shell of the anisotropic and inhomogeneous medium can enable perfect cloaking of an object located inside the spherical cavity. Note that conversely, the cloaked object inside the spherical cavity can never communicate with the external world because all the fields emitted by a source inside would be mapped onto regions within the cloak only. Thus the invisible object is also thus blind and secluded from the rest of the electromagnetic world at that frequency.

The main requirements of an invisibility cloak, namely, the anisotropy and inhomogeneity, are both well within technological reach. An experimental demonstration of cloaking of a cylinder using a system of split ring resonators was given in Schurig et al. (2006) where it was shown that the scattering from a cloaked cylindrical cavity was much reduced. Invisibility by a similar cloak was investigated in Zolla et al. (2007) using the finite element method and the response to a closely placed radiating line source was studied. It was found that the primary effect due to the invisibility cloak was that the objects near the cloak would appear slightly shifted from their original positions – in the manner of a mirage.

We also note that due to the necessary dispersion in the material parameters of metamaterials, the invisibility is principally functional only within a narrow band of frequencies. Another important limitation of such invisibility devices stems from the fact that in order to make the cloak layer substantially thinner than the cloaked region, materials with very large constitutive parameters are required. While obtaining such characteristics is feasible by working close to resonances, important challenges may arise, typically associated with dissipation.

10.3   Cloaking with localized resonances

A somehow natural idea when trying to reduce the scattering cross-section of an object is to look for a coating material that exhibits a compensating effect. For example, for a dielectric core inducing a positive polarizability, one could look for a shell that induces a negative polarizability. While this is a straightforward outcome in the context of complementary optical layers (see Section 8.2), it is not immediately straightforward in higher dimensions. By defining a proper metric (typically the scattering cross-section) and outlining a design procedure, one could adjust the parameters of the shell so as to cancel as much as possible the scattering of the core. Against intuition, one would thus realize a larger object, thus exhibiting a larger geometric cross-section, but a lower scattering cross-section. This is precisely the approach proposed by Alù and Engheta (2005).

Let us simplify the conceptual discussion to spherical geometries, for which the exact scattering field is known to be governed by the Mie theory (see Section 5.2.5 on 198). The scattering cross-section normalized to the geometric cross-section for a sphere of radius a is given by (Tsang et al. 2000a)

$$\sigma_{scat} = \frac{2}{(|k|a)^2} \sum_{m=1}^{\infty} (2m+1)(|a_m|^2 + |b_m|^2), \qquad (10.4)$$

where k is the wavenumber of the surrounding medium and (a_m, b_m) are the coefficients of the external field expanded in the spherical coordinate system, as derived in Section 5.2.5. Minimizing σ_{scat} is seen to be a non-linear problem that does not offer a trivial solution. Let us therefore assume that the sphere is very small so that the quasi-static is applicable. Under this approximation, the small sphere responds primarily as a point source and thus re-radiates as a dipole. Under a $\hat{z}$ polarized incident field propagating along the $\hat{x}$ direction, the electric and magnetic fields can be written as (Kong 2000)

$$\mathbf{E}(\mathbf{r}) = -\frac{i\omega\mu I\ell e^{ikr}}{4\pi} \left[\hat{r} \left[\left(\frac{i}{kr}\right)^2 + \frac{i}{kr} \right] 2\cos\theta + \hat{\theta} \left[\left(\frac{i}{kr}\right)^2 + \frac{i}{kr} + 1 \right] \sin\theta \right],$$
$$(10.5a)$$

$$\mathbf{H}(\mathbf{r}) = -\hat{\phi} \frac{ik I\ell e^{ikr}}{4\pi r} \left(\frac{i}{kr} + 1 \right) \sin\theta, \qquad (10.5b)$$

where the dipole moment $I\ell$ can be determined by matching the boundary conditions. Under the quasi-static approximation, which is equivalent to a near-field approximation ($kr \ll 1$), we see that the electric field dominates over the magnetic field. Consequently, in order to reduce the scattering cross-section of such sphere, one only needs to design a shield that compensates for the first electric dipole term, all the other terms being negligible already.[‡]

[‡]Note that if the dipole term is canceled, the higher order terms become dominant and cannot be neglected anymore in the computation of the scattering cross-section. However, this scattering is already much weaker than the original one, which therefore goes along the purpose of reducing σ_{scat}.

Within the Mie theory, the small sphere regime, also known as the Rayleigh regime, corresponds to reducing the sum to the first term $m = 1$ only. Under this approximation, it is seen that the TM term b_1' is dominant, confirming the fact that the response is primarily of electric nature (the prime indicates that the b_1' term is a generalization of the b_1 term of Eq. (5.33) to a two-layer spherical coating geometry). The reduction of the scattering cross-section is therefore reduced to the minimization of the b_1' term. This minimum has been shown to be exactly zero within the small sphere approximation if the ratio between the core radius r_c and the shell external radius r_s is (Alù and Engheta 2005)

$$\frac{r_c}{r_s} = \sqrt[2m+1]{\frac{(\varepsilon_c - \varepsilon_0)[(m+1)\varepsilon_c + n\varepsilon]}{(\varepsilon_c - \varepsilon)[(m+1)\varepsilon_c + n\varepsilon_0]}}, \qquad (10.6)$$

where ε, ε_c, and ε_0 are the permittivities of the core, the shell, and the outside free-space, respectively. Upon studying this relation, it has been shown that in order to reduce the scattering cross-section of a regular dielectric core, one has to resort to shell permittivities that are lower than those of the outside medium (typically free-space), possibly taking negative values. This is in good agreement with the intuitive view of polarization compensation. As a matter of fact, the polarizations of the core and the shell can be written as $\mathbf{P}_1 = (\varepsilon - \varepsilon_0)\mathbf{E}$ and $\mathbf{P}_2 = (\varepsilon_c - \varepsilon_0)\mathbf{E}$, respectively, so that one is positive and the other one negative if $\varepsilon_c < \varepsilon_0 < \varepsilon$. In addition, although this condition has been derived within the Rayleigh approximation, it has been shown to still yield reasonably low scattering cross-sections for radii up to $\lambda/5$, i.e. away from the Rayleigh regime already.

Naturally, the same principle can be applied to the magnetic dipole, which can be reduced or even canceled by the proper choice of the permeability of the shell as function of the one of the core. In fact, in view of Eqs. (10.5), the magnetic dipole moment is the next important term after the electric dipole effect has been canceled. Consequently, further reduction of the scattering cross-section can be achieved by lowering the permeability of the shell simultaneously to lowering its permittivity.

Remarkably, this phenomenon of cancellation of scattering cross-section is not associated with a resonance of the Mie coefficients, which would instead yield a strongly enhanced scattering effect. Such resonances being nonetheless present, one has to be careful at not designing a configuration where the condition of zero and infinite σ_{scat} are too close to each other, in which case the sensitivity between transparency and strong scattering would be too large. In fact, whenever possible, it is judicious to design a configuration where no resonances are excited, which is possible by using only positive permittivities, with the internal one being lower than the external one (Alù and Engheta 2005).

The transparency achieved using this method presents one major advantage and one major disadvantage. The advantage is that since the cancellation of the scattering cross-section does not rely on any resonance, it is a robust effect to various perturbations. In particular, variations in sizes and material parameters, including losses, have been shown to have a moderate effect on the increase of the scattering cross-section, whereas variations due to geometrical imprecision in the realization of the optimized split-rings and rods (necessary to obtain the required effective constitutive parameters) have been speculated to have a limited impact as well, for the same reason. This weak sensitivity can also be translated into a robustness of the configuration to frequency changes, so that the technique presented here can be expected to perform relatively well over a frequency band surrounding the ideal situation.

This configuration, however, presents an important drawback which is that it is specifically optimized for a given core. In other words, the design and the physical implementation of the shell need to be repeated anew for each core property.

A

The Fresnel coefficients for reflection and refraction

In this appendix, which is primarily meant for the uninitiated reader, we derive the Fresnel coefficients for the reflected and refracted radiation across an interface between two isotropic media with dielectric permittivities ε_1 and ε_2 and magnetic permeabilities μ_1 and μ_2. Consider Fig. A.1 for the geometry involved (the case of P-polarization is considered there).

The wave vectors in the two media then obey the dispersion (assuming that the plane of incidence is the x-z plane)

$$k_{1x}^2 + k_{1z}^2 = \varepsilon_1 \mu_1 \frac{\omega^2}{c^2}, \tag{A.1}$$

$$k_{2x}^2 + k_{2z}^2 = \varepsilon_2 \mu_2 \frac{\omega^2}{c^2}. \tag{A.2}$$

For proper continuity of the electromagnetic fields across the interface at all times, we require the frequency ω to be unchanged. Similarly, $k_{1x} = k_{2x}$ which is required to match the phases at all points along the x direction as the system is invariant along the x direction. This enables us to obtain the Snells law for the angle of incidence and refraction as

$$\pm\sqrt{\varepsilon_1 \mu_1} \sin\theta_1 = \pm\sqrt{\varepsilon_2 \mu_2} \sin\theta_2. \tag{A.3}$$

A proper choice for the sign of the square should be taken as per the discussion in Chapter 5. Now k_{1z} and k_{2z} can be obtained from the above dispersion relations.

For the P-polarized radiation, it is easier to work with the magnetic fields and for S-polarization it is easier to work with the electric fields. Note that the Fresnel coefficient would then differ by a factor of $\sqrt{\mu/\varepsilon}$ which is the impedance of the medium in the two cases. In the case of P-polarization, the magnetic field is normal to the plane of incidence. Hence, **H** has only a y component. Matching the tangential component of **H** across the interface, we have

$$H_{iy} + H_{ry} = H_{ty}, \tag{A.4}$$

where the subscripts i, r, t stand for the incident, reflected and transmitted fields. The Maxwell equation,

$$\nabla \times \mathbf{H} = \frac{\partial \mathbf{D}}{\partial t}, \tag{A.5}$$

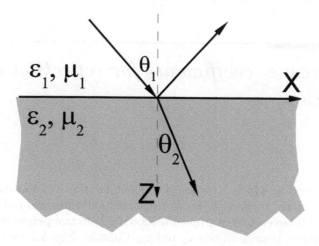

Figure A.1   The reflection and refraction of light across an interface between two isotropic media with arbitrary dielectric permittivity and magnetic permeability.

should be used to obtain the associated electric field. For a plane harmonic wave, we obtain

$$\mathbf{k} \times \mathbf{H} = -\omega\varepsilon\mathbf{E}, \tag{A.6}$$

from which we obtain

$$E_z = -\frac{k_x}{\omega\varepsilon}H_y, \tag{A.7}$$

$$E_x = \frac{k_z}{\omega\varepsilon}H_y. \tag{A.8}$$

The continuity of the tangential electric field (E_x) implies that

$$\frac{k_{iz}H_{iy}}{\omega\varepsilon_1} + \frac{k_{rz}H_{ry}}{\omega\varepsilon_1} = \frac{k_{tz}H_{ty}}{\omega\varepsilon_2}. \tag{A.9}$$

Noting $k_{rz} = -k_{iz}$ and eliminating H_{ry} from the two equations, we get

$$T = \frac{H_{ty}}{H_{iy}} = \frac{2k_{iz}/\varepsilon_1}{k_{tz}/\varepsilon_2 + k_{iz}/\varepsilon_1} \tag{A.10}$$

and similarly eliminating H_{ty} we obtain

$$R = \frac{H_{ry}}{H_{iy}} = \frac{k_{tz}/\varepsilon_2 - k_{iz}/\varepsilon_1}{k_{tz}/\varepsilon_2 + k_{iz}/\varepsilon_1}. \tag{A.11}$$

A similar analysis can be made for the S-polarized light. The results are similar with only ε being replaced by the corresponding μ everywhere and the coefficients relate the electric fields across the interface in this case.

B

The dispersion and Fresnel coefficients for a bianisotropic medium

We will present in this appendix, principally for the benefit of the less experienced reader, a derivation of the dispersion in a bianisotropic medium and the Fresnel coefficients across the interface between an isotropic medium and a bianisotropic medium. This is primarily to only demonstrate the manner of calculating them. By following the procedure the reader should be able to derive all the cases (anisotropic, bi-isotropic or bianisotropic) that (s)he would confront in the book. We will consider the case when the constitutive relations for the bianisotropic medium are given by

$$\mathbf{D} = \varepsilon_0 \begin{pmatrix} \varepsilon_x & 0 & 0 \\ 0 & \varepsilon_y & 0 \\ 0 & 0 & \varepsilon_z \end{pmatrix} \mathbf{E} + \frac{1}{c} \begin{pmatrix} 0 & i\xi_{xy} & 0 \\ 0 & 0 & 0 \\ 0 & i\xi_{zy} & 0 \end{pmatrix} \mathbf{H}, \tag{B.1}$$

$$\mathbf{B} = \frac{1}{c} \begin{pmatrix} 0 & 0 & 0 \\ i\zeta_{yx} & 0 & i\zeta_{yz} \\ 0 & 0 & 0 \end{pmatrix} \mathbf{E} + \mu_0 \begin{pmatrix} 1 & 0 & 0 \\ 0 & \mu_y & 0 \\ 0 & 0 & 1 \end{pmatrix} \mathbf{H}. \tag{B.2}$$

These relations are seen to be reciprocal if $\zeta_{yx} = -\xi_{xy}$ and $\zeta_{yz} = -\xi_{zy}$. An array of split-ring resonators with their axes along the y direction (or cylinders along the y directions) would be well described by these constitutive relations (see Chapter 3). This is a special case of a bianisotropic medium where the modes in the medium are linearly polarized.

We will consider the plane of incidence to be the x-z plane (zero k_y component) and the case of TM modes (The magnetic field, H_y, is normal to the plane of incidence). Hence we have

$$H_x = H_z = E_y = 0. \tag{B.3}$$

The constituent relations can be rewritten in their components:

$$D_x = \varepsilon_0 \varepsilon_x E_x + i\frac{\xi_{xy}}{c} H_y, \tag{B.4}$$

$$D_z = \varepsilon_0 \varepsilon_z E_z + i\frac{\xi_{zy}}{c} H_y, \tag{B.5}$$

$$B_y = \mu_0 \mu_y H_y - i\frac{\xi_{xy}}{c} E_x - i\frac{\xi_{zy}}{c} E_z. \tag{B.6}$$

Now consider the Maxwell equations for plane waves

$$\mathbf{k} \times \mathbf{E} = \omega \mathbf{B}, \qquad \mathbf{k} \times \mathbf{H} = -\omega \mathbf{D}. \tag{B.7}$$

Assuming a complete y-invariance and since we have $k_y = 0$, we can expand these equations in the bianisotropic medium as

$$k_z E_x - k_x E_z = +\omega B_y = \omega [\mu_0 \mu_y H_y - i \frac{\xi_{xy}}{c} E_x - i \frac{\xi_{zy}}{c} E_z], \tag{B.8}$$

$$-k_z H_y = -\omega D_x = -\omega [\varepsilon_0 \varepsilon_x E_x + i \frac{\xi_{xy}}{c} H_y], \tag{B.9}$$

$$k_x H_y = -\omega D_z = -\omega [\varepsilon_0 \varepsilon_z E_z + i \frac{\xi_{zy}}{c} H_y], \tag{B.10}$$

from which we obtain

$$E_z = \frac{1}{-\varepsilon_0 \varepsilon_z} [\frac{k_x}{\omega} + i \frac{\xi_{zy}}{c}] H_y, \tag{B.11}$$

$$E_x = \frac{1}{\varepsilon_0 \varepsilon_x} (\frac{k_z}{\omega} - i \frac{\xi_{xy}}{c}) H_y, \tag{B.12}$$

$$(\frac{k_z}{\omega} + i \frac{\xi_{xy}}{c}) E_x = \left[\mu_0 \mu_y - \frac{1}{\varepsilon_0 \varepsilon_z} (\frac{k_x^2}{\omega^2} + \frac{\xi_{zy}^2}{c^2}) \right] H_y. \tag{B.13}$$

For a non-trivial solution, we will require that the determinant for the last two homogeneous equations (for E_x and H_y) vanishes. This gives us the dispersion of the waves in this medium:

$$k_z^2 = \left(\varepsilon_x \mu_y - \xi_{xy}^2 - \frac{\varepsilon_x}{\varepsilon_z} \xi_{zy}^2 \right) \frac{\omega^2}{c^2} - \frac{\varepsilon_x}{\varepsilon_z} k_x^2, \tag{B.14}$$

or to put it more symmetrically

$$\varepsilon_x k_x^2 + \varepsilon_z k_z^2 + \varepsilon_z \xi_{xy}^2 \omega^2 / c^2 + \varepsilon_x \xi_{zy}^2 \omega^2 / c^2 = \varepsilon_x \varepsilon_z \mu_y \omega^2 / c^2. \tag{B.15}$$

Here we can see that $n_{\text{eff}} = \pm \sqrt{\varepsilon_x \mu_y - \xi_{xy}^2 - \xi_{zy}^2}$ acts like the refractive index when $\varepsilon_x = \varepsilon_z$. The sign of the square root will be determined by whether the real part of the quantity inside the square root is positive or negative (see Chapter 5 for details).

Now we will calculate the Fresnel coefficients for the reflected and transmitted wave amplitudes. Let us consider the reflection or transmission of a plane wave (moving along $+z$ direction) incident from vacuum onto the bianisotropic medium. Since we consider here P-polarised light, we can write for the incident, reflected and the transmitted fields, and the corresponding wave-vectors:

$$H_{iy} = e^{i(k_x x + k_{1z} z)}, \tag{B.16}$$

$$H_{ry} = r_{\text{mp}}\, e^{i(k_x x - k_{1z} z)}, \tag{B.17}$$

$$H_{ty} = t_{\text{pp}}\, e^{i(k_x x + k_{2z} z)}, \tag{B.18}$$

$$k_{1z}^2 = \frac{\omega^2}{c^2} - k_x^2, \tag{B.19}$$

$$k_{2z}^2 = [\varepsilon_x \mu_y - \xi_{xy}^2 - \xi_{zy}^2]\frac{\omega^2}{c^2} - k_x^2. \tag{B.20}$$

The requirement of continuity of the parallel component of $\mathbf{H} \Rightarrow k_x = \text{const}$ across the interface and that

$$1 + r_{\text{mp}} = t_{\text{pp}}, \tag{B.21}$$

where the subscripts "mp" imply reflection of a positive-going wave into the negative direction and the subscripts "pp" imply the transmission into the positive direction of a wave initially moving in the positive direction. Noting that

$$E_{2x} = \frac{1}{\varepsilon_0 \varepsilon_x}\left(\frac{k_{2z}}{\omega} - i\frac{\xi_{xy}}{c}\right)H_{2y}, \tag{B.22}$$

$$E_{1x} = \frac{k_{1z}}{\varepsilon_0 \varepsilon_1 \omega} H_{1y} \tag{B.23}$$

in the bianisotropic and isotropic medium respectively, the requirement of continuity of the parallel components of the electric field (E_x) yields

$$\frac{k_{z1}}{\varepsilon_0 \varepsilon_1 \omega} + \frac{-k_{z1}}{\varepsilon_0 \varepsilon_1 \omega}r_{\text{mp}} = \frac{1}{\varepsilon_0 \varepsilon_x}\left(\frac{k_{2z}}{\omega} - i\frac{\xi_{xy}}{c}\right)t_{\text{pp}}. \tag{B.24}$$

Hence, we have for the reflection coefficient

$$r_{\text{mp}} = \frac{k_{z1}/\varepsilon_1 - (k_{z2}/\varepsilon_1 - i\xi_{xy}\omega/c)/\varepsilon_x}{k_{z1} + (k_{z2} - i\xi_{xy}\omega/c)/\varepsilon_x}. \tag{B.25}$$

Similarly, the transmission coefficient across the interface is obtained as

$$t_{\text{pp}} = \frac{2k_{z1}/\varepsilon_1}{k_{z1}/\varepsilon_1 + (k_{z2} - i\xi_{xy}\omega/c)/\varepsilon_x}. \tag{B.26}$$

Next, let us calculate the Fresnel coefficients for a wave incident on an isotropic dielectric medium with $(\varepsilon_1 \mu_1)$ from the bianisotropic medium. The bianisotropic medium is now assumed to occupy the negative half-space ($z < 0$ and the vacuum the positive half-space ($z > 0$). Once again, using the continuity of the tangential components of the magnetic and electric fields, we have

$$1 + r'_{\text{mp}} = t'_{\text{pp}}, \tag{B.27}$$

$$\frac{1}{\varepsilon_0 \varepsilon_x}\left(\frac{k_{z2}}{\omega} - i\frac{\xi_{xy}}{c}\right) + \frac{1}{\varepsilon_0 \varepsilon_x}\left(-\frac{k_{z2}}{\omega} - i\frac{\xi_{xy}}{c}\right)r'_{\text{mp}} = \frac{k_{z1}}{\varepsilon_0 \varepsilon_1 \omega}t'_{\text{pp}}, \tag{B.28}$$

from which we obtain

$$r'_{mp} = \frac{(k_{z2} - \mathrm{i}\,\xi_{xy}\omega/c)/\varepsilon_x - k_{z1}/\varepsilon_1}{(k_{z2} + \mathrm{i}\,\xi_{xy}\omega/c)/\varepsilon_x + k_{z1}/\varepsilon_1}, \tag{B.29}$$

$$t'_{pp} = \frac{2k_{z2}/\varepsilon_x}{(k_{z2} + \mathrm{i}\,\xi_{xy}\omega/c)/\varepsilon_x + k_{z1}/\varepsilon_1} \tag{B.30}$$

We do not simply have $r_{mp} = r'_{mp}$ as in the case of an isotropic medium. In fact, we also do not have the same Fresnel coefficients if the bianisotropic medium occupied the positive half-space ($z > 0$), the isotropic medium occupied the negative-half space, and the wave were incident in the negative direction. This stems about from the fact that the impedance in a bianisotropic medium depends on the direction of propagation. We can call

$$\eta_p = \frac{\sqrt{\varepsilon_x\mu_y - \xi_{xy}^2 - \xi_{zy}^2 - (k_x/k_0)^2} - \mathrm{i}\xi_{xy}}{\varepsilon_x}, \tag{B.31}$$

$$\eta_m = \frac{-\sqrt{\varepsilon_x\mu_y - \xi_{xy}^2 - \xi_{zy}^2 - (k_x/k_0)^2} - \mathrm{i}\xi_{xy}}{\varepsilon_x} \tag{B.32}$$

as the effective impedances for the positive- and negative-going waves in the bianisotropic medium (where $k_0 = \omega/c$). Also defining

$$\eta_i = \frac{\sqrt{\varepsilon_1\mu_1 - (k_x^2/k_0^2)}}{\varepsilon_1} \tag{B.33}$$

for the isotropic medium, we can write all the Fresnel coefficients for the different cases when the wave is incident from the positive or negative directions as

$$r_{mp} = \frac{\eta_i - \eta_p}{\eta_i + \eta_p}, \qquad t_{pp} = \frac{2\eta_i}{\eta_i + \eta_p}, \tag{B.34}$$

$$r'_{mp} = -\frac{\eta_i - \eta_p}{\eta_i - \eta_m}, \qquad t'_{pp} = \frac{\eta_p - \eta_m}{\eta_i - \eta_m}, \tag{B.35}$$

$$r_{pm} = \frac{\eta_i + \eta_m}{\eta_i - \eta_m}, \qquad t_{mm} = \frac{2\eta_i}{\eta_i - \eta_m}, \tag{B.36}$$

$$r'_{pm} = -\frac{\eta_i + \eta_m}{\eta_i + \eta_p}, \qquad t'_{mm} = \frac{\eta_p - \eta_m}{\eta_i + \eta_p}. \tag{B.37}$$

Here the primed coefficients indicate the cases where the wave is incident from the bianisotropic medium onto the isotropic medium and the unprimed coefficients are for the case when the wave is incident from the isotropic medium onto the bianisotropic medium.

The derivation of the dispersion and the Fresnel coefficients for bianisotropic media (even where the two polarizations are not linearly polarized) can be performed in an analogous manner and is left as an exercise to the reader.

C

The reflection and refraction of light across a material slab

In this Appendix, we will calculate the transmission or reflection coefficient from a slab of any arbitrary material. The material can be isotropic, anisotropic or bianisotropic. All that information is included in the Fresnel coefficients for reflection and transmission across the interfaces comprising the slab, the wave impedance in the different directions within the slab, and the dispersion relation within the medium via the component of the wave vector normal to the interfaces. The wave impedance is the same for an isotropic medium in all directions, but different for waves moving along the positive and negative directions as seen in Appendix B in a bianisotropic medium.

We will perform the calculations in three conceptually differen ways: (i) by the usual way of matching the tangential components of the fields inside and outside the slab, (ii) by using the fact that we have one upgoing and one downgoing wave in the medium, and (iii) by the multiple scattering method. The three methods obviously lead to the same final result which is a matter of consistency. But each of these methods can give rise to better understanding of the concepts involved and the resulting phenomena.

Method 1: Matching fields across the interfaces

Consider the material slab to occupy the region between the planes $z = 0$ and $z = d$ and the region outside to be vacuum. Consider a plane wave incident from the left to be incident on the slab $\exp(ik_{z1}z$. The magnetic fields inside the slab medium can be written as

$$H_y = A\exp(+ik_{z2}z) + B\exp(-ik_{z2}z). \tag{C.1}$$

At $z = 0$ and $z = d$, we have the conditions of continuity for the tangential **E** and **H** fields as usual:

$$1 + R = A + B, \tag{C.2}$$
$$1 - R = \eta_p\, A + \eta_m\, B, \tag{C.3}$$
$$Te^{ik_{z1}d} = Ae^{ik_{z2}d} + Be^{-ik_{z2}d}, \tag{C.4}$$
$$Te^{ik_{z1}d} = \eta_p\, Ae^{ik_{z2}d} + \eta_m\, Be^{-ik_{z2}d}. \tag{C.5}$$

Solving the above, we have for the reflection (R) and transmission (T) coefficients of the slab

$$Te^{ik_{z1}d} = \frac{2(\eta_p - \eta_m)e^{ik_{z2}d}}{(1+\eta_p)(1-\eta_m) - (1+\eta_m)(1-\eta_p)e^{2ik_{z2}d}}, \tag{C.6}$$

$$R = \frac{(1-\eta_p)(1-\eta_m)(1-e^{2ik_{z2}d})}{(1+\eta_p)(1-\eta_m) - (1+\eta_m)(1-\eta_p)e^{2ik_{z2}d}}. \tag{C.7}$$

In the case of an isotropic medium, we have $\eta_p = -\eta_m = \eta_i$ as defined in Appendix B. It can be easily verified that we have an invariance of these expressions when $k_{z2} \to -k_{z2}$ as we then also have $\eta_p \to \eta_m$ simultaneously. So the choice of the sign of the wave-vector inside the medium does not manifest in any measurable effect with a symmetric slab.

Method 2: Coupling waves across the interfaces

We note that we have the positive and negative waves coupled via the Fresnel coefficients to the incident, reflected and transmitted waves across the two interfaces at $z = 0$ and $z = d$ as

$$A = t_{pp} + r'_{pm}B, \tag{C.8}$$

$$Be^{-ik_{z2}d} = r'_{mp}Ae^{ik_{z2}d}, \tag{C.9}$$

$$Te^{ik_{z1}d} = t'_{pp}Ae^{ik_{z2}d}, \tag{C.10}$$

$$R = r_{mp} + t'_{mm}B. \tag{C.11}$$

Solving these equations for T and R, we get

$$Te^{ik_{z1}d} = \frac{t_{pp}t'_{pp}e^{ik_{z2}d}}{1 - r'_{pm}r'_{mp}e^{2ik_{z2}d}}, \tag{C.12}$$

$$R = r_{mp} + \frac{t'_{mm}r'_{mp}t_{pp}e^{2ik_{z2}d}}{1 - r'_{pm}r'_{mp}e^{2ik_{z2}d}}, \tag{C.13}$$

which can be seen to be identical to the earlier one by substituting in the values of the Fresnel coefficients in terms of η_p, η_m and η_i presented before in Appendix B.

Method 3: The multiple scattering technique

Now let us consider the typical multiple scattering method in Born and Wolf (1999). The transmission and reflection can be written as a sum of infinite waves in terms of the partial Fresnel scattering coefficients:

$$T = t_{pp}t'_{pp}e^{ik_{z2}d} + t_{pp}r'_{mp}r'_{pm}t'_{pp}e^{3ik_{z2}d} + \cdots , \tag{C.14}$$

$$R = r_{mp} + t_{pp}r'_{mp}e^{2ik_{z2}d}t'_{mm} + \cdots . \tag{C.15}$$

These geometric series can be summed to yield exactly the expressions given by Eq. (C.12) and Eq. (C.13), respectively, obtained in method 2. There is a crucial assumption that the optical path-lengths for forward-going and backward-going waves are the same. This is also true in the bianisotropic case, although the wave impedances in both directions are not the same.

References

M. Abraham. Zur Elektrodynamik bewegter Körper. *Rend. Circ. Mat. Palermo*, 28 (1), 1909.

V. M. Agranovich and V. L. Ginzburg. *Spatial Dispersion in Crystal Optics and the Theory of Excitons*. NY Wiley-Interscience, 1966.

V. M. Agranovich, Yu. N. Gartstein, and A. A. Zakhidov. Negative refraction in gyrotropic media. *Phys. Rev. B*, 73:045114, 2005.

Y. Akahane, T. Asano, and S. Noda. High Q photonic crystal optical cavities. *Nature*, 425:944–947, 2003.

A. Alù, A. Salandrino, and N. Engheta. Negative effective permeability and left-handed materials at optical frequencies. *Opt. Express*, 14:1557–1566, 2006a.

Andrea Alù and Nader Engheta. Three-dimensional nanotransmission lines at optical frequencies: A recipe for broadband negative-refraction optical metamaterials. *Phys. Rev. B*, 75(2):024304, January 2007.

Andrea Alù and Nader Engheta. Achieving transparency with plasmonic and metamaterial coatings. *Phys. Rev. E*, 72:016623, 2005.

Andrea Alù, Nader Engheta, and Richard W. Ziolkowski. Finite-difference time-domain analysis of the tunneling and growing exponential in a pair of epsilon-negative and mu-negative slabs. *Phys. Rev. E*, 74(016604), 2006b.

P. W. Anderson. Plasmons, gauge invariance, and mass. *Phys. Rev.*, 130:439–442, 1963.

Xianyu Ao and Sailing He. Negative refraction of left-handed behavior in porous alumina with infiltrated silver at an optical wavelength. *Appl. Phys. Lett.*, 87: 101112, 2005.

G. Arfken. *Mathematical Methods for Physicists*. Academic Press, San Diego, 1985.

K. Artmann. Berechnung der Seitenversetzung des totalreflektierten Strahles. *Ann. Phys. (Leipzig)*, 2:87–102, 1948.

Neil W. Ashcroft and N. David Mermin. *Solid State Physics*. Saunders College, Philadelphia, 1976. ISBN 0-03-049346-3.

J. Askne and B. Lind. Energy of electromagnetic waves in the presence of absorption and dispersion. *Phys. Rev. A*, 2(6):2235–2340, December 1970.

K. Aydin, Zhaofeng Li, M. Hudlicka, S. A. Tretyakov, and Ekmel Ozbay. Transmission characteristics of bianisotropic metamaterials based on omega shaped metallic inclusions. *New J. Phys.*, 9(326), 2007.

J. Baker-Jarvis, E. J. Vanzura, and W. A. Kissick. Improved technique for determining complex permittivity with the transmission/reflection method. *IEEE Trans. Microwave Theory Tech.*, 38(8):1096–1103, August 1990.

W. L. Barnes, A. Dereux, and T. W. Ebbesen. Surface plasmon subwavelength optics. *Nature*, 424:824–830, August 2003.

P. A. Belov, S. I. Maslovsky, and S. A. Tretyakov. Wire media with negative effective permittivity: a quasi-static model. *Microwave Opt. Tech. Lett.*, 35:47–51, 2002.

P. A. Belov, R. Marques, S. I. Maslovski, I. S. Nefedov, M. Silveirinha, C. R. Simovski, and S. A. Tretyakov. Strong spatial dispersion in wire media in the very large wavelength limit. *Phys. Rev. B*, 67:113103, 2003.

J. P. Berenger. A perfectly matched layer for the absorption of electromagnetic waves. *J. Comput. Phys.*, 114:185–200, 1994.

Dwight W. Berreman. Optics in stratified and anisotropic media: 4×4-matrix formulation. *J. Opt. Soc. Am.*, 62(4):502–510, April 1972.

H. A. Bethe. Theory of diffraction by small holes. *Phys. Rev.*, 66:163–182, 1944.

A. D. Boardman, N. King, and L. Velasco. Negative refraction in perspective. *cond-mat/0508501*, 2005.

Allan D. Boardman and Kiril Marinov. Electromagnetic energy in a dispersive metamaterial. *Phys. Rev. B*, 73:165110, 2006.

C. F. Bohren. Applicability of effective medium theories to problems of scattering and absorption by nonhomogeneous atmospheric particles. *J. Atmos. Sci.*, 43: 468–475, 1986.

C. F. Bohren and D. R. Huffman. *Absorption and Scattering of Light by Small Particles*. John Wiley & Sons, New York, 1983.

E. L. Bolda, J .C. Garrison, and R. Y. Chiao. Optical pulse propagation at negative group velocities due to a nearby gain line. *Phys. Rev. A*, 49:2938, 1994.

M. Born and E. Wolf. *Principles of Optics*. Cambridge University Press, Cambridge, UK, 7th edition, 1999.

Robert W. Boyd. *Nonlinear Optics*. Academic Press, San Diego, 2003.

A M Bratkovsky, A. Cano, and A. P. Levanyuk. Strong effect of surfaces on resolution limit of negative-index superlens. *Appl. Phys. Lett.*, 87:103507, 2005.

L. Brillouin. *Wave Propagation and Group Velocity*. Academic Press, New York, 1960.

R. F. J. Broas, D. Sievenpiper, and E. Yablonovitch. A high impedance ground plane applied to a cellphone handset geometry. *IEEE Trans. Microwave Theory Tech.*, 49(7):1262–1265, July 2001.

J. Brown. Artificial dielectrics. *Prog. Dielectr.*, 2:194–225, 1960.

J. A. Buck. *Fundamentals of Optical Fibres*. John Wiley & Sons, Hoboken, NJ, 2nd edition, 2004.

Irfan Bulu, Humeyra Caglayan, and Ekmel Ozbay. Experimental demonstration of labyrinth-based left-handed metamaterials. *Opt. Exp.*, 13(25):10238–10247, December 2005.

M. Büttiker and H. Thomas. Front propagation in evanescent media. *Superlattices Microstruct.*, 23:781, 1998. Also at ArXiv:quant-ph/9711019.

C. Caloz and T. Itoh. *Electromagnetic Metamaterials:Transmission Line Theory and Microwave Applications*. John Wiley & Sons, Hoboken, NJ, 2005.

C. Caratheodory. The most general transformations of plane regions which transform circles into circles. *Bull. Amer. Math. Soc.*, 43:573–579, 1937.

D. Cassagne. Matériaux à bandes interdites photoniques. *Ann. Phys. Fr.*, 23(4), 1998.

Dan Censor. Scattering of a plane wave at a plane interface separating two moving media. *Radio Sci.*, 4(11):1079–1088, November 1969.

P. A. Čerenkov. Visible radiation produced by electrons moving in a medium with velocities exceeding that of light. *Phys. Rev.*, 52:378–379, 1934.

H. Chen, L. Ran, J. Huangfu, T. M. Grzegorczyk, and Jin Au Kong. Equivalent circuit model for left-handed metamaterials. *J. Appl. Phys.*, 100:024915, 2003.

H. Chen, L. Ran, D. Wang, J. Huangfu, Q. Jiang, and J. A. Kong. Metamaterial with randomized patterns for negative refraction of electromagnetic waves. *Appl. Phys. Lett.*, 88:031908, 2006a.

Hongsheng Chen, Lixin Ran, Jiangtao Huangfu, Xianmin Zhang, Kangsheng Chen, Tomasz M. Grzegorczyk, and Jin Au Kong. Left-handed metamaterials composed of only S-shaped resonators. *Phys. Rev. E*, 70(057605), 2004a.

Hou-Tong Chen, Willie J. Padilla, Joshua M. O. Zide, Arthur C. Gossard, Antoinette J. Taylor, and Richard D. Averitt. Active terahertz metamaterial devices. *Nature*, 444:597–600, November 30, 2006b.

Jianbing J. Chen, Tomasz M. Grzegorczyk, Bae-Ian Wu, and Jin Au Kong. Role of evanescent waves in the positive and negative Goos-Hänchen shifts with left-handed material slabs. *J. Appl. Phys.*, 98:094905, 2005a.

Jianbing J. Chen, Tomasz M. Grzegorczyk, Bae-Ian Wu, and Jin Au Kong. Limitation of FDTD in simulation of perfect lens imaging system. *Opt. Express*, 13(26): 10840–10845, 2005b.

Jianbing J. Chen, Tomasz M. Grzegorczyk, Bae-Ian Wu, and Jin Au Kong. Imaging properties of finite-size left-handed material slabs. *Phys. Rev. E*, 74:046615, 2006c.

Long Chen, Sailing He, and Linfang Shen. Finite-size effects of a left-handed material slab on the image quality. *Phys. Rev. Lett.*, 92:107404, 2004b.

X. L. Chen, Ming He, YinXiao Du, W. Y. Wang, and D. F. Zhang. Negative refraction: An intrinsic property of uniaxial crystals. *Phys. Rev. B*, 72:113111,

2005c.

Xudong Chen, Tomasz M. Grzegorczyk, Bae-Ian Wu, Joe Pacheco Jr., and Jin Au Kong. Robust method to retrieve the constitutive effective parameters of metamaterials. *Phys. Rev. E*, 70:016608, 2004c.

Xudong Chen, Bae-Ian Wu, Jin Au Kong, and Tomasz M. Grzegorczyk. Retrieval of the effective constitutive parameters of bianisotropic metamaterials. *Phys. Rev. E*, 71:046610, 2005d.

Xudong Chen, Tomasz M. Grzegorczyk, and Jin Au Kong. Optimization approach to the retrieval of the constitutive parameters of a slab of general bianisotropic medium. *Prog. in Electromagnetic Res.*, 60:1–18, 2006d.

Xudong Chen, Bae-Ian Wu, Jin Au Kong, and Tomasz M. Grzegorczyk. Erratum: Retrieval of the effective constitutive parameters of bianisotropic metamaterials. *Phys. Rev. E*, 73:019905(E), 2006e.

Weng Cho Chew. *Waves and Fields in Inhomogeneous Media*. Van Nostrand Reinhold, 1990. ISBN 0-442-23816-9.

R. Y. Chiao and A. M. Steinberg. Tunneling times and superluminality. In E. Wolf, editor, *Prog. Opt.*, volume 37, pages 347–406. Elsevier, Amsterdam, 1997.

Dmitry N. Chigrin. Radiation pattern of a classical dipole in a photonic crystal: Photon focusing. *Phys. Rev. E*, 70:056611, 2004.

Dmitry N. Chigrin, Stefan Enoch, Clivia M. Sotomayor Torres, and Gerard Tayeb. Self-guiding in two-dimensional photonic crystals. *Opt. Express*, 11(10):1203–1211, May 2003.

Robert E. Collin. *Field Theory of Guided Waves*. IEEE Press, New York, 2nd edition, 1990. ISBN 0-87942-237-8.

Tie Jun Cui and Jin Au Kong. Time-domain electromagnetic energy in a frequency-dispersive left-handed medium. *Phys. Rev. B*, 70:205106, November 2004.

S. Cummer. Simulated causal subwavelength focusing by a negative refractive index slab. *Appl. Phys. Lett.*, 82(10):1503–1505, March 2003.

T. Decoopman, G. Tayeb, S. Enoch, D. Maystre, and B. Gralak. Photonic crystal lens: From negative refraction and negative index to negative permittivity and permeability. *Phys. Rev. Lett.*, 97:073905, August 2006.

V. Dmitriev. Constitutive tensors of omega- and chiroferrites. *Microwave Opt. Tech. Lett.*, 29:201–205, 2001.

G. Dolling, C. Enrich, M. Wegener, J. F. Zhou, C. M. Soukoulis, and S. Linden. Cut-wire pairs and plate pairs as magnetic atoms for optical metamaterials. *Opt. Lett.*, 30(23), December 2005.

Gunnar Dolling, Christian Enrich, Martin Wegener, Costas M. Soukoulis, and Stefan Linden. Low-loss negative-index metamaterial at telecommunication wavelengths. *Opt. Lett.*, 31(12), 2006a.

Gunnar Dolling, Christian Enrich, Martin Wegener, Costas M. Soukoulis, and Stefan Linden. Simultaneous negative phase and group velocity of light in a meta-material. *Science*, 312:892–894, May 2006b.

William T. Doyle. Optical properties of a suspension of metal spheres. *Phys. Rev. B*, 39:9852, 1989.

YinXiao Du, Ming He, X. L. Chen, W. Y. Wang, and D. F. Zhang. Uniaxial crystal slabs as amphoteric-reflecting media. *Phys. Rev. B*, 73:245110, 2006.

Thomas Dumelow, Jose Alzamir Pereira da Costa, and Valder Nogueira Freire. Slab lenses from simple anisotropic media. *Phys. Rev. B*, 72:235115, 2005.

S. Duttagupta, R. Arun, and G. S. Agarwal. Subluminal to superluminal propagation in a left-handed medium. *Phys. Rev. B*, 69:113104, 2004.

T. W. Ebbesen, H. J. Lezec, T. Thio, and P. A. Wolff. Extraordinary optical transmission through sub-wavelength hole arrays. *Nature*, 391:667–669, 1998.

A. Einstein. Die Grundlage der allgemeinen Relativitätstheorie. *Ann. Phys.*, 49, 1916.

George V. Eleftheriades and A. Grbic. Experimental verification of backward-wave radiation from a negative refractive index metamaterial. *J. Appl. Phys.*, 92:5930–5935, 2002.

George V. Eleftheriades, Ashwin K. Iyer, and Peter C. Kremer. Planar negative refractive index media using periodically L-C loaded transmission lines. *IEEE Trans. Microwave Theory Tech.*, 50(12):2702–2712, December 2002.

N. Engheta. Circuits with light at nanoscales: Optical circuits with light at nanoscales: Optical. *Science*, 317:1698, 2007.

N. Engheta, D. L. Jaggard, and M. W. Kowarz. Electromagnetic waves in a Faraday chiral medium. *IEEE Trans. Antennas Propagat.*, 40:367–374, 1992.

P. P. Ewald. Die Berechnung Optischer und Elektrostatischen Gitterpotentiale. *Ann. Phys.*, 64:253–258, 1921.

Nicholas Fang and Xiang Zhang. Imaging properties of a metamaterial superlens. *Appl. Phys. Lett.*, 82:161–163, 2003.

Nicholas Fang, Hyesog Lee, Cheng Sun, and Xiang Zhang. Subdiffraction-limited optical imaging with a silver superlens. *Science*, 308(5721):534–537, April 22, 2005.

Michael W. Feise, P. J. Bevelacqua, and J. B. Schneider. Effects of surface waves on the behaviour of perfect lenses. *Phys. Rev. B*, 66:035113, 2002.

S. Foteinopoulou and C. M. Soukoulis. Electromagnetic wave propagation in two-dimensional photonic crystals: A study of anomalous refractive effects. *Phys. Rev. B*, 72:165112, 2005.

S. Foteinopoulou, E. N. Economou, and C. M. Soukoulis. Refraction in media with a negative refractive index. *Phys. Rev. Lett.*, 90:107402, 2003.

I. M. Frank and I. G. Tamm. Coherent visible radiation of fast electrons passing through matter. *Comp. Rend. (Dokl)*, 14:109–114, 1937.

N. Garcia and M. Nieto-Vesperinas. Left-handed materials do not make a perfect lens. *Phys. Rev. Lett.*, 88(20):207403, May 2002.

C. G. Garrett and D. E. McCumber. Propagation of a Gaussian light pulse through an anomalous dispersion medium. *Phys. Rev. A*, 1:305, 1970.

Philippe Gay-Balmaz and Olivier J. F. Martin. Efficient isotropic magnetic resonators. *Appl. Phys. Lett.*, 81(5):939–941, 29 July 2002.

V. L. Ginzburg. *The Propagation of Electromagnetic Waves in Plasmas*. Pergamon Press, Oxford, 1970.

V. L. Ginzburg. O zakone sokhraneniya i vyrazhenii dlya plotnosti energii v elektrodinamike pogloshchayushchei dispergiruyushchei sredy. *Izvestiya vysshikh uchebnykh zavedenii: Radiofizika*, 4:74–88, 1961.

S. Glasgow, M. Ware, and J. Peatross. Poyntings theorem and luminal total energy transport in passive dielectric media. *Phys. Rev. E*, 64(4):046610, October 2001.

G. Gomez-Santos. Universal features of the time evolution of evanescent modes in a left-handed perfect lens. *Phys. Rev. Lett.*, 90:077401, 2003.

James P. Gordon. Radiation forces and momenta in dielectric media. *Phys. Rev. A*, 8(1):14–21, July 1973.

Anthony Grbic and George V. Eleftheriades. Overcoming the diffraction limit with a planar left-handed transmission-line lens. *Phys. Rev. Lett.*, 92:117403, 2004.

R. B. Greegor, C. G. Parazzoli, K. Li, and M. H. Tanielian. Origin of dissipative losses in negative index of refraction materials. *Appl. Phys. Lett.*, 82(14):2356–2358, April 7 2003.

R. B. Greegor, C. G. Parazzoli, J. A. Nielsen, M. A. Thompson, M. H. Tanielian, and D. R. Smith. Simulation and testing of a graded negative index of refraction lens. *Appl. Phys. Lett.*, 87:091114, 2005.

A. N. Grigorenko. Negative refractive index in artificial metamaterials. *Opt. Lett.*, 31(16), August 2006.

A. N. Grigorenko, A. K. Geim, H. F. Gleeson, Y. Zhang, A. A. Firsov, I. Y. Khrushchev, and J. Petrovic. Nanofabricated media with negative permeability at visible frequencies. *Nature*, 438:17–20, November 2005.

Tomasz M. Grzegorczyk and Jin Au Kong. Electrodynamics of moving media inducing positive and negative refraction. *Phys. Rev. B*, 74:033102, 2006.

Tomasz M. Grzegorczyk, Xudong Chen, Joe Pacheco, Jr., Jianbing Chen, Bae-Ian Wu, and Jin Au Kong. Reflection coefficients and Goos-Hänchen shifts in anisotropic and bianisotropic left-handed metamaterials. *Prog. in Electromagnetic Res.*, 51:83–113, 2005a.

Tomasz M. Grzegorczyk, Christopher D. Moss, Jie Lu, Xudong Chen, Joe Pacheco, Jr., and Jin Au Kong. Properties of left-handed metamaterials: transmission, backward phase, negative refraction, and focusing. *IEEE Trans.*

Microwave Theory Tech., 53(9):2956–2967, 2005b.

Tomasz M. Grzegorczyk, Madhusudhan Nikku, Xudong Chen, Bae-Ian Wu, and Jin Au Kong. Refraction laws for anisotropic media and their application to left-handed metamaterials. *IEEE Trans. Microwave Theory Tech.*, 53(4):1443–1450, April 2005c.

Tomasz M. Grzegorczyk, Zachary Thomas, and Jin Au Kong. Inversion of critical angle and brewster angle in anisotropic left-handed metamaterials. *Appl. Phys. Lett.*, 86:251909, 2005d.

S. Guenneau, B. Gralak, and J. B. Pendry. Perfect corner reflector. *Opt. Lett.*, 30: 1204–1206, 2005a.

S. Guenneau, Amar C. Vutha, and S. A. Ramakrishna. Negative refraction in 2D checkerboards related by mirror anti-symmetry and 3D corner lenses. *New J. Phys.*, 7(164), 2005b.

K. Guven and E. Ozbay. A plain photonic crystal for generating directional radiation from embedded sources. *J. Opt. A: Pure Appl. Opt.*, 9:239–242, 2007.

F. D. M. Haldane. Electromagnetic surface modes at interfaces with negative refractive index make a 'not-quite-perfect' lens. *ArXiv:condmat0206420*, 2002. Accessible at http://arxiv.org.

Roger F. Harrington. *Field Computation by Moment Method*. Wiley-IEEE Press, New York, 1993. ISBN 0-7803-1014-4.

T. E. Hartman. Tunneling of a wave packet. *J. Appl. Phys.*, 33:3427, 1962.

Sailing He, Yi Jin, Zhichao Ruan, and Jinguo Kuang. On subwavelength and open resonators involving metamaterials of negative refraction index. *New J. Phys.*, 7: 210, 2005.

K. M. Ho, C. J. Chan, and C. M. Soukoulis. Existence of a photonic gap in dielectric structures. *Phys. Rev. Lett.*, 65:3152–3155, 1990.

C. L. Holloway, E. F. Kuester, J. Baker-Jarvis, and P. Kabos. A double negative (DNG) composite medium composed of magnetodielectric spherical particles embedded in a matrix. *IEEE Trans. Antennas Propagat.*, 51:2596–2603, 2003.

A. Ishimaru, S. W. Lee, Y. Kuga, and V. Jandhyala. Generalized constitutive relations for metamaterials based on the quasi-static lorentz theory. *IEEE Trans. Antennas Propagat.*, 51:2550–2557, 2003.

M. S. Islam and V. J. Logeeswaran. Microfabrication of self-assembling elements for 3D negative-index materials. *SPIE News Room*, 2006.

Ashwin Iyer, Peter Kremer, and George V. Eleftheriades. Experimental and theoretical verification of focusing in a large, periodically loaded transmission line negative refractive index metamaterial. *Opt. Express*, 11:696–708, 2003.

J. D. Jackson. *Classical Electrodynamics*. John Wiley & Sons, New York, 1999.

Zubin Jacob, Leonid V. Alekseyev, and Evgenii Narimanov. Optical hyperlens: Far-field imaging beyond the diffraction limit. *Opt. Express*, 14(18):8247–8256,

2006.

Japanese   scientist   invents   'invisibility   coat',   2003.
http://news.bbc.co.uk/2/2777111.stm.

Y. Jin and S. He. Focusing by a slab of chiral medium. *Optics Express*, 13(13):
4974–4979, June 2005.

John D. Joannopoulos, Robert D. Meade, and Joshua N. Winn. *Photonic Crystals,
Molding the Flow of Light.* Princeton University Press, Princeton, NJ, 1995. ISBN
0-691-03744-2.

Sajeev John. Strong localization of photons in certain disordered dielectric superlattices. *Phys. Rev. Lett.*, 58(23):2486–2489, June 1987.

P. B. Johnson and R. W. Christy. Optical constants of the noble metals. *Phys. Rev.
B*, 6:4370, 1972.

M. K. Karkkainen and S. A. Maslovski. Wave propagation, refraction and focusing
phenomena in Lorentzian double negative materials: A theoretical and numerical
study. *Microwave Opt. Tech. Lett.*, 37:4–7, 2002.

A. V. Kats, Sergey Savel'ev, V. A. Yampol'skii, and Franco Nori. Left-handed
interfaces for electromagnetic surface waves. *Phys. Rev. Lett.*, 98:073901, 2007.

G. Kaupp. *Atomic Force Microscopy, Scanning Nearfield Optical Microscopy and
Nanoscratching.* Springer, Berlin, 2006. ISBN: 978-3-540-28405-5.

Brandon A. Kemp, Tomasz M. Grzegorczyk, and Jin Au Kong. Optical momentum
transfer to absorbing mie particles. *Phys. Rev. Lett.*, 97:133902, 29 September
2006.

Brandon A. Kemp, Jin Au Kong, and Tomasz M. Grzegorczyk. Reversal of wave
momentum in isotropic left-handed media. *Phys. Rev. A*, 75:053810, 2007.

Kyoung-Youm Kim. Goos-Hänchen shifts via composite layers of negative- and
positive-index media. *Jpn. J. Appl. Phys.*, 44(8):6295–6296, 2005.

R. J. King, D. V. Thiel, and K. S. Park. The synthesis of surface reactance using
and artificial dielectric. *IEEE Trans. Antennas Propagat.*, AP31:471–475, 1983.

C. Kittel, S. Fahy, and S. G. Louie. Magnetic screening by a thin superconducting
surface layer. *Phys. Rev. B*, 37:642, 1988.

Charles Kittel. *Introduction to Solid State Physics.* John Wiley & Sons, New York,
7th edition, 1996. ISBN 0-471-11181-3.

M. W. Klein, C. Enrich, M. Wegener, and S. Linden. Second-harmonic generation
from magnetic metamaterials. *Science*, 313:502–504, 2006a.

M. W. Klein, C. Enrich, M. Wegener, C. M. Soukoulis, and S. Linden. Single-slit
split-ring resonators at optical frequencies: Limits of size scaling. *Opt. Lett.*, 31
(9), May 2006b.

Pavel Kolinko and David Smith. Numerical study of electromagnetic waves interacting with negative index materials. *Opt. Express*, 11:640–648, 2003.

Jin Au Kong. *Electromagnetic Wave Theory*. EMW Cambridge, MA, 2000. ISBN 0-9668143-9-8.

Th. Koschny, P. Markoš, E. N. Economou, D. R. Smith, D. C. Vier, and C. M. Soukoulis. Impact of inherent periodic structure on effective medium description of left-handed and related metamaterials. *Phys. Rev. B*, 71:245105, 2005.

G. J. Kovacs and G. D. Scott. Optical excitation of surface plasma waves in layered media. *Phys. Rev. B*, 16:1297, 1977.

A. Krishnan, T. Thio, T. J. Kim, H. J. Lezec, T. W. Ebbesen, P. A. Wolff, J. Pendry, L. Martin-Moreno, and F. J. Garcia-Vidal. Evanescently coupled resonance in surface plasmon enhanced transmission. *Opt. Commun*, 200:1–7, 2001.

N. Kumar. In P. R. Rao, M. Doyama, G. Sundararajan, Y. R. Mahajan, and K. Oda, editors, *Asia Academic Seminar on New Materials*, pages 33–39. International Centre for Powder Metallurgy and New Materials, and the Japan Society for Promotion of Science, 2002.

A. N. Lagarkov and V. N. Kissel. Near-perfect imaging in a focusing system based on a left-handed-material plate. *Phys. Rev. Lett.*, 92(7):077401, 2004.

A. N. Lagarkov and A. K. Sarychev. Electromagnetic properties of composites containing elongated conducting inclusions. *Phys. Rev. B*, 53:6318, 1996.

L. D. Landau, E. M. Lifschitz, and L. P. Pitaeskii. *Electrodynamics of Continuous Media*. Butterworth-Heinemann, Oxford, 1984.

R. Landauer and Th. Martin. Barrier interaction time in tunneling. *Rev. Mod. Phys.*, 66:217–228, 1994.

M. Lapine and M. Gorkunov. Three wave coupling of microwaves in metamaterial with nonlinear resonant conductive elements. *Phys. Rev. E*, 70:066601, 2004.

M. Lapine, M. Gorkunov, and K. H. Ringhofer. Nonlinearity of a metamaterial arising from diode insertions into resonant conductive elements. *Phys. Rev. E*, 67:065601, 2003.

U. Leonhardt. Notes on conformal invisibility devices. *New J. Phys.*, 8:118, July 2006a.

Ulf Leonhardt. Optical conformal mapping. *Science*, 312(5781):1777–1780, June 2006b.

Henri Lezec and T. Thio. Diffracted evanescent wave model for enhanced and suppressed optical transmission through subwavelength hole arrays. *Opt. Express*, 12:3629–3651, 2004.

J. Li, L. Zhou, C. T. Chan, and P. Sheng. Photonic band gap from a stack of positive and negative index materials. *Phys. Rev. Lett.*, 90:083901, 2003a.

Z. W. Li, L. Chen, Xuesong Rao, and C. K. Ong. Subwavelength imaging by a left-handed material superlens. *Phys. Rev. B*, 67:055409, 2003b.

Zhi-Yuan Li and Lan-Lan Lin. Evaluation of lensing in photonic crystal slabs exhibiting negative refraction. *Phys. Rev. B*, 68:245110, 2003.

Peng Liang, Ran Lixin, Chen Hongsheng, Zhang Haifei, Jin Au Kong, and Tomasz M. Grzegorczyk. Experimental observation of left-handed behavior in an array of standard dielectric resonators. *Phys. Rev. Lett.*, 98:157403, 2007.

I. V. Lindell, A. H. Sihvola, S. A. Tretyakov, and A. J. Viitanen. *Electromagnetic Waves in Chiral and Bi-Isotropic Media*. Artech House, Boston, 1994.

I. V. Lindell, S. A. Tretyakov, K. I. Nikoskinen, and S. Ilvonen. BW media – media with negative parameters, capable of supporting backward waves. *Microwave Opt. Tech. Lett.*, 31(2):129–133, 20 October 2001.

Stefan Linden, Christian Enkrich, Martin Wegener, Jiangfeng Zhou, Thomas Koschny, and Costas M. Soukoulis. Magnetic response of metamaterials at 100 terahertz. *Science*, 306:1351–1353, November 19, 2004.

Z. Liu, H. Lee, Y. Xiong, C. Sun, and X. Zhang. Far-field optical hyperlens magnifying sub-diffraction-limited objects. *Science*, 315:1686, March 2007.

P. F. Loschialpo, D. L. Smith, D. W. Forester, F. J. Rachford, and J. Schelleng. Electromagnetic waves focused by a negative-index planar lens. *Phys. Rev. E*, 67: 025602, 2003.

R. Loudon. The propagation of electromagnetic energy through an absorbing dielectric. *J. Phys. A: Gen. Phys.*, 3:233–245, 1970.

R. Loudon, L. Allen, and D. F. Nelson. Propagation of electromagnetic energy and momentum through an absorbing dielectric. *Phys. Rev. E*, 55(1):1071–1085, January 1997.

Rodney Loudon, Stephen M. Barnett, and C. Baxter. Radiation pressure and momentum transfer in dielectrics: The photon drag effect. *Phys. Rev. A*, 71:063802, 2005.

Jie Lu, Tomasz M. Grzegorczyk, Yan Zhang, Joe Pacheco, Jr., Bae-Ian Wu, and Jin. A. Kong. Čerenkov radiation in materials with negative permittivity and permeability. *Opt. Express*, 11(7), April 7, 2003.

Jie Lu, Tomasz M. Grzegorczyk, Bae-Ian Wu, Joe Pacheco, Jr., Min Chen, and Jin Au Kong. Effect of poles on the sub-wavelength focusing by an LHM slab. *Microwave Opt. Tech. Lett.*, 45(1):49–53, April 5, 2005a.

W. T. Lu, J. B. Sokoloff, and S. Sridhar. Comment on 'wave refraction in negative-index materials: Always positive and very inhomogeneous'. *ArXiv:cond-mat/0207689*, 2002.

W. T. Lu, J. B. Sokoloff, and S. Sridhar. Refraction of electromagnetic energy for wave packets incident on a negative-index medium is always negative. *Phys. Rev. E*, 69:026604, 2004.

Zhaolin Lu, Janusz A. Murakowski, Christopher A. Schuetz, Shouyuan Shi, Garrett J. Schneider, and Dennis W. Prather. Three-dimensional subwavelength imaging by a photonic-crystal flat lens using negative refraction at microwave frequencies. *Phys. Rev. Lett.*, 95:153901, 2005b.

Chiyan Luo, Steven G. Johnson, J. D. Joannopoulos, and J. B. Pendry. All-angle negative refraction without negative effective index. *Phys. Rev. B*, 65:201004, May 2002.

Chiyan Luo, Mihai Ibanescu, Steven G. Johnson, and J. D. Joannopoulos. Cerenkov radiation in photonic crystals. *Science*, 299:368–371, 17 January 2003.

Stefan Alexander Maier. *Plasmonics: Fundamentals and Applications*. Springer, Berlin, 2007.

L. I. Mandelshtam. Lectures on some problems of the theory of oscillations (1944). In *Complete Collection of Works*, volume 5, pages 428–467. Academy of Science, Moscow, 1950.

L. I. Mandelshtam. Group velocity in crystal lattice. *Zhurnal Experimentalnoi i Teoreticheskoi FizikiJournal of Experimental and Theoretical Physics (JETP)*, 15:475, 1945. (lecture notes in Russian).

Ricardo Marques, Francisco Medina, and Rachid Rafii-El-Idrissi. Role of bian-isotropy in negative permeability and left-handed metamaterials. *Phys. Rev. B*, 65:144440, April 2002.

Th. Martin and R. Landauer. Time delay of evanescent electromagnetic waves and the analogy to particle tunneling. *Phys. Rev. A*, 45:2611–2617, 1992.

Alejandro Martinez and Javier Marti. Negative refraction in two-dimensional photonic crystals: Role of lattice orientation and interface termination. *Phys. Rev. B*, 71:235115, 2005.

J. C. Maxwell-Garnett. Colors in metal glasses and in metallic films. *Phil. Trans. Roy. Soc. A*, 203:385–420, 1904.

M. McCall, W. S. Weiglhoffer, and A. Lakhtakia. The negative index of refraction demystified. *Eur. J. Phys.*, 23:353, 2002.

A. R. Melnyk and M. J. Harrison. Theory of optical excitation of plasmons in metals. *Phys. Rev. B*, 2:835–850, 1970.

David O. S. Melville and Richard J. Blaikie. Super-resolution imaging through a planar silver layer. *Opt. Express*, 13(6):2127–2134, March 2005.

David O. S. Melville and Richard J. Blaikie. Experimental comparison of resolution and pattern fidelity in single- and double-layer planar lens lithography. *J. Opt. Soc. Am. B*, 23:461–467, 2006.

R. Merlin. Analytical solution of the almost-perfect-lens problem. *Appl. Phys. Lett.*, 84(8):1290–1292, February 23 2004.

Krzysztof A. Michalski. Extrapolation methods for Sommerfeld integral tails. *IEEE Trans. Antennas Propagat.*, 46(10):1405–1418, October 1998.

P. W. Milonni. *Fast Light, Slow Light and Left-Handed Light*. Institute of Physics Publishing, Bristol, UK, 2005.

G. W. Milton. *The Theory of Composites*. Cambridge University Press, Cambridge, 2002.

H. Minkowski. Die Grundgleichungen für die elektromagnetischen Vorgänge in bewegten Körpern. *Nachrichten der K. Gesellschaft der Wissenschaften zu Göttingen, mathem.-physik. Klasse*, 1908.

Cesar Monzon and D. W. Forester. Negative refraction and focusing of circularly polarized waves in optically active media. *Phys. Rev. Lett.*, 95:123904, 2005.

Cesar Monzon, Donald W. Forester, and Peter Loschialpo. Exact solution to plane-wave scattering by an ideal "left-handed" wedge. *J. Opt. Soc. Am. A*, 23:339–348, 2005.

Cesar Monzon, P. Loschialpo, D. Smith, F. Rachford, P. Moore, and D. W. Forester. Three-dimensional focusing of broadband microwave beams by a layered photonic structure. *Phys. Rev. Lett.*, 96:207402, May 2006.

R. Moussa, S. Foteinopoulou, Lei Zhang, G. Tuttle, K. Guven, E. Ozbay, and C. M. Soukoulis. Negative refraction and superlens behavior in a two-dimensional photonic crystal. *Phys. Rev. B*, 71:085106, 2005.

P. Muhlschlegel, H. J. Eisler, O. J. F. Martin, B. Hecht, and D. W. Pohl. Resonant optical antennas. *Science*, 308:1607, 2005.

A. I. Nachman. Reconstructions from boundary measurements. *Ann. Math.*, 128: 531–576, 1988.

L. Nanda and S. A. Ramakrishna. Time for pulse traversal through slabs of dispersive and negative (ε, μ) materials. *Phys. Rev. A*, 76:063807, 2007.

L. Nanda, A. Basu, and S. A. Ramakrishna. Delay times and detector times for optical pulses traversing plasmas and negative refractive media. *Phys. Rev. E*, 74:036601, 2006.

A. M. Nicolson and G. F. Ross. Measurement of the intrinsic properties of materials by time-domain techniques. *IEEE Trans. Inst. Meas.*, IM-19:377–382, November 1970.

M. A. Noginov, G. Zhu, M. Bahoura, J. A. Adegoke, C. Small, B. A. Ritzo, V. P. Drachev, and V. M. Shalaev. The effect of gain and absortion of surface plasmons in metal nanoparticles. *Appl. Phys. B*, 86(3):455–460, 2006.

M. A. Noginov, V. A. Podolskiy, G. Zhu, M. Mayy, M. Bahoura, J. A. Adegoke, B. A. Ritzo, and K. Reynolds. Compensation of loss in propagating surface plasmon polariton by gain in adjacent gain medium. *arXiv:0704.1513*, 2007.

M. Norgen. General optimization approach to a frequency-domain inverse problem of a stratified bianisotropic slab. *J. Electromagn. Waves Applicat.*, 11:515–546, 1997.

M. Notomi. Theory of light propagation in strongly modulated photonic crystals: Refractionlike behavior in the vicinity of the photonic band gap. *Phys. Rev. B*, 62(16):10696–10705, October 2000.

M. Notomi. Negative refraction in photonic crystals. *Opt. Quant. Electr.*, 34:133–143, 2002.

S. O'Brien and J. B. Pendry. Magnetic activity at infrared frequencies in structured metallic photonic crystals. *J. Phys.: Condens. Matter*, 14(25), June 2002.

S. O'Brien and J. B. Pendry. Photonic band-gap effects and magnetic activity in dielectric composites. *J. Phys.: Condens. Matt.*, 14(15):4035–4044, April 2002.

S. O'Brien, D. MacPeake, S. A. Ramakrishna, and J. B. Pendry. Near infrared photonic bandgaps and non-linear effects in negative magnetic materials. *Phys. Rev. B*, 69:241101, 2004.

K. Ohtaka and Y. Tanabe. Photonic band using vector spherical waves. 1. various properties of Bloch electric fields and heavy photons. *J. Phys. Soc. Jpn.*, 65:2265, 1996.

A. A. Oliner and T. Tamir. Backward waves on isotropic plasma slabs. *J. Appl. Phys.*, 33(1):231, January 1962.

Ouail Ouchetto, Cheng-Wei Qiu, Sad Zouhdi, Le-Wei Li, and Adel Razek. Homogenization of 3-D periodic bianisotropic metamaterials. *IEEE Trans. Microwave Theory Tech.*, 54(11):3893–3898, November 2006.

J. Pacheco, Jr., T. M. Grzegorczyk, B.-I. Wu, Y. Zhang, and J. A. Kong. Power propagation in homogeneous isotropic frequency dispersive left-handed media. *Phys. Rev. Lett.*, 89(25):7401, 16 December 2002.

Joe Pacheco, Jr. *Theory and Application of Left-Handed Metamaterials*. PhD thesis, Massachusetts Institute of Technology, February 2004.

Michael A. Paesler and Patrick J. Moyer. *Near-Field Optics: Theory, Instrumentation, and Applications*. John Wiley & Sons, Hoboken, NJ, June 1996. ISBN: 978-0-471-04311-9.

L. V. Panina, A. N. Grigorenko, and D. P. Makhnovskiy. Optomagnetic composite medium with conducting nanoelements. *Phys. Rev. B*, 66:155411, 2002.

C. G. Parazzoli, R. B. Greegor, K. Li, B. E. C. Koltenbah, and M. Tanielian. Experimental verification and simulation of negative index of refraction using Snell's law. *Phys. Rev. Lett.*, 90(10):107401, March 2003.

P. V. Parimi, W. T. Lu, P. Vodo, J. Sokoloff, J. S. Derov, and S. Sridhar. Negative refraction and left-handed electromagnetism in microwave photonic crystals. *Phys. Rev. Lett.*, 92(12):127401, March 2004.

W. F. Parks and J. T. Dowell. Fresnel drag in uniformly moving media. *Phys. Rev. A*, 9(1):565–567, January 1974.

J. Paul, C. Christopoulos, and D. W. P. Thomas. Time-domain modelling of negative refractive index material. *Electron. Lett.*, 37(14):912–913, 2001.

J. Peatross, S. A. Glasgow, and M. Ware. Average energy flow of optical pulses in dispersive media. *Phys. Rev. Lett.*, 84:2370–2373, 2000.

J. B. Pendry. Negative refraction makes a perfect lens. *Phys. Rev. Lett.*, 85:3966–3969, October 2000.

J. B. Pendry. A chiral route to negative refraction. *Science*, 306:1353–1355, November 2004a.

J. B. Pendry. Perfect cylindrical lenses. *Opt. Express*, 11:755–760, 2003.

J. B. Pendry. Negative refraction. *Contemporary Physics*, 45:191–202, 2004b.

J. B. Pendry. Photonic band structures. *J. Modern Opt.*, 41(2):209–229, February 1994.

J. B. Pendry and S. A. Ramakrishna. Focusing light using negative refraction. *J. Phys.: Condens. Matter*, 15:6345–6364, 2003.

J. B. Pendry and S. A. Ramakrishna. Near-field lenses in two dimensions. *J. Phys.: Condens. Matter*, 14:8463–8479, 2002.

J. B. Pendry and D. R. Smith. Reversing light with negative refraction. *Phys. Today*, 57:37–43, June 2004.

J. B. Pendry, A. J. Holden, W. J. Stewart, and I. Youngs. Extremely low frequency plasmons in metallic mesostructures. *Phys. Rev. Lett.*, 76(25):4773–4776, June 1996.

J. B. Pendry, A. J. Holden, D. J. Robbins, and W. J. Stewart. Low frequency plasmons in thin-wire structures. *J. Phys.: Condens. Matter*, 10:4785–4809, 1998.

J. B. Pendry, A. J. Holden, D. J. Robbins, and W. J. Stewart. Magnetism from conductors and enhanced nonlinear phenomena. *IEEE Trans. Microwave Theory Tech.*, 47(11):2075–2084, November 1999.

J .B. Pendry, L. Martin-Moreno, and F. J. Garcia-Vidal. Mimicking surface plasmons with structured surfaces. *Science*, 305:847, 2004.

J. B. Pendry, D. Schurig, and D. R. Smith. Controlling electromagnetic fields. *Science*, 312:1780–1782, June 2006.

Paul Penfield and Hermann A. Haus. *Electrodynamics of Moving Media*. MIT Press, Cambridge, MA, 1967. Research Monograph no. 40.

Y. I. Perelman. *Physics Can Be Fun*. Firebird Publications, 1913. ISBN-13: 978-0828528948.

J. M. Pitarke, F. J. Garcia-Vidal, and J. B. Pendry. Effective electronic response of a system of metallic cylinders. *Phys. Rev. B*, 57:15261, 1998.

A. K. Popov and V. M. Shalaev. Negative index metamaterials: Second harmonic generation, Manley-Rowe relations and parametric amplification. *Appl. Phys. B*, 84:131–137, 2006.

David M. Pozar. *Microwave Engineering*. John Wiley & Sons, New York, 3rd edition, 2005. ISBN 0-471-44878-8.

J. C. Quail, J. G. Rako, H. J. Simon, and R. T. Deck. Optical second-harmonic generation with long-range surface plasmons. *Phys. Rev. Lett.*, 50:1987–1990, 1981.

Heinz Raether. *Surface Plasmons on Smooth and Rough Surfaces and Gratings*. Springer-Verlag, Berlin, 1986.

S. A. Ramakrishna. Physics of negative refractive index materials. *Rep. Prog. Phys.*, 68:449–521, 2005.

S. A. Ramakrishna and A. D. Armour. Propagating and evanescent waves in absorbing media. *Am. J. Phys.*, 71:562, 2003.

S. A. Ramakrishna and N. Kumar. Correcting the quantum clock: conditional sojourn times. *Europhys. Lett.*, 60:491–497, 2002.

S. A. Ramakrishna and J. B. Pendry. Removal of absorption and increase in resolution in a near-field lens via optical gain. *Phys. Rev. B*, 67:201101, 2003.

S. A. Ramakrishna, J. B. Pendry, D. Schurig, D. R. Smith, and S. Schultz. The asymmetric lossy near-perfect lens. *J. Modern Opt.*, 49(10):1747–1762, August 2002.

S. A. Ramakrishna, J. B. Pendry, M. C. K. Wiltshire, and W. J. Stewart. Imaging the near-field. *J. Modern Opt.*, 50:1419–1430, 2003.

S. A. Ramakrishna, S. Chakrabarti, and O. J. F. Martin. Metamaterials with negative refractive index at optical frequencies. *arXiv:physics/0703003*, 2007a.

S. A. Ramakrishna, S. Guenneau, S. Enoch, G. Tayeb, and B. Gralak. Confining light with negative refraction in checkerboard metamaterials and photonic crystals. *Phys. Rev. A*, 75:063830, 2007b.

S. Anantha Ramakrishna and J. B. Pendry. Spherical perfect lens: Solutions of Maxwell's equations for spherical geometry. *Phys. Rev. B*, 69:115115, March 2004.

X. S. Rao and C. K. Ong. Amplification of evanescent waves in a lossy left-handed material slab. *Phys. Rev. B*, 68:113103, 2003.

R. Rashed. A pioneer in anaclastics: Ibn Sahl on burning mirrors and lenses. *Isis*, 81:464–491, 1990.

W. Rayleigh. The propagation of waves through a medium endowed with a periodic structure. *Philos. Mag.*, 24:145–159, 1887.

R. H. Ritchie. Plasma losses by fast electrons in thin films. *Phys. Rev.*, 106:874, 1957.

C. Rockstuhl, F. Lederer, C. Etrich, T. Pertsch, and T. Scharf. Design of an artificial three-dimensional composite metamaterial with magnetic resonances in the visible range of the electromagnetic spectrum. *Phys. Rev. Lett.*, 99:017401, 2007.

W. Rotman. Plasma simulation by artificial dielectrics and parallel-plate media. *IRE Trans. Antennas Propag.*, AP10:82–85, 1962.

R. Ruppin. Non-local optics of the near-field lens. *J. Phys.: Condens. Matter*, 17:1803–1810, 2005.

R. Ruppin. Evaluation of extended Maxwell-Garnett theories. *Opt. Comm.*, 182(4):273–279, August 2000a.

R. Ruppin. Surface polaritons of a left-handed medium. *Phys. Lett. A*, 277(1),

November 2000b.

M. M. I. Saadoun and N. Engheta. A reciprocal phase shifter using novel pseudochiral or Omega medium. *Microwave Opt. Tech. Lett.*, 5(4):184–188, April 1992.

M. H. Saffouri. Treatment of Čerenkov radiation from electric and magnetic charges in dispersive and dissipative media. *Nuovo Cimento*, 3D:589–622, 1984.

K. Sakoda. *Optical Properties of Photonic Crystals.* Springer, Berlin, 2005.

Dror Sarid. Long-range surface-plasma waves on very thin metal films. *Phys. Rev. Lett.*, 47:1927–1930, 1981.

Andrey K. Sarychev and Vladimir M. Shalaev. Electromagnetic field fluctuations and optical nonlinearities in metal-dielectric composites. *Phys. Rep.*, 335(275), 2000.

B. Sauviac, C. R. Simovski, and S. A. Tretyakov. Double split-ring resonators: Analytical modelling and numerical simulations. *Electromagnetics*, 24(5):317–338, 2004.

D. Schurig, J. J. Mock, B. J. Justice, S. A. Cummer, J. B. Pendry, A. F. Starr, and D. R. Smith. Metamaterial electromagnetic cloak at mircowave frequencies. *Science*, 314:977–980, November 2006.

A. S. Schwanecke, A. Krasavin, D. M. Bagnall, A. Potts, A. V. Zayats, and N. I. Zheludev. Broken time reversal of light interaction with planar chiral nanostructures. *Phys. Rev. Lett.*, 91(247404), 2003.

M. O. Scully and M. S. Zubairy. *Quantum Optics.* Cambridge University Press, Cambridge, U.K., 1997.

J. Seidel, S. Grafstrom, and L. Eng. Stimulated emission of surface plasmons at the interface between a silver film and an optically pumped dye solution. *Phys. Rev. Lett.*, 94:177401, 2005.

L. C. Sengupta. Reports on leading edge engineering in 1996 NAE Symp. on Frontiers of Engineering. In *1996 NAE Symp. on Frontiers of Engineering*, Washington, D.C., 1997. National Academy Press.

Ilya V. Shadrivov, Andrey A. Sukhorukov, and Yuri S. Kivshar. Beam shaping by a periodic structure with negative refraction. *Appl. Phys. Lett.*, 82(22):3820–3822, June 2 2003.

Ilya V. Shadrivov, Andrey A. Sukhorukov, Yuri S. Kivshar, Alexander A. Zharov, Allan D. Boardman, and Peter Egan. Nonlinear surface waves in left-handed materials. *Phys. Rev. E*, 69:016617, 2004.

Ilya V. Shadrivov, Andrey A. Sukhorukov, and Yuri S. Kivshar. Complete band gaps in one-dimensional left-handed periodic structures. *Phys. Rev. Lett.*, 95:193903, 2005.

V. M. Shalaev. Optical negative-index metamaterials. *Nature Photonics*, 1:41–48, 2007.

V. M. Shalaev, Wenshan Cai, Uday K. Chettiar, Hsiao-Kuan Yuan, Andrey K.

Sarychev, Vladimir P. Drachev, and Alexander V. Kildishev. Negative index of refraction in optical metamaterials. *Opt. Lett.*, 30(24), December 2005.

E. Shamonina, V. A. Kalinin, K. H. Ringhofer, and L. Solymar. Imaging, compression and poynting vector streamlines with negative permittivity materials. *Electron. Lett.*, 37:124, 2001.

R. A. Shelby, D. R. Smith, S. C. Nemat-Nasser, and S. Schultz. Microwave transmission through a two-dimensional, isotropic, left-handed metamaterial. *Appl. Phys. Lett.*, 78(4):489–491, January 22 2001a.

R. A. Shelby, D. R. Smith, and S. Schultz. Experimental verification of a negative index of refraction. *Science*, 292:77–79, April 6 2001b.

J. T. Shen and P. M. Platzman. Near field imaging with negative dielectric constant lenses. *Appl. Phys. Lett.*, 80:3286–3288, 2002.

Zhongyan Sheng and Vasundara V. Varadan. Tuning the effective properties of metamaterials by changing the substrate properties. *J. Appl. Phys.*, 101(014909), Jan. 1 2007.

H. Shin and S. H. Fan. All-angle negative refraction for surface plasmon waves using a metal-dielectric-metal structure. *Phys. Rev. Lett.*, 96:073907, February 2006.

Jonghwa Shin and Shanhui Fan. Conditions for self-collimation in three-dimensional photonic crystals. *Opt. Lett.*, 30(18):2397–2399, 2005.

D. Sievenpiper, L. Zhang, R. F. J. Braos, N. J. Alexopolous, and E. Yablonovitch. High impedance electromagnetic surface with a forbidden frequency band. *IEEE Trans. Microwave Theory Tech.*, 47:2059–2074, 1999.

D. F. Sievenpiper, M. E. Sickmiller, and E. Yablonovitch. 3D wire mesh photonic crystal. *Phys. Rev. Lett.*, 76:2480–2483, 1996.

Ari Sihvola. *Electromagnetic Mixing Formulae and Applications*. IEE Electromagnetic Waves Series 47, May 2000. ISBN 0852967721.

Peter P. Silvester and Ronald L. Ferrari. *Finite Elements for Electrical Engineers*. Cambridge University Press, Cambridge, UK, 3rd edition, 1996.

D. R. Smith and J. B. Pendry. Homogenization of metamaterials by field averaging. *J. Opt. Soc. Am. B*, 23:391–403, 2006.

D. R. Smith and D. Schurig. Electromagnetic wave propagation in media with indefinite permittivity and permeability tensors. *Phys. Rev. Lett.*, 90(7):077405, February 2003.

D. R. Smith, Willie J. Padilla, D. C. Vier, S. C. Nemat-Nasser, and S. Schultz. Composite medium with simultaneously negative permeability and permittivity. *Phys. Rev. Lett.*, 84(18):4184–4187, May 1 2000.

D. R. Smith, S. Schultz, P. Markos, and C. M. Soukoulis. Determination of effective permittivity and permeability of metamaterials from reflection and transmission coefficients. *Phys. Rev. B*, 65:195104, 2002a.

D. R. Smith, D. Schurig, and J. B. Pendry. Negative refraction of modulated electromagnetic waves. *Appl. Phys. Lett.*, 81:2713–2715, 2002b.

D. R. Smith, D. Schurig, M. Rosenbluth, S. Schultz, S. A. Ramakrishna, and John B. Pendry. Limitations on subdiffraction imaging with a negative refractive index slab. *Appl. Phys. Lett.*, 82(10):1506–1508, March 2003.

D. R. Smith, P. Kolinko, and D. Schurig. Negative refraction in indefinite media. *J. Opt. Soc. Am. B*, 21(5), May 2004a.

D. R. Smith, D. Schurig, J. J. Mock, P. Kolinko, and P. Rye. Partial focusing of radiation by a slab of indefinite media. *Appl. Phys. Lett.*, 84(13), March 2004b.

D. R. Smith, D. C. Vier, Th. Koschny, and C. M. Soukoulis. Electromagnetic parameter retrieval from inhomogeneous metamaterials. *Phys. Rev. E*, 71:036617, 2005.

I. I. Smolyaninov, A. V. Zayats, A. Gungor, and C. C. Davis. Single photon tunnelling via localized surface plasmons. *Phys. Rev. Lett.*, 88:187402, 2002.

Daniel D. Stancil, Benjamin E. Henty, Ahmet G. Cepni, and J. P. Van't Hof. Observation of an inverse Doppler shift from left-handed dipolar spin waves. *Phys. Rev. B*, 74:060404(R), 2004.

Allen Taflove and Susan C. Hagness. *Computational Electrodynamics: The Finite-Difference Time-Domain Method*. Artech House, Boston, 3rd edition, 2005.

Chen-To Tai. *Dyadic Green Functions in Electromagnetic Theory*. IEEE Press, 2nd edition, 1993.

H. Takeda and K. Yoshino. Tunable photonic band schemes in two-dimensional photonic crystals composed of copper oxide high-temperature superconductors. *Phys. Rev. B*, 67:245109, 2003.

A. I. Talukder, T. Haruta, and M. Tomita. Measurement of net group and reshaping delays for optical pulses in dispersive media. *Phys. Rev. Lett.*, 94:054502, 2005.

A. I. Talukder, S. Kawakita, and M. Tomita. Propagation of the centroid of arbitrary pulses through angularly dispersive systems. *J. Opt. Soc. Am. B*, 24:1406–1409, 2007.

G. Tayeb and D. Maystre. Rigorous theoretical study of finite-size two-dimensional photonic crystals doped by microcavities. *J. Opt. Soc. Am. A*, 14:3323–3332, 1997.

Zachary M. Thomas, Tomasz M. Grzegorczyk, and Jin A. Kong. Design and measurement of a four-port device using metamaterials. *Opt. Express*, 13(12):4737–4744, 13 June 2005.

S. A. Tretyakov. Electromagnetic field energy density in artificial microwave materials with strong dispersion and loss. *Phys. Lett. A*, 343:231–237, August 2005.

S. A. Tretyakov. Meta-materials with wideband negative permittivity and permeability. *Microwave Opt. Tech. Lett.*, 31:163–165, 2001.

S. A. Tretyakov, F. Mariotte, C. R. Simovski, T. G. Kharina, and J. P. Heliot. Analytical antenna model for chiral scatterers: Comparison with numerical and experimental data. *IEEE Trans. Antennas Propagat.*, 44:1006–1014, 1996.

S. A. Tretyakov, I. Nefedov, A. Sihvola, S. Maslovski, and C. Simovski. Waves and energy in chiral nihility. *J. Electromagn. Wave Appl.*, 17:65–706, 2003.

S. A. Tretyakov, A. Sihvola, and L. Jylha. Backward-wave regime and negative refraction in chiral composites. *Photon. Nanostruct.: Fund. Appl.*, 3:107–115, 2005.

S. A. Tretyakov, C. R. Simovski, and M. Hudlička. Bianisotropic route to the realization and matching of backward-wave metamaterial slabs. *Phys. Rev. B*, 75: 153104, 2007.

L. Tsang, J. A. Kong, and K. H. Ding. *Scattering of Electromagnetic Waves: Theories and Applications*. Wiley, New York, 2000a.

L. Tsang, J. A. Kong, K. H. Ding, and C. O. Ao. *Scattering of Electromagnetic Waves: Numerical Simulations*. Wiley, New York, 2000b.

P. M. Valanju, R. M. Walser, and A. P. Valanju. Wave refraction in negative-index media: Always positive and very inhomogeneous. *Phys. Rev. Lett.*, 88(18):187401, May 2002.

V. G. Veselago. The electrodynamics of substances with simultaneously negative values of ϵ and μ. *Sov. Phys. USPEKHI*, 10(4):509–514, January-February 1968.

R. M. Walser. Metamaterials: An introduction. In W. S. Weiglhofer and A. Lakhtakia, editors, *Introduction to Complex Mediums for Optics and Electromagnetics*, volume PM123. SPIE Press, Bellingham, WA, 2003.

L. J. Wang, A. Kuzmich, and A. Dogariu. Gain-assisted superluminal light propagation. *Nature*, 406:277–279, July 2000.

Li-Gang Wang and Shi-Yao Zhu. Large positive and negative Goos-Hänchen shifts from a weakly absorbing left-handed slab. *J. Appl. Phys.*, 98(043522), 2005.

X. Wang, Z. F. Ren, and K. Kemp. Unrestricted superlensing in a triangular two-dimensional photonic crystal. *Opt. Express*, 12(13):2919–2924, June 2004.

A. J. Ward and J. B. Pendry. Refraction and geometry in Maxwell's equations. *J. Modern Opt.*, 43:773, 1996.

Werner S. Weiglhofer. *Constitutive Characterization of Simple and Complex Mediums*, chapter 2, pages 27–61. SPIE Press, Bellingham, WA, 2003. W. S. Weiglhofer and A. Lakhtakia editors.

H. G. Wells. *The Invisible Man*. Signet Classic, A division of Penguin Group Inc., New York, new edition, 2002. Available at http://www.online-literature.com/wellshg/invisible/.

E. P. Wigner. Lower limit for the energy derivative of the scattering phase shift. *Phys. Rev.*, 98:145–147, 1955.

M. C. K. Wiltshire, J. B. Pendry, I. R. Young, D. J. Larkman, D. J. Gilderdale, and J. V. Hajnal. Microstructured magnetic materials for RF flux guides in magnetic resonance imaging. *Science*, 291:849–851, 2001.

M. C. K. Wiltshire, J. V. Hajnal, J. B. Pendry, and D. J. Edwards. Metamaterial endoscope for magnetic field transfer: Near field imaging with magnetic wires. *Opt. Express*, 11:709, 2003a.

M. C. K. Wiltshire, J .B. Pendry, J. V. Hajnal, and D. J. Edwards. Swiss roll metamaterials: An effective medium with strongly negative permeability. In *IEE Seminar on Metamaterials for Microwave and (Sub)Millimeter Wave Applications*, London, November 2003b. Paper 13.

M. C. K. Wiltshire, J. B. Pendry, and J. V. Hajnal. Subwavelength imaging at radio frequency. *J. Phys. Condens. Matter*, 18:L315–L321, 2006.

E. Wolf and T. Habashy. Invisible bodies and uniqueness of the inverse scattering problem. *J. Modern. Opt.*, 40:785–792, 1993.

B. Wood, J. B. Pendry, and D. P. Tsai. Directed subwavelength imaging using a layered metal-dielectric system. *Phys. Rev. B*, 74:115116, 2006.

Bae-Ian Wu, Tomasz M. Grzegorczyk, Y. Zhang, and J. A. Kong. Guided modes with imaginary transverse wavenumber in a slab waveguide with negative permittivity and permeability. *J. Appl. Phys.*, 93(11):9386–9388, June 2003.

Eli Yablonovitch. Inhibited spontaneous emission in solid-state physics and electronics. *Phys. Rev. Lett.*, 58(20):2059–2062, May 1987.

Eli Yablonovitch, T. J. Gmitter, and K. M. Leung. Photonic band structure: The face-centred cubic case employing non-spherical atoms. *Phys. Rev. Lett.*, 67(17): 2295–2298, October 1991.

Vassilios Yannopapas and Alexander Moroz. Negative refractive index metamaterials from inherently non-magnetic materials for deep infrared to terahertz frequency ranges. *J. Phys.: Cond. Matt*, 17, June 2005.

T. J. Yen, W. J. Padilla, N. Fang, D. C. Vier, D. R. Smith, J. B. Pendry, D. N. Basov, and X. Zhang. Terahertz magnetic response from artificial materials. *Science*, 303, March 2004.

Shuang Zhang, Wenjun Fan, K. J. Malloy, S. R. Brueck, N. C. Panoiu, and R. M. Osgood. Near-infrared double negative metamaterials. *Opt. Express*, 13:4922–4930, 2005a.

Shuang Zhang, Wenjun Fan, B. K. Minhas, Andrew Frauenglass, K. J. Malloy, and S. R. J. Brueck. Mid-infrared resonant magnetic nanostructures exhibiting a negative permeability. *Phys. Rev. Lett.*, 94:037402, January 2005b.

Alexander A. Zharov, Ilya V. Shadrivov, and Yuri S. Kivshar. Nonlinear properties of left-handed metamaterials. *Phys. Rev. Lett.*, 91(3), July 2003.

J. Zhou, Th. Koschny, M. Kafesaki, E. N. Economou, J. B. Pendry, and C. M. Soukolis. Saturation of the magnetic response of split-ring resonators at optical frequencies. *Phys. Rev. Lett.*, 95:223902, 2005.

R. W. Ziolkowski and A. Erentok. Metamaterial-based efficient electrically small antennas. *IEEE Trans. Antennas Propagat.*, 54:2113–2130, 2006.

R. W. Ziolkowski and E. Heyman. Wave propagation in media having negative permittivity and permeability. *Phys. Rev. E*, 64:056625, 2001.

F. Zolla, S. Guenneau, A. Nicolet, and J. B. Pendry. Electromagnetic analysis of cylindrical invisibility cloaks and the mirage effect. *Opt. Lett.*, 32(9):1069–1071, May 2007.

Index

Milton Keynes UK
Ingram Content Group UK Ltd.
UKHW021837071024
449327UK00021B/1512

9 780367 577483